Christian Sardet

CELLS

The Illustrated Story *of* Life

Foreword by Eric Karsenti

THE EXPERIMENT

NEW YORK

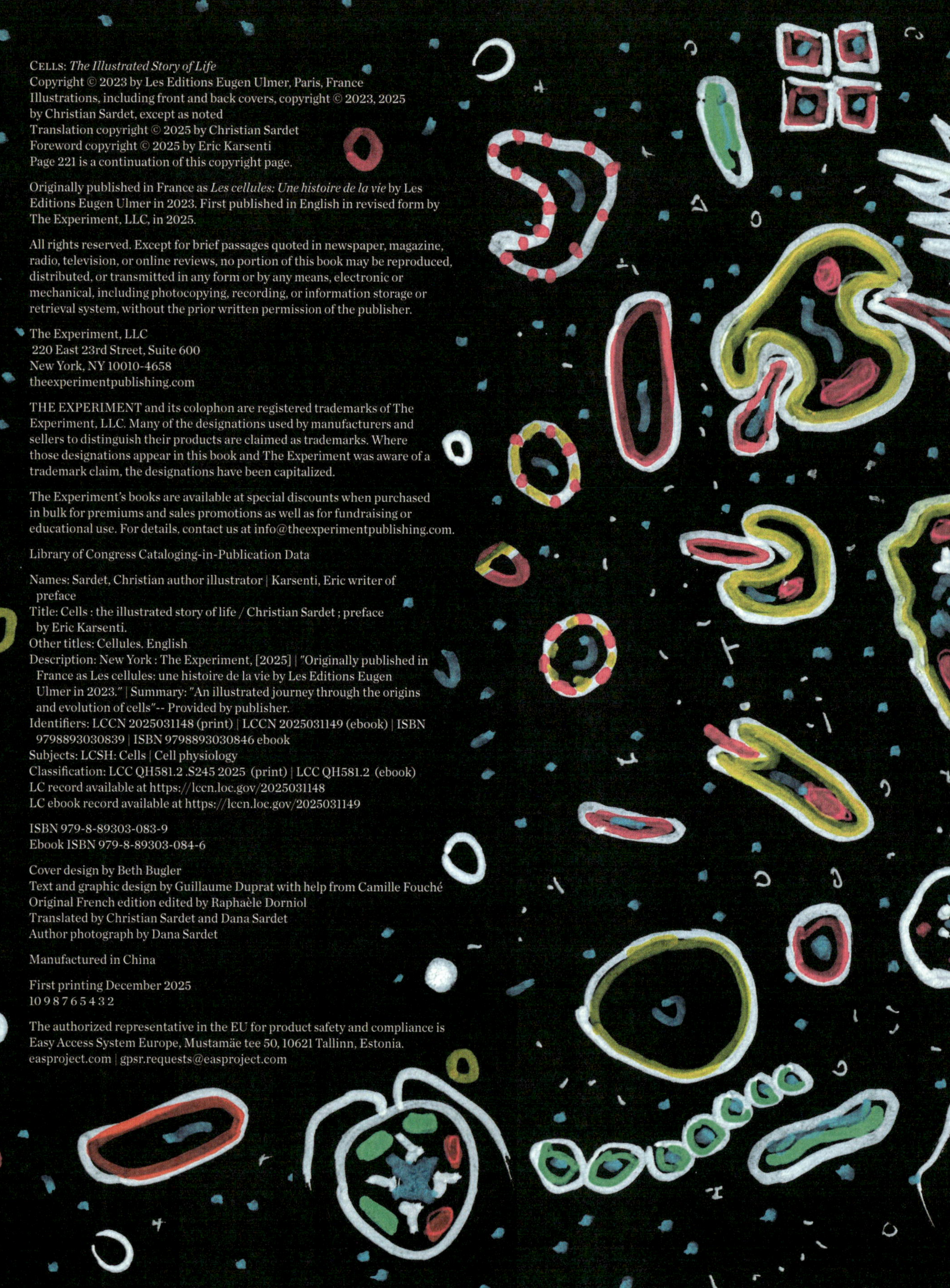

The Experiment, LLC
220 East 23rd Street, Suite 600
New York, NY 10010-4658
theexperimentpublishing.com

THE EXPERIMENT and its colophon are registered trademarks of The
Experiment, LLC. Many of the designations used by manufacturers and
sellers to distinguish their products are claimed as trademarks. Where
those designations appear in this book and The Experiment was aware of a
trademark claim, the designations have been capitalized.

The Experiment's books are available at special discounts when purchased
in bulk for premiums and sales promotions as well as for fundraising or
educational use. For details, contact us at info@theexperimentpublishing.com.

Library of Congress Cataloging-in-Publication Data

Names: Sardet, Christian author illustrator | Karsenti, Eric writer of
  preface
Title: Cells : the illustrated story of life / Christian Sardet ; preface
  by Eric Karsenti.
Other titles: Cellules. English
Description: New York : The Experiment, [2025] | "Originally published in
  France as Les cellules: une histoire de la vie by Les Editions Eugen
  Ulmer in 2023." | Summary: "An illustrated journey through the origins
  and evolution of cells"-- Provided by publisher.
Identifiers: LCCN 2025031148 (print) | LCCN 2025031149 (ebook) | ISBN
  9798893030839 | ISBN 9798893030846 ebook
Subjects: LCSH: Cells | Cell physiology
Classification: LCC QH581.2 .S245 2025 (print) | LCC QH581.2 (ebook)
LC record available at https://lccn.loc.gov/2025031148
LC ebook record available at https://lccn.loc.gov/2025031149

ISBN 979-8-89303-083-9
Ebook ISBN 979-8-89303-084-6

Cover design by Beth Bugler
Text and graphic design by Guillaume Duprat with help from Camille Fouché
Original French edition edited by Raphaèle Dorniol
Translated by Christian Sardet and Dana Sardet
Author photograph by Dana Sardet

Manufactured in China

First printing December 2025
10 9 8 7 6 5 4 3 2 1

The authorized representative in the EU for product safety and compliance is
Easy Access System Europe, Mustamäe tee 50, 10621 Tallinn, Estonia.
easproject.com | gpsr.requests@easproject.com

This book would never have seen the light of day without the Ulmer publishing team's enthusiasm for the original French edition. With Antoine Isambert and Guillaume Duprat, we had already lived a great adventure together with *Plankton: Wonders of the Drifting World*, published in French in 2013 and in English in 2015. The book was born out of our desire to share our explorations of the planktonic ecosystem, following the *Tara Oceans* expedition and our *Plankton Chronicles* films, undertaken with Véronique Kleiner, my son Noé Sardet, and his Montreal partner Sharif Mirshak. This film series attracted the attention of our friend Cédric Pollet, whom I would like to thank for introducing me to Editions Eugen Ulmer in Paris, who beautifully published my first book and oversaw its translation into several languages. When I assembled my first series of drawings and writings on cells, Antoine and Guillaume encouraged me and patiently waited for the project to mature. The final assembly into a richly illustrated book of art and science was a feast of exchanged designs and ideas, and a real pleasure. The book's superb design and layout owe a lot to Guillaume Duprat's artistic choices. And I am grateful to Antoine Isambert for having believed in this unique project. Additional thanks go to Emmanuelle Christophe, who succeeded Antoine as head of Ulmer, and to Anthony Gachet, who liaised with foreign publishers interested in translating this book. Working on the English edition with The Experiment team was a real pleasure— notably with Karen Giangreco, whose rigorous and creative editorial work reflected a longstanding connection to biology, and publisher Matthew Lore, whom we had the pleasure of meeting in New York. I say "we" because this translation and editing journey would not have been possible without my spouse, Dana Sardet, a native New Yorker, whose contribution was invaluable every step of the way.

The desire to draw and write about cells, all cells, came a few years after my retirement as a researcher in biology. I felt an irrepressible need to return to the sources and share my love of biology. This project seemed a natural continuation of the *Tara Oceans* adventure, which was imagined and accomplished with Eric Karsenti and a great team of plankton explorers. Thank you Eric for your enthusiastic foreword, and to the Taranauts for broadening my horizons regarding the diversity of life and human beings. Thanks also to my colleagues and friends for their encouragement, particularly to Catherine Jessus for her comments and invaluable suggestions, as well as Vincent Laudet, Bertie Goïc, Eric Scarfone, and the Villefranche-sur-Mer biologists for their feedback. Colleagues at the Villefranche Marine Station and collaborators around the world have nourished my passion for cells, embryos, and organisms for more than forty years.

I am very grateful to the Centre National de la Recherche Scientifique and Sorbonne Université for their continual support. I have been able to benefit from their infrastructure, databases, and friendly laboratory atmosphere during my emeritus tenure, including for this project. My early drawings and drafts did not amount to a book, so I owe a debt of gratitude to my screenwriter friend Peter Lauridsen. He encouraged me to tell the story of life using the evolution of cells and life-forms over the past 4 billion years as a common thread. My writing skills are greatly improvable, and I owe a lot to Dana Sardet, Françoise Dalban, and Claire Denis, who read, reread, corrected, and criticized various versions that were difficult for non-biologists to understand. They helped me simplify the text and thus be able to share the metamorphoses of this project with family and friends. Many researchers and biologist-artists, notably David Goodsell, Janet Iwasa, and Gaël McGill, generously made available their remarkable representations of cells and molecules. I sincerely thank them. My hope is that the photographic and digital works in this book, combined with my writings and drawings, will allow as many people as possible to understand and appreciate the cells that constitute life in all its diversity.

membrane
Chromosome
condensin
organelle
Chromosome
Chromosome
Chromosome
Chromosome
Chromosome
condensin
organelle

CELLULAR CALLIGRAMS

A bacteria ◯ or an archaea ◯.
This prokaryotic cell is surrounded
by a membrane ▮ and a cell wall.
It contains a single chromosome ▮,
as well as biomolecular
condensates in its cytoplasm.

A larger and more complex animal
cell. It is also surrounded by a
membrane ▮. This eukaryotic cell is
characterized by the presence of a
nucleus ⬤ in its center, containing
several linear chromosomes ▮ and a
biomolecular condensate. In
addition to the nucleus, delimited
by its intracellular membrane, the
cell contains other organelles and
biomolecular condensates, as well
as networks of microtubules ▮ and
microfilaments ▮, which make up
the cytoskeleton.

The background features different
viruses.

CALLIGRAMS
My drawings of the cell, using
words, are inspired by Guillaume
Apollinaire's "calligrams," which
are a form of visual poetry.
Here is what Apollinaire wrote in
his *Calligrams: Poems of Peace and
War, 1913–1916*: "As for the
calligrams, they are an example of
free-verse poetry using creative
typography and layout, at the dawn
of new means of expression such as
the cinema and the phonograph."

FOREWORD
BY ERIC KARSENTI

European Molecular Biology Laboratory
& *Tara Oceans*

We are all composed of billions of cells, very different from each other, that begin with the encounter between a father's spermatozoon and a mother's egg. This process, both fascinating and mysterious, often makes us wonder, how did such a complex system of reproduction evolve? By what miracle did we come to exist? For centuries, our ancestors grappled with these profound questions, often invoking religious explanations, which all turned out to be extremely naive and unsatisfactory.

Since the 18th century, advances in rational scientific inquiry have provided answers to many of our questions, even as the ultimate question of life's origin remains unsolved. Over the last five decades, however, we've made remarkable progress in understanding the increasing complexity of the living world. These insights have led us, *Homo sapiens*, to contemplate the vastness of the universe in ways that far exceed the boundaries of our lifespan. In doing so, we are gradually forming a comprehensive view of life's position and intricacy among the cosmos. We're even beginning to grasp how the complex molecules of life might have emerged from the atoms produced within those extraordinary cauldrons: the stars.

The different types of cells that exist today: bacteria, archaea, and the cells that make us up—the eukaryotes—all share a number of common features: They are bounded by a membrane composed of lipids (fat molecules, including cholesterol), and they contain DNA, proteins, and all the complex molecules that enable them to function, reproduce, and proliferate. The difficulty lies in imagining how all these molecular components were first encapsulated in a lipid bubble, and how they functioned to produce the first cell capable of reproduction and proliferation.

The cell is considered the irreducible structure of life. Many complex molecules exist in the universe, but without a favorable environment, there can be no cells and no life. Viruses often raise the question of what constitutes "life," since no virus can reproduce itself without infecting a cell. A virus is thus dependent upon the cell, but can also exist on its own—without the ability to self-reproduce.

We and our cells are "self-organizing" beings

One of the characteristics of cells, and therefore of life, is that they function far from thermodynamic equilibrium. Unlike thermodynamically stable molecules, which require no external energy, living systems must continually absorb energy to maintain their integrity.

Photosynthetic organisms draw their energy from the photons provided by sunlight; some bacteria draw energy from chemistry; and some organisms—called *heterotrophs*—eat other organisms. This constant energy flow underpins the dynamics of life, enabling the molecules of living beings to interact and form ever-more-complex structures with novel functions. This is why a virus outside its cell is not alive. It's a kind of dormant "germ" that only comes back to life as a parasite when it is able to harness the energy of the cell it infects. As soon as the energy flow stops, the cell stops and "dies." Similarly, understanding the origins of life involves understanding how molecular interactions within lipid environments gave rise to the first metabolic processes capable of sustaining a cell. There are many examples of self-organizing cellular structures in this book, such as biomolecular condensates (aggregates of molecules that acquire functions), the mitotic spindle, and, in fact, virtually all dynamic cellular elements.

Writing a book about the cell for the general public is an arduous task. Biology is complex, and it seems very difficult for a scientist to talk about life, as today's cellular and molecular biologists imagine it, without using complicated jargon. It's also risky for an author, because you can't write about life without dealing with its origin and place in the universe and on our planet (and possibly others).

I met Christian a long time ago, and our paths have crossed on many occasions, whether during our time as cell biologists or during the *Tara Oceans* adventure. We share the same desire: to help as many people as possible understand what life is, and its relationship to the environment, to the evolution of our planet, the solar system, and the universe in general. We can't talk about the cell without examining its chemical components and the physical forces at play, as well as the extraordinary complexification and evolution of the living world. There's also an important aesthetic component to living things. We've always admired the beauty of plants, animals, and men and women. We experience this same sense of wonder when we observe cells under the microscope, including our own. We admire the profusion of cellular forms that make up plankton, as well as the beauty of the molecules of life that we can now "see" using sophisticated physical methods and microscopes.

In this book, Christian has masterfully combined art, the most recent scientific knowledge, and ease of understanding. The choice of illustrations, including his own superb drawings, allows us to follow the odyssey of life with real pleasure, from its beginnings to the present day. This book also reveals the complexity of cells' architecture, diversity, and function. Other books about the cell do already exist, but, in my opinion, most are too academic. Christian's book is different! I sincerely hope that readers who found biology boring in school will rediscover its beauty and philosophical importance. For many of us, learning about our innermost nature may seem impossibly complex and ultimately irrelevant to everyday life. Nothing could be further from the truth. Understanding where we come from and how we function allows us to live more serenely, and even to approach our end, death, with tranquility and wisdom, by placing us firmly within the universe.

ERIC KARSENTI holds a PhD in immunology and cell biology from the Pasteur Institute in Paris. After working as a postdoctoral fellow in Dr. Marc Kirschner's laboratory at UCSF, he went on to establish his own group at the European Molecular Biology Laboratory, which made significant contributions to our understanding of the cell-cycle clock, mitotic spindle assembly, and cell morphogenesis. At EMBL, he established the Cell Biology and Biophysics unit, an interdisciplinary team of cell biologists, geneticists, physicists, and imaging developers. He created the Tara Oceans program, which began as an around-the-world expedition to map the plankton ecosystem to a depth of 1000 meters, and directed the Tara Oceans Consortium until 2021. Karsenti is emeritus research director at the CNRS, visiting senior scientist at EMBL, and a member of the European Molecular Biology Organization, the French Academy of Sciences, and the board of the Tara Oceans foundation.

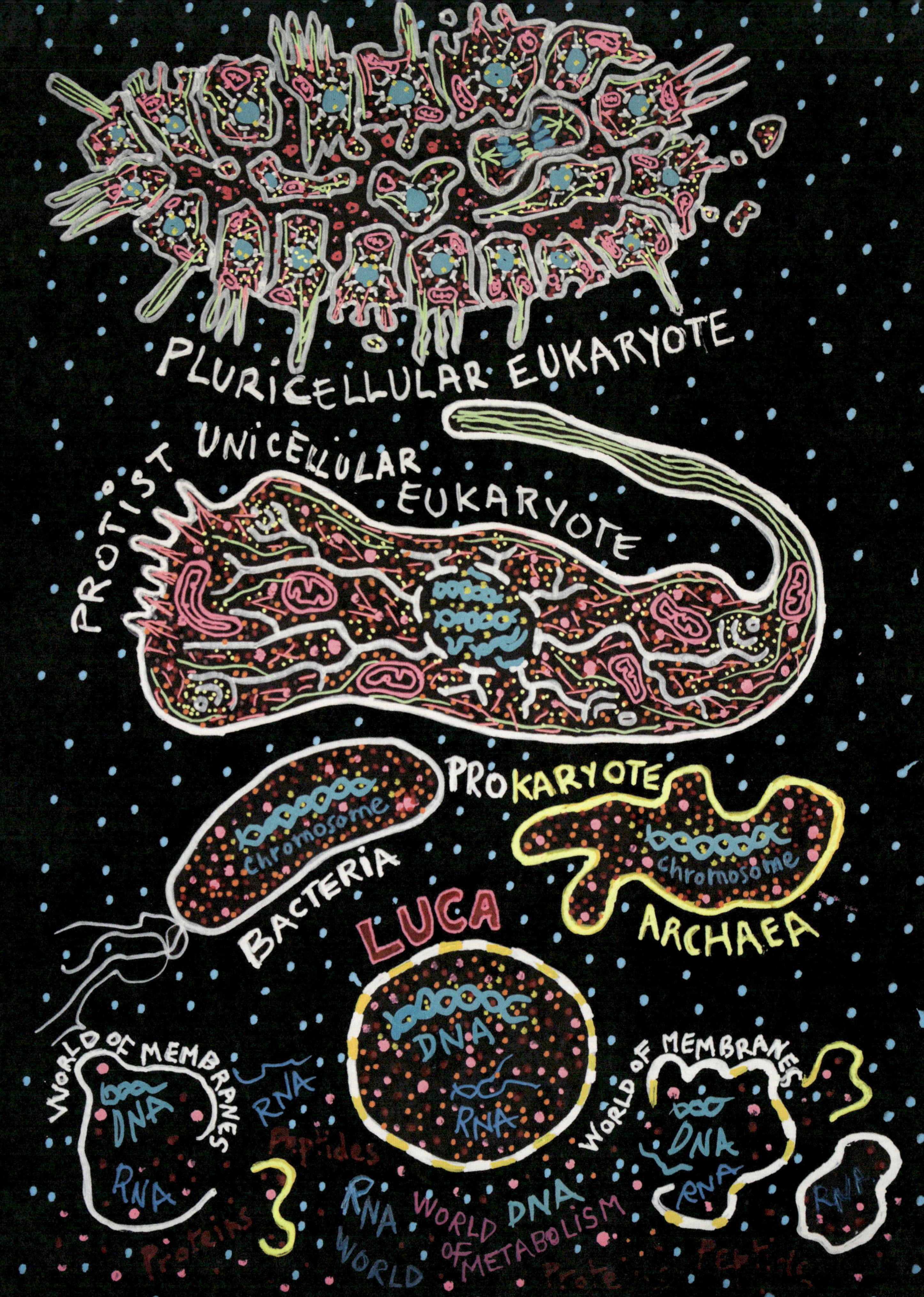

PLURICELLULAR EUKARYOTE
PROTIST UNICELLULAR EUKARYOTE
PROKARYOTE
chromosome
BACTERIA
chromosome
ARCHAEA
LUCA
DNA
RNA
WORLD OF MEMBRANES
DNA
RNA
RNA
RNA
Peptides
Proteins
RNA WORLD
WORLD OF METABOLISM
DNA
WORLD OF MEMBRANES
DNA
RNA
RNA
Proteins
Peptides

CHAPTER I

Cells & evolution

THE DISCOVERY OF CELLS AND EVOLUTION

What is life? Until the 17th century, this question belonged to the realms of philosophy and religion. Three hundred years ago, the mystery of life took on a material dimension. While telescopes were revolutionizing our understanding of the universe, microscopes revealed a world of tiny organisms and the building blocks of life called *cells*.

Two centuries of meticulous observation of plants, animals, and single-celled organisms culminated in the formulation of the cell theory. It states that all living things are composed of cells which reproduce and proliferate through division.

In the mid-1800s, Charles Darwin introduced a temporal dimension to this material view of life. He proposed that natural selection was the driving force of evolution. Darwin hypothesized that all living organisms descended from a common ancestor and that all species were connected in a tree of life. Ernst Haeckel expanded this vision, placing unicellular organisms—bacteria and protists—at the tree's root. At the end of the 19th century, it was understood that the evolution of these ancestral single cells into multicellular organisms led to the appearance of our planet's amazing diversity of animals, plants, algae, and fungi.

The discovery of cells as the fundamental units of life did not immediately dispel ancient beliefs in spontaneous generation—the idea that life can arise spontaneously from non-living matter. It wasn't until Louis Pasteur's experiments on microorganisms in the 1860s that these beliefs were seriously challenged. Today, these ideas persist in new forms, as current hypotheses on the origins of life propose that cells arose spontaneously through harnessing energy and the natural selection of molecules.

Finally, the dynamic and energetic dimensions of life became apparent in the late 19th century. The emerging field of chemistry unveiled the physiological and molecular principles underlying respiration, fermentation, photosynthesis, and all forms of metabolism—the chemical transformations that fuel all living things.

Three centuries have passed since the Dutch scientist Nicolaas Hartsoeker observed a human sperm cell through one of the first optical microscopes and envisioned a miniature human, a *homunculus*, nestled within.

Since then, observations with optical and electron microscopes have revealed that each spermatozoon is surrounded by a membrane ❚ and that it moves thanks to a flagellum which propels it to meet and fertilize an oocyte.

The purpose of fertilization is to inject the male's chromosomes ❚, located in the head of the sperm, into the oocyte and transform it into an embryo that will become an adult. The energy needed for the sperm to swim is provided by an organelle (a mitochondrion ●) located at the base of the flagellum, which propels the spermatozoon thanks to the microtubules ❚ of the cytoskeleton.

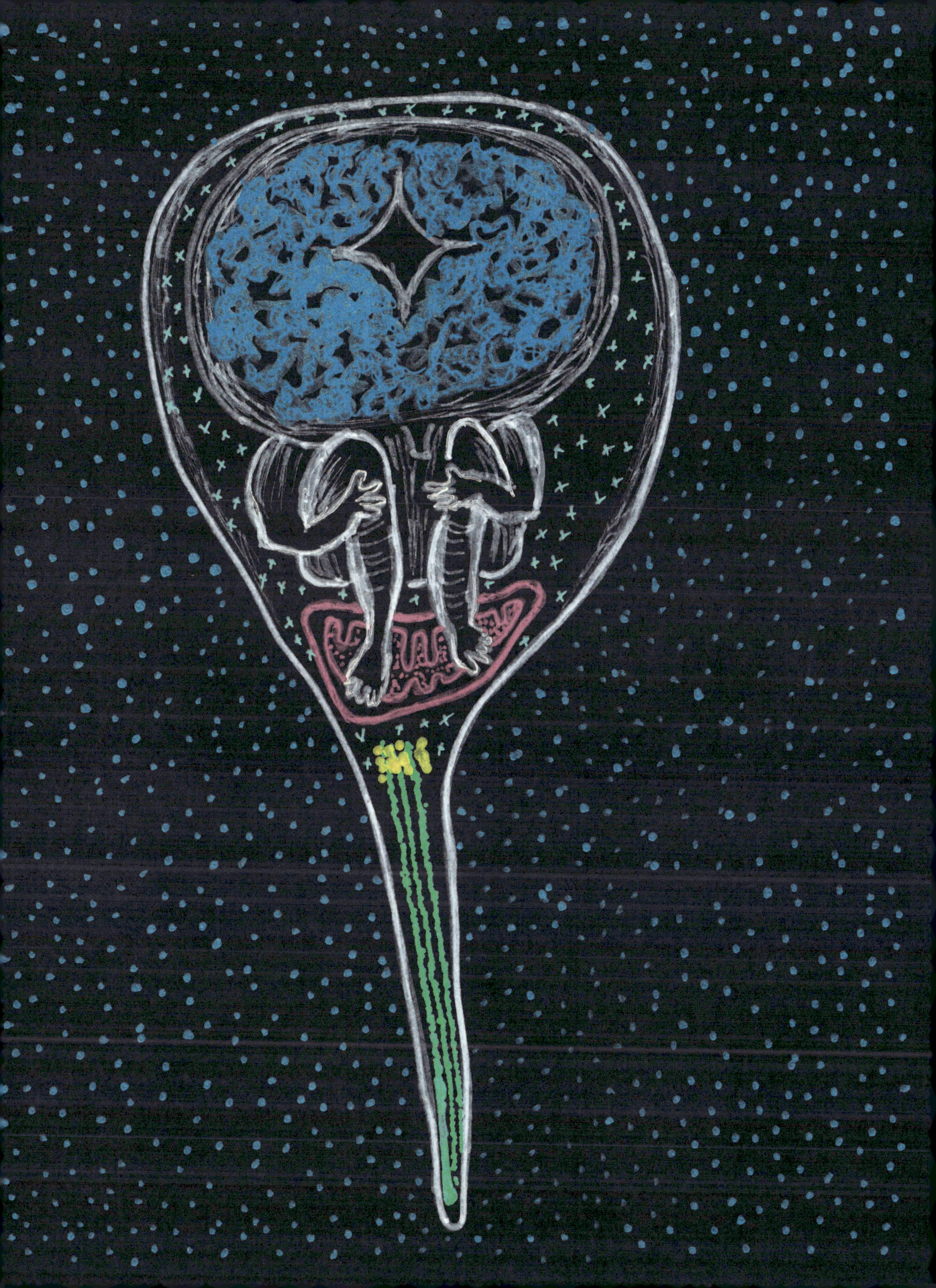

1660: The invention of the word *cell* and the discovery of a microscopic world

Inspired by the cells in which monks and prisoners lived, the English scientist Robert Hooke had the idea of using the word *cell* to describe the microscopic compartments visible in a thin slice of cork. He drew these first cells in his book *Micrographia*. His successors adopted the term, and *cells* came to mean the elementary units making up all living beings.

Dutch craftsmen mastered the art of making magnifying lenses in the mid-17th century. Through the lenses of his tiny microscopes, Antonie Van Leeuwenhoek saw minute creatures he called *animalcules* in all liquids and bodily fluids. At the same time, the Dutch biologist and physicist Nicolaas Hartsoeker imagined that, in seminal fluid, each spermatozoon carried a miniature man, which he dubbed a *homunculus* (see drawing on previous page).

ANTONIE VAN LEEUWENHOEK REVEALS THE EXISTENCE OF MICROSCOPIC LIFE

Van Leeuwenhoek, holding his tiny microscope: portrait by Ernest Boar, commissioned by H. S. Wellcome in 1913 to celebrate the history of medicine. Antonie Van Leeuwenhoek, a cloth merchant and later an employee of the Dutch town of Delft, perfected lens polishing. From the 1660s onward, he produced the best microscopes of the time. Magnifying 200 to 300 times, they allowed bacteria to be seen for the first time. Van Leeuwenhoek examined a wide range of organisms under his microscope, including bacteria (2) and protists (10 to 19). He even studied animal spermatozoa (29 to 30) and, remarkably, those of a recently deceased human (31 to 40). Van Leeuwenhoek's discoveries were initially overlooked by the Royal Society of England, but they were eventually published after he sent 190 letters in Dutch. Despite long being considered an amateur, Van Leeuwenhoek was finally recognized and became a member of learned societies in England and France, receiving visits from the kings and queens of his time.

Left

ROBERT HOOKE INVENTS THE WORD *CELL* BASED ON HIS MICROSCOPIC OBSERVATIONS

In *Micrographia*, his masterly treatise on microscopy published in 1664, Robert Hooke introduced his double-lens microscope, inspired by the telescopes of the time. The microscope magnified the sample some twenty times, enabling Hooke to distinguish partitions in a thin slice of cork. He named these structures *cells*. These were in fact the cellulose walls between dead cork cells, emptied of their contents.

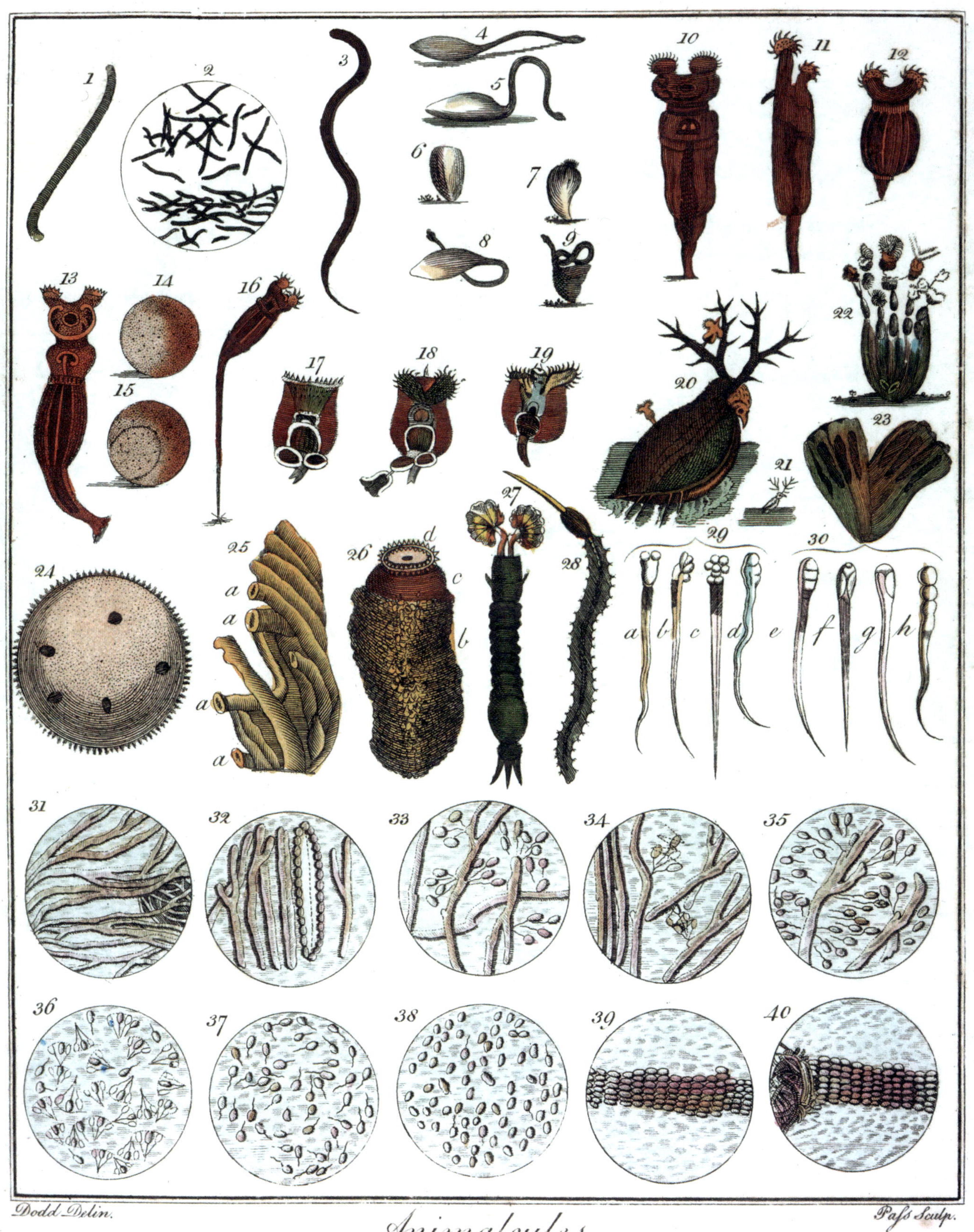

This image was inspired by Van Leeuwenhoek's black-and-white sketches, as transcribed a few decades later by English naturalist Henry Baker. His representations were then assembled in a poster and colorized by an unidentified artist.

1839: Animals and plants are made up of cells with nuclei

Legend has it that the idea that all plants and animals are made up of the same types of cells was born in 1837 at a dinner in Berlin between two young biologists, the German botanist Matthias Schleiden and the Belgian zoologist Theodor Schwann. The discussion revolved around the fact that Schwann had discovered every animal cell to have at its center a microscopic globule (the nucleus), an organelle whose presence Schleiden recognized as characteristic of all plant cells.

On the basis of this similarity—the systematic presence of a nucleus—in 1839 Schleiden and Schwann set out the premises of the cell theory: Animal and plant cells are of the same nature, and all living things are made up of cells. In 1857, Rudolf Virchow followed in their footsteps with the famous *omnis cellula e cellula* ("every cell comes from another cell").

The maturation of a fairly complete and coherent cell theory actually took about a hundred years between 1780 and 1880. It was the collective work of biologists from half a dozen European countries, who benefited from advances in microscopes and methods of fixing and staining cells. These scientists exchanged ideas, samples, and discoveries through letters and visits. They also benefited from a network of national academies, emerging scientific journals, and increasingly better-equipped and -organized laboratories.

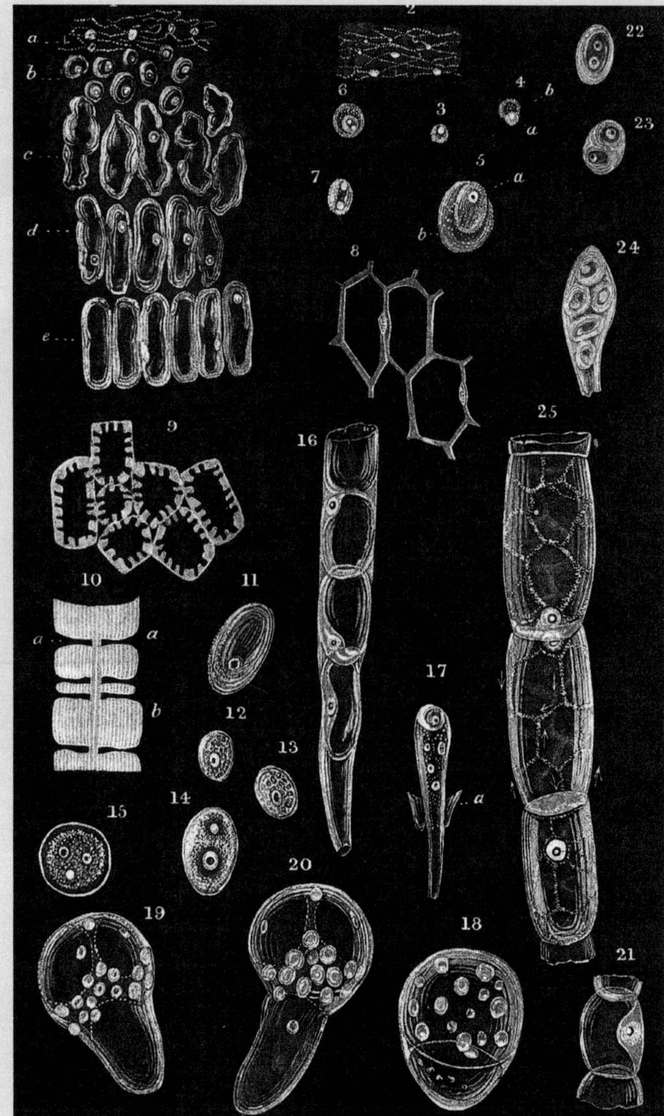

ALL PLANT CELLS HAVE A NUCLEUS

These cells were drawn in 1838 by Matthias Schleiden, a professor at the University of Jena, based on his observations of the embryonic sacs of plants as diverse as the dwarf palm, the green orchid, and the fountain liverwort. In all cells, he represents one or more globules, which are the nuclei discovered in 1831 by the Scotsman Robert Brown. In his book *Contributions to Our Knowledge of Phytogenesis* (1938), Schleiden stated that nuclei are at the origin of cell genesis, an erroneous notion that was soon corrected.

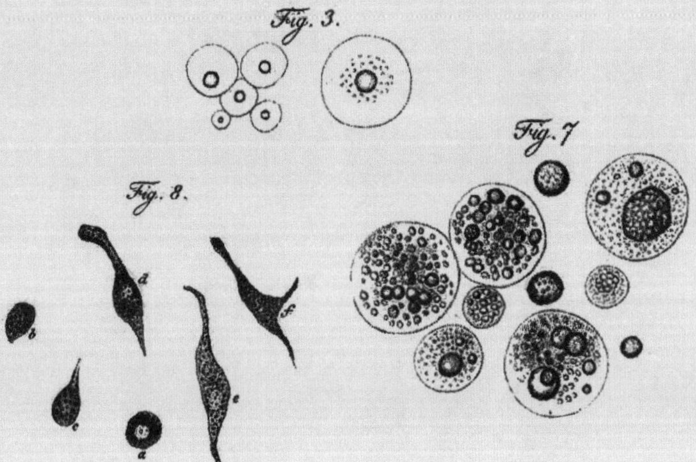

Left

ALL ANIMAL CELLS ALSO HAVE A NUCLEUS

As a young researcher in his mentor Johannes Müller's laboratory in Berlin, Theodor Schwann examined a vast number of animal tissues—cartilage taken from tadpole gills; bones, eggs, teeth, feathers, tendons, muscles, and nerves. Schwann found that all animal cells contain a globular nucleus, as do all plant cells. In 1839, in his book *Microscopical Researches into the Accordance in the Structure and Growth of Animals and Plants*, he enuciated the first principles of a cell theory.

1860: From spontaneous generation to fertilization

The concept of spontaneous generation is an ancient one. The Greek philosopher Aristotle (384–322 BCE) defined it as the idea that all life can emerge from inanimate matter if the material contains pneuma (the vital breath). This belief was based on the observation that vermin naturally appeared on rotting meat, frogs emerged from mud, and mice from piles of rags or grain. The same must have been true for the microscopic world of animalcules. This idea persisted, even among scientists, well into the 19th century, despite challenges from Francesco Redi and Lazzaro Spallanzani in the 17th and 18th centuries. Their experiments showed that the appearance of maggots and tadpoles was due to the reproductive capabilities of flies and frogs, which laid eggs on meat or in ponds. Spallanzani even dressed the male frogs in small pants to demonstrate that sperm production is necessary for oocytes to be fertilized and become embryos and larvae.

It would take another century and the observations of Hermann Fol in 1870 to understand that reproduction in animals depends on the fusion of a sperm cell with an oocyte. This fusion between two reproductive cells is the first step of development in animals and plants. Shortly before this discovery, in 1860, Louis Pasteur, drawing on his solid knowledge of bacteria and yeasts and of methods for sterilizing culture media, demonstrated that spontaneous generation does not exist in microbes.

Even if spontaneous generation in its original sense is an obsolete idea, researchers working on the origins of life believe that a primordial cell could have arisen spontaneously from inanimate matter through self-organization of the molecules of life: proteins, RNA, DNA, lipids, etc. As we'll discuss in later chapters, this might have happened 4 billion years ago on Earth, or perhaps even longer ago if life originated in the cosmos. This is a major quest of space exploration!

View *The Extraordinary Life of Hermann Fol* by Yannick Mahé, and Christian and Noé Sardet

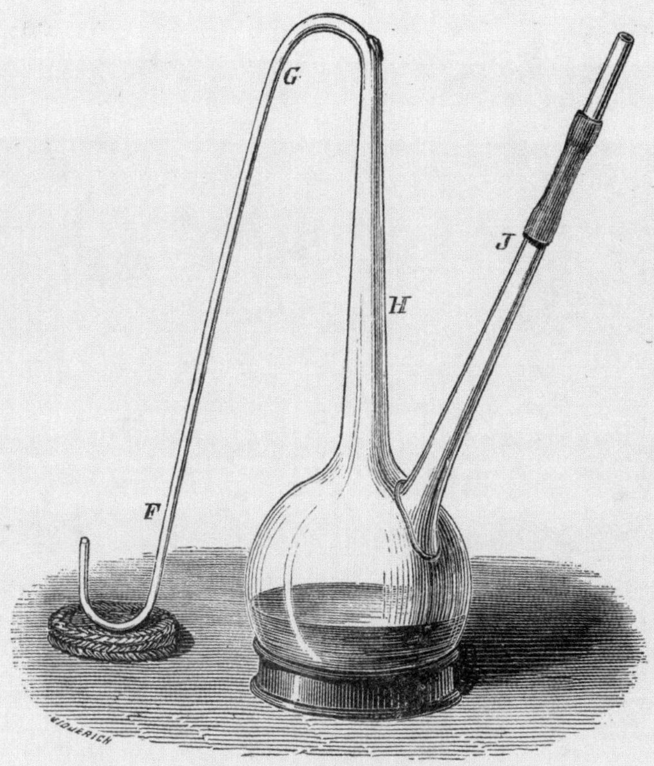

LOUIS PASTEUR'S ELEGANT EXPERIMENTS

In the mid-19th century, proponents and opponents of spontaneous generation clashed. In Paris, the Académie des Sciences decided to offer a prize for solving the problem. Louis Pasteur participated, using swan-neck flasks to definitively demonstrate that spontaneous generation did not exist in microorganisms. The culture medium in the flask was first boiled for a few minutes, until steam escaped through the open end of the bottle (J). Pasteur then allowed the sterilized liquid to cool. During cooling, the air entering the bottle deposited dust and microbes in the first bend (F). Because microbes could not pass through, the liquid remained sterile, even though it was in contact with the outside air. However, when Pasteur cut the bottle at the points marked G or H, air containing microbes reached the liquid. The microbes then proliferated in the culture medium, contaminating the liquid. Pasteur won the prize.

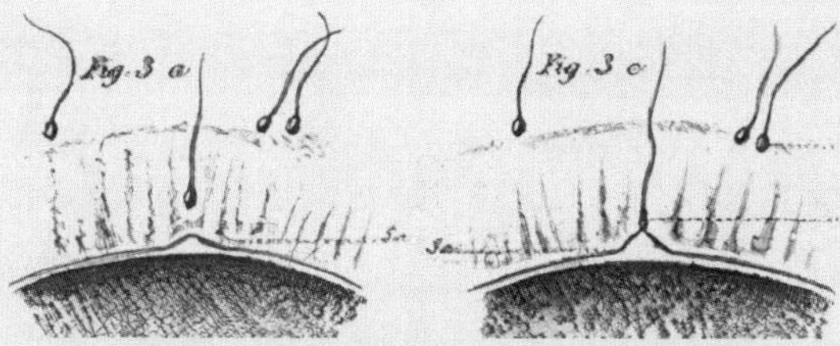

HERMANN FOL SHOWS THAT REPRODUCTION STARTS WITH FERTILIZATION

In the late 1870s, the young Hermann Fol set up a small laboratory in Messina, Sicily, at his own expense. He studied the fertilization and development of the many local aquatic fauna, including starfish, sea urchins, and mollusks. Fol discovered that a spermatozoon fuses with an oocyte, and that the nuclei of the two cells meet and fuse. The fertilized oocyte becomes an embryo and then a larva. This important discovery, rapidly extended to all animals, contributed to our understanding of heredity. Knowledge of the principles of sexual reproduction in animals helped discredit spontaneous generation.

LIFE HAS EVOLVED ON EARTH FOR 4 BILLION YEARS

For over 4 billion years, life has been unfolding on Earth. In the late 19th century, inquiries into life's origins and evolution gained momentum, informed by revelations about Earth's history and the fossilized remnants of extinct organisms.

Our planet emerged approximately 4.5 billion years ago from monumental cosmic collisions, events that also led to the formation of the Moon. One theory proposes that early cellular prototypes—protocells—arrived from space, transported by icy meteorites, and seeded life on Earth. This notion, known as *panspermia*, remains a topic of scientific exploration, especially in the quest for extraterrestrial life. However, the predominant scientific consensus holds that the first cells originated on Earth itself. These protocells likely emerged hundreds of millions of years after Earth's formation, in energy-rich environments abundant with complex molecules, some of them undoubtedly delivered from space. We will explore this further in pages 60–64.

The prevailing theory suggests that a primordial cell, known as LUCA (the last universal common ancestor), gave rise to a continuous lineage, shaping all past and present life on Earth. LUCA is estimated to have existed around 4.2 billion years ago, already possessing a complex genome. Initially, a vast community of microorganisms thrived in the nascent Earth's extreme environments, vastly different from those of today. Over millions to billions of years, the proliferation of bacteria and archaea—prokaryotic cells—produced gases such as hydrogen, methane, and oxygen, which transformed the planet and its atmosphere. This paved the way for the emergence of the first complex single-celled organisms—protists—the ancestors of all eukaryotic cells. Within a billion years, life flourished, spreading from the oceans to submerged lands, giving rise to fungi, algae, plants, and animals. Biodiversity exploded, shaped by periods of rapid evolution punctuated by mass extinctions.

By studying the planet's geochemical transformations, fossilized traces of ancient life, and the evolution of organisms, we can imagine the deep past and intertwined histories of life and Earth from their very beginnings.

Cells are enclosed by membranes **ǁ** and protective walls. They contain genetic material in the form of one or more chromosomes **ǀ** carrying DNA. After 2 billion years of evolution, cells became more complex, acquiring a nucleus ● containing several chromosomes **ǀ**, internal membranes **ǀ**, and organelles, including energy powerhouses: mitochondria ● and chloroplasts ●. Various particles and viruses made of nucleic acids and proteins have bathed all living organisms since their origins.

3 – Pluricellular organisms
Multicellular organisms

2 – Microorganisms

1 – The earliest protocells

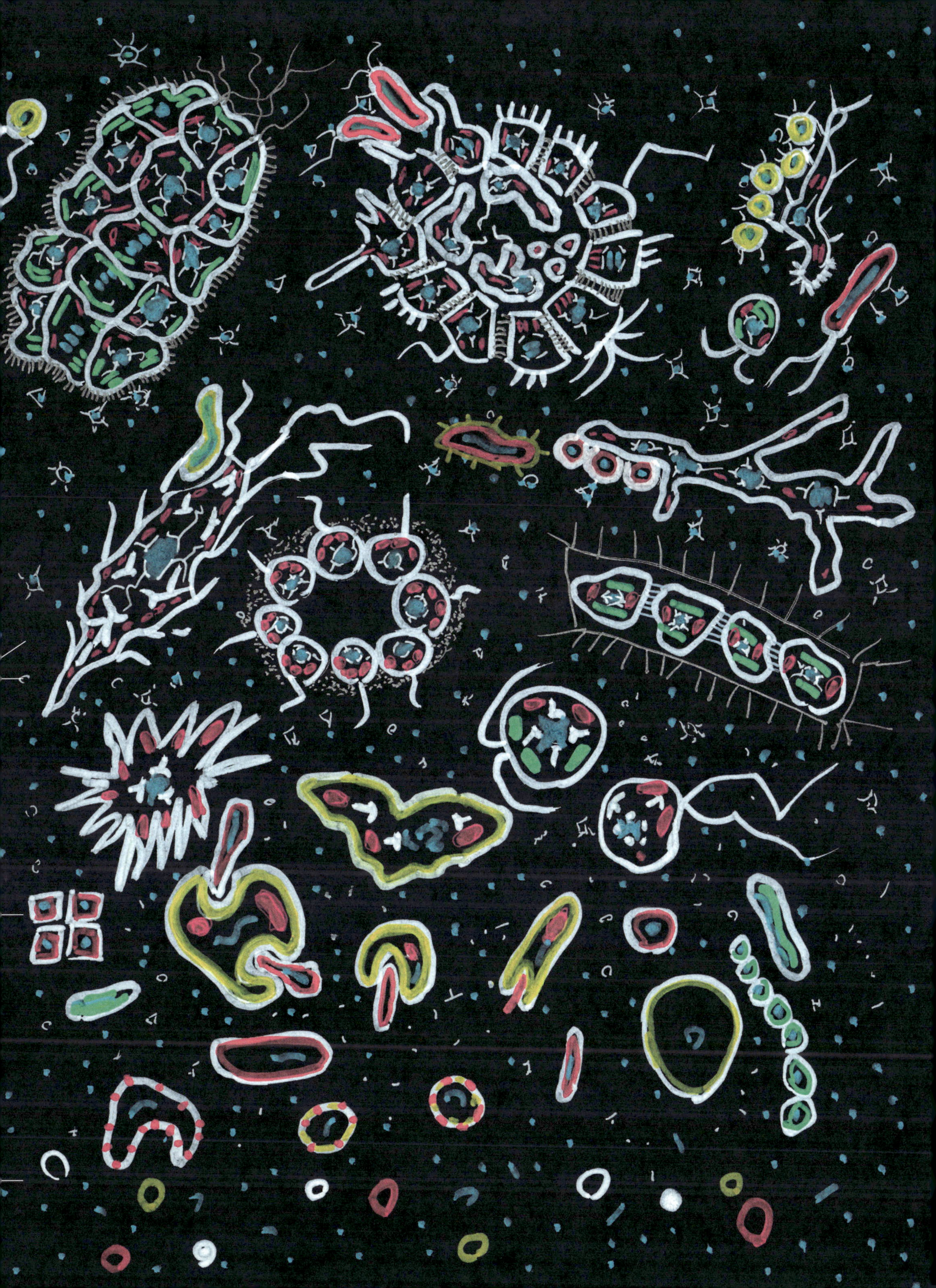

TWO TYPES OF CELLS— PROKARYOTES AND EUKARYOTES

Before delving into the step-by-step evolution of cells and life, and how it relates to the transformations of the planet, it helps to know that there is not just one type of cell on Earth, but two. Life is made up of cells without a nucleus (prokaryotes) and cells with a nucleus (eukaryotes)—two categories defined and named by Édouard Chatton in the 1920s.

Prokaryotes—cells without a nucleus—encompass two distinct groups of single-celled organisms: bacteria and archaea, both of which have been evolving for over 3.5 billion years. The first eukaryotes—cells with a nucleus—emerged more than 1.5 billion years ago, as single-celled protists. These more complex and voluminous cells might have resembled amoebas, paramecia, and microalgae, the types of protist which still exist today. Then, over 600 million years ago, some unicellular protists evolved into eukaryotic organisms, whose cells began forming colonies. This transition led to multicellular organisms (composed of identical cells) and pluricellular organisms (composed of several cell types) which eventually became the ancestors of animals, plants, algae, and fungi.

Prokaryotes (bacteria and archaea) contain a single, circular chromosome within their cytoplasm—an essential distinction from eukaryotic cells, which house multiple, linear chromosomes inside a nucleus enclosed by a double membrane. Besides the nucleus, eukaryotic cells possess a wealth of membrane-bound organelles—such as mitochondria, chloroplasts, the Golgi apparatus, and lysosomes—as well as a structural skeleton called the *cytoskeleton*. These features are largely absent in bacteria and archaea, with rare exceptions. In addition, bacteria and archaea are encased in thick, rigid cell walls and are equipped with unique appendages, distinct from the typical cilia and flagella which eukaryotic cells use to propel themselves.

Eukaryotic cells are particularly diverse in terms of their organelles and appendages. For example, all plant cells are characterized by photosynthetic organelles called *chloroplasts*, and animal cells by the presence of specific organelles and appendages that enable motility, defense, and communication. We will devote a later chapter to this morphological diversity of cells *(see chap. IV)*.

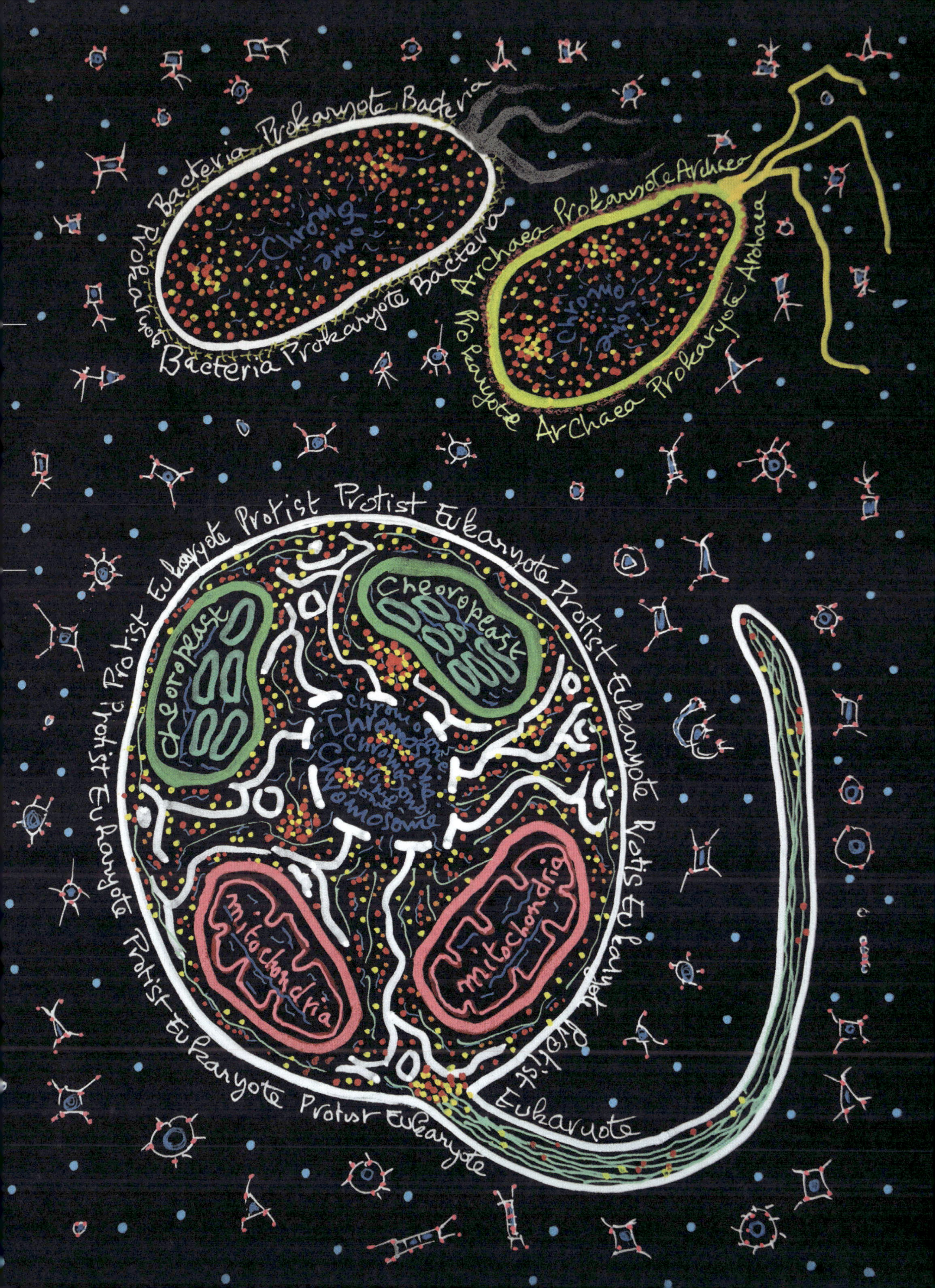

Bacteria Prokaryote Bacteria
Prokaryote
Bacteria Prokaryote Bacteria
Chromosome
Bacteria Prokaryote Bacteria
Prokaryote Archaea
Chromosome
Prokaryote
Archaea Prokaryote Archaea
Protist Protist Eukaryote
Protist Eukaryote
Protist Eukaryote
Protist Eukaryote
Chloroplast
Chloroplast
Chromosome
Chromosome
Chromosome
Chromosome
mitochondria
mitochondria
Eukaryote
Protist Eukaryote
Protist Eukaryote Protist Eukaryote

THE INTERTWINED EVOLUTIONS OF EARTH AND LIFE

Let's envision the Earth's origins, 4.5 billion years ago. As the planet cools, its atmosphere consists of carbon dioxide (CO_2), water vapor (H_2O), nitrogen (N_2), methane (CH_4), and sulfurous gases (H_2S). A billion years later, the proliferation of the first microorganisms has already begun to modify the atmosphere. Photosynthetic bacteria called *cyanobacteria* produce abundant quantities of oxygen (O_2), dramatically altering geochemical equilibriums on the planet. Soluble iron (Fe^{2+}) in the oceans oxidizes, forming rust. Vast deposits of banded iron layers turn the planet red.

The rise in oxygen levels, along with shifts in carbon dioxide (CO_2) and methane (CH_4) concentrations, triggered a cataclysmic event—a total glaciation of the planet's surface, 2.4 billion years after its formation. On this inhospitable "Snowball Earth," early microbial life was drastically reduced. Then, over hundreds of millions of years, the CO_2 released by volcanic activity gradually warmed the planet, allowing the surviving microorganisms to adapt. As the atmosphere and once-tropical climate stabilized, conditions became favorable for the emergence of more complex cells—the precursors of eukaryotes—born from symbiotic associations among microorganisms *(see chap. IV)*.

Cyanobacteria produced so much oxygen that it caused a major oxygenation event around 2.4 billion years ago, known as the GOE (Great Oxygenation Event). It was followed 1.7 billion years later by another surge in oxygen production, the NOE (Neoproterozoic Oxygenation Event). Beginning 750 million years ago, the NOE triggered two global ice ages, ultimately raising atmospheric oxygen to near-present levels while reducing carbon dioxide and methane, leading to climate stabilization. At the same time, an ozone layer formed, shielding life from harmful UV radiation. These conditions paved the way for an explosion of multicellular and macroscopic life, first in the oceans and later on land.

Over the past 500 million years, a vast diversity of microorganisms, fungi, animals, algae, and plants have filled every ecological niche. Yet, this evolutionary success has been punctuated by five mass extinctions caused by meteorite impacts, volcanic eruptions, and climate shifts. Today, a sixth extinction is underway, this time caused by human activities.

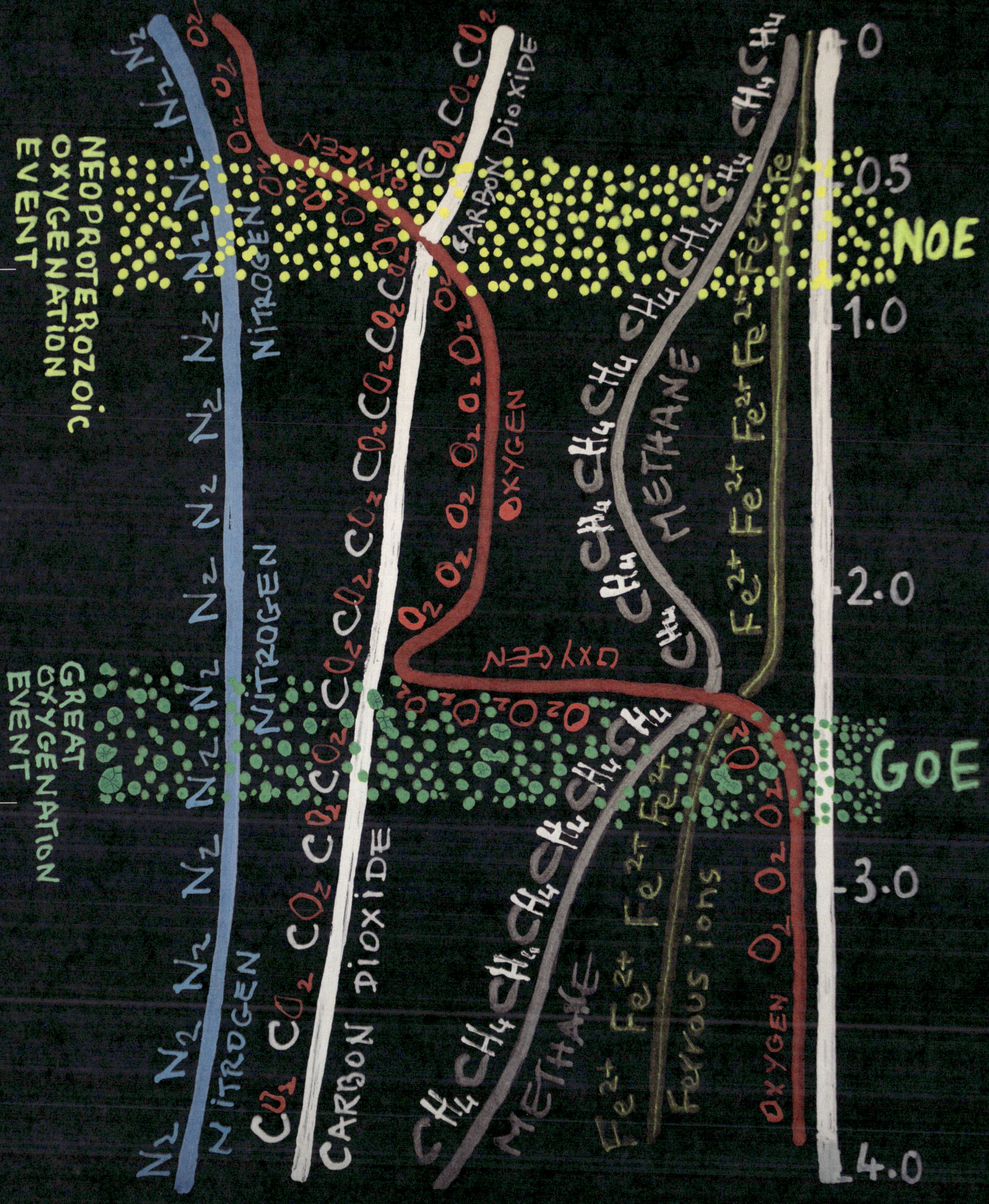

NEOPROTEROZOIC OXYGENATION EVENT
GREAT OXYGENATION EVENT
N_2 N_2 N_2 N_2 N_2 N_2
NITROGEN
O_2 O_2 O_2 O_2
OXYGEN
CO_2 CO_2 CO_2 CO_2 CO_2 CO_2 CO_2
CARBON DIOXIDE
CH_4 CH_4 CH_4 CH_4 CH_4 CH_4
METHANE
Fe^{2+} Fe^{2+} Fe^{2+} Fe^{2+} Fe^{2+}
Ferrous ions
O_2 O_2 O_2 O_2 O_2
OXYGEN
O_2
OXYGEN
NOE
GOE
0
0.5
1.0
2.0
3.0
4.0

VERY ANCIENT COMPLEX ORGANISMS GONE EXTINCT
In 2010 in Gabon, Africa, numerous fossils measuring just a few centimeters each were discovered in sediments dating back 2.1 billion years. Microtomography techniques enable reconstruction of their external (*left*) and internal (*right*) morphology. These intriguing fossils contain sulfur, iron, zinc, and organic carbon, indicative of life. These fossils likely represent ancestral forms of macroscopic life. This type of ancient and probably multicellular organism could have disappeared following a drop in the environment's oxygen concentration, approximately 1.9 to 2.1 billion years ago.

© Courtesy Abderrazak El Albani, CNRS, University of Poitiers, France

Traces of ancestral life and the first Great Oxygenation Event (GOE)

Less than a billion years after Earth's formation, microorganisms were already thriving. Among them, cyanobacteria had pioneered oxygenic photosynthesis—a process that harnesses solar energy to produce chemical energy, while releasing oxygen as a by-product. These cyanobacteria proliferated in shallow aquatic environments, associating with other microorganisms to form biofilms, the fossilized traces of which can be found in different regions of the planet.

The large quantities of oxygen produced by cyanobacteria were responsible for the planet's first major oxygenation event, known as GOE. Cataclysmic planetary transformations followed, in particular, the massive glaciation episodes of more than 2 billion years ago.

Today, descendants of the original cyanobacteria continue to proliferate in all sunny wetlands and aquatic areas. The greenish filaments formed by cyanobacteria (blue-green algae) invade lakes, coasts, and beaches where phosphates and nitrates are abundant. These photosynthetic bacteria are now the most numerous cells on the planet. They generate one fifth of the atmosphere's oxygen, while absorbing large quantities of CO_2.

We can date cyanobacteria and other types of prokaryotic microorganisms to the first billion years of the planet's existence, but we do not know exactly when the first, more complex and voluminous eukaryotic lifeforms appeared. Some macroscopic life-forms date back over 2 billion years, for example those found fossilized in sediments in Gabon. There may even have been earlier, multicellular life-forms which disappeared during the cataclysms following the first Great Oxygenation Event.

CYANOBACTERIA, THE OLDEST TRACES OF FOSSIL LIFE
Stromatolites (shown here in Shark Bay, Australia) are stratified rocks containing traces of mineralized and fossilized cyanobacterial films dating back more than 3 billion years. No indisputable trace of life exists before 3.7 billion years ago, the estimated age of the most ancient stromatolites.

A second oxygenation event (NOE) sets off an explosion of complex life

Large, multicellular life-forms appeared over periods spanning tens of millions of years (Ma), beginning in the Ediacaran period (–635 to –538 Ma) and continuing into the Cambrian period (–538 to –485 Ma). Since the fossils from these two periods are quite different, some researchers believe that two successive explosions of life took place. In any case, the fauna fossilized during the more recent period (the Cambrian) include plausible ancestors of organisms whose descendants are still living—cnidarians, ctenophores, sponges, arthropods, echinoderms, and mollusks. These multicellular creatures depend on respiration, and their emergence is considered a likely consequence of the substantial increase in oxygen concentration in the atmosphere and oceans.

This abundant presence of oxygen for over 600 million years followed the second major oxygenation event, the NOE (Neoproterozoic Oxygenation Event, –750 Ma). The causes of this second massive oxygenation are uncertain, but the photosynthetic organisms of this period (cyanobacteria and phytoplanktonic protists) probably produced enormous quantities of oxygen. What's more, this period (the end of the Proterozoic era) was characterized by glaciations and dramatic atmospheric and climatic disturbances linked to tectonic plate shifts

ANIMALS AND PLANTS OF THE CAMBRIAN PERIOD
Some life-forms dating back to the Cambrian clearly display characteristics of animals, plants, algae and fungi, which have evolved into the life-forms we see today. One such creature, Kylinxia—whose name comes from the kylin, a chimeric creature from Chinese mythology—was discovered in Chengjiang, China. Dating back 520 million years, this marine organism had a segmented body, five eyes, and two claws. The top image shows an artistic representation of Kylinxia.
FOSSIL PHOTO AND ARTISTIC REPRESENTATION OF THE CAMBRIAN ENVIRONMENT
© Courtesy Diying Huang and Fangchen Zhao, Nanjing, Institute of Geology and Paleontology, China

and intense volcanic activity. These profound disruptions seem to have played a crucial role in both the rise and extinction of complex multicellular life, ultimately shaping the evolutionary trajectory of organisms whose descendants persist to this day.

Mass extinctions sculpt the tree of life

We have described how cataclysmic events around the great oxygenation periods (GOE −2,500 Ma and NOE −750 Ma) decimated certain organisms and paved the way for the rise of others. The importance of these oxygenation events due to microorganisms is a recent discovery, but periods of extinction of living organisms have been known since the beginnings of paleontology in the 19th century.

Jean-Baptiste Lamarck, at the end of the 18th century, was the first naturalist to perceive the continuity of life. At the beginning of the 19th century, Georges Cuvier held the chair of comparative anatomy at the National Museum of Natural History (MNHN) in Paris. Inspired by George Buffon, Cuvier perceived through the study of fossils that some animal and plant species had disappeared, and that others had appeared, during catastrophes that he called "global revolutions."

Mass extinctions are caused by planetary catastrophes: climatic cataclysms, meteorite collisions, and/or volcanic eruptions. They are defined as major extinctions when they wipe out more than 75 percent of animals and plants in a short period of time. That said, regional and localized extinctions are commonplace and affect all species.

Since the Cambrian, five major mass extinctions have been recognized (−65, −201, −252, −365, and −436 Ma). These extinctions have reshuffled the phyla and the tree of life. For example, the trilobites, the dominant arthropod marine animals at the beginning of the Paleozoic era (−538 Ma), disappeared completely during the Permian extinction (−252 Ma), the most severe extinction of all, during which 90 percent of living organisms disappeared. In the absence of trilobites, other arthropods—ancestors of crustaceans, arachnids, and insects—were able to evolve, diversify, and occupy most ecological niches; today they represent 80 percent of known animal species.

In a similar way, the non-avian dinosaurs, descendants of the gigantic predators of the Jurassic (−201 to −145 Ma), disappeared in history's most recent mass extinction (the Cretaceous extinction, −65 Ma). However, some avian dinosaurs survived this extinction, to become the ancestors of modern birds.

We are currently experiencing a new global extinction—the sixth, known as the Holocene extinction—which is driven by human activities. One in eight bird species, one in four mammals, and 60 percent of all plants are endangered and threatened with total extinction by the end of the century.

TRILOBITES THRIVED, THEN DISAPPEARED DURING THE PERMIAN EXTINCTION (−252 Ma)

Trilobites were a predominant group of arthropods during the Paleozoic (−538 to −252 Ma). Their name comes from the three lobes that make up a trilobite's body, covered with a chitin carapace that is easily mineralized into calcium carbonate. This explains the profusion of fossils from more than 20,000 species, ranging in size from 2 to 75 cm. In southern Morocco, we have even found fossils dating back 480 million years of trilobites (*Ampyx priscus*) in single file. This suggests collective migration or reproduction behavior reminiscent of that observed in other arthropods, such as lobsters.

© Courtesy Jean Vannier, CNRS and University of Lyon 1,
and Muriel Vidal, University of Brest, France

Opposite
IN 1900, HEINRICH HARDER POPULARIZED EXTINCT, PRIMITIVE FAUNA

Harder prepared sixty lithographs for a series of cards titled "Animals of the Primitive World" which accompanied chocolates made by the Reichardt company. These illustrations depict dinosaurs that became extinct 65 million years ago, after a giant meteorite collided with Earth off the coast of present-day Yucatan. Harder also painted trilobites which became extinct 252 million years ago, following gigantic volcanic eruptions.

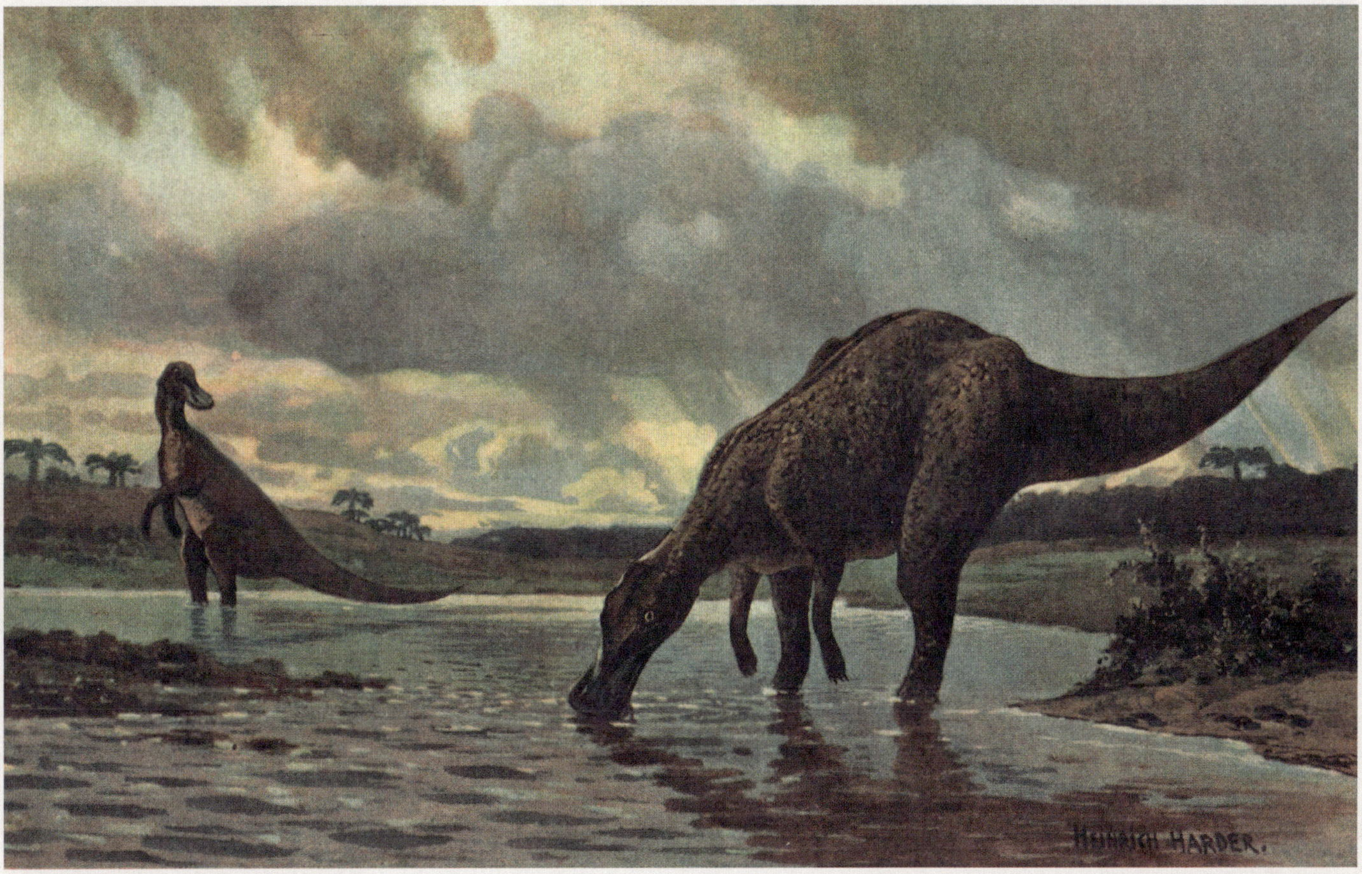
HEINRICH HARDER.

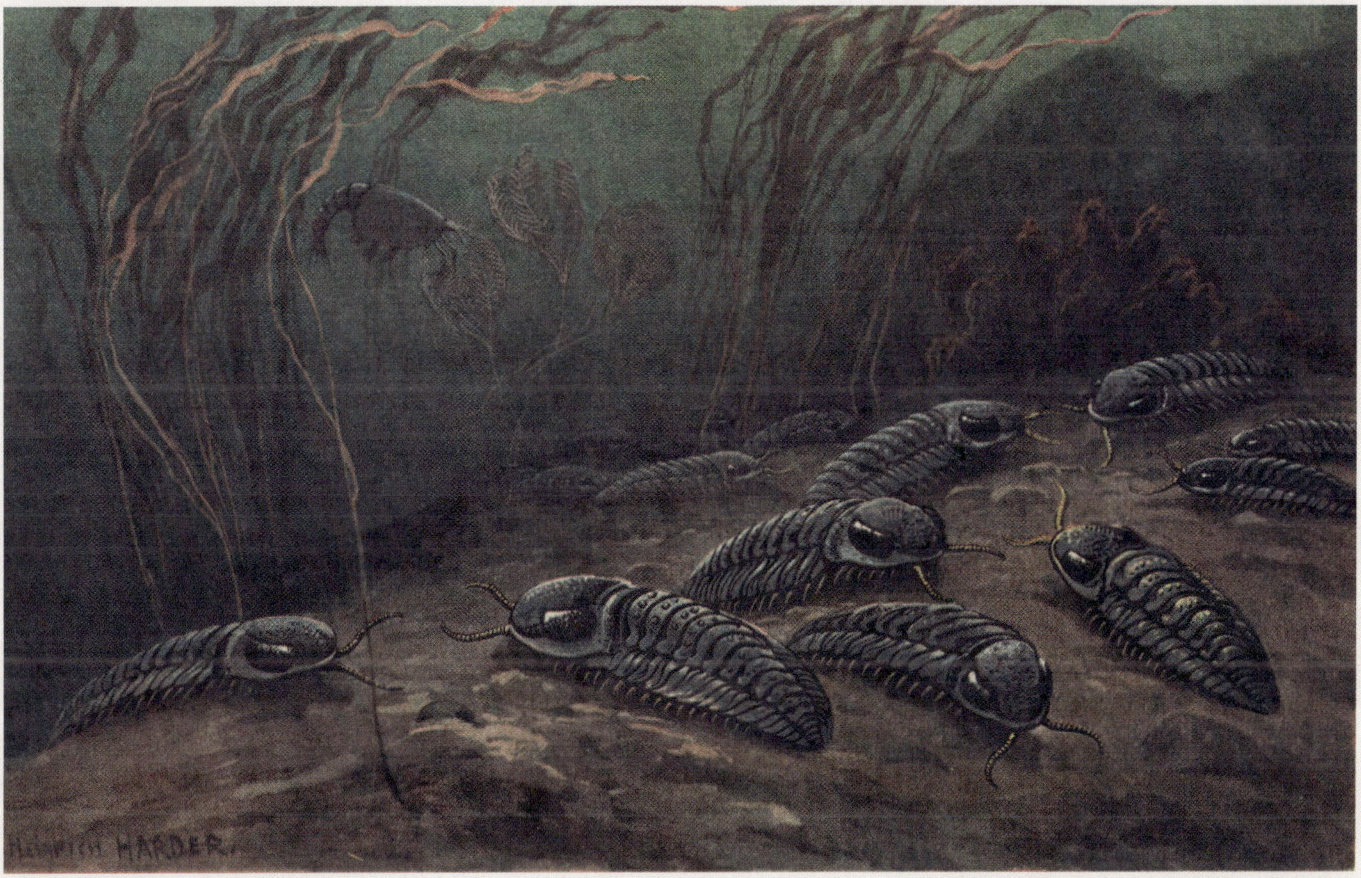
HEINRICH HARDER.

TREES OF LIFE— EVOLUTION AND BIODIVERSITY

In the 19th century, Charles Darwin and Ernst Haeckel invented and popularized a tree-shaped representation of the evolution of life. At the base of the trunk are the microorganisms of the origins, and the branches show the diversification of the different forms of life that followed. Every species evolves at its own pace, whether unicellular or multicellular.

The single-celled prokaryotes—bacteria and archaea—lie at the root of the tree of life. These microorganisms have been evolving for nearly 4 billion years, and their descendants survive to this day—exploring, exploiting, and shaping all available environments and resources. Bacteria and archaea thrive in diverse ecosystems, from soil and water to rock, sediment, and even the upper atmosphere and the deepest drill holes. By forming intricate biofilms and colonies, countless species interact and collaborate, creating resilient microbial communities. As they colonize and permeate all life-forms, bacteria and archaea constitute a living thread that connects all ecosystems.

One to 2 billion years ago, more advanced unicellular organisms emerged called *protists*, a type of eukaryote. They were chimeric cells with a nucleus, which resulted from symbiosis between bacteria and archaea. Since those primordial times, various single-celled entities—bacteria, archaea, and protists—have engaged in intricate interactions: merging, parasitizing, or appropriating components from one another. Dynamic gene exchanges have been further intensified by the presence of viruses, contributing to the rich tapestry of microbial evolution *(see chap. VI on viruses and p. 198)*. Finally, between 600 and 700 million years ago, protist-like organisms made up of different cell types appeared. These first multicellular organisms heralded the explosion in biodiversity of animals, plants, algae, and fungi and their associations with each other and with microorganisms.

For the past fifty years, researchers have been exploring the molecular information inscribed in genes to determine who is descended from whom, who has exchanged with whom, and when. These genetic analyses are producing ever-finer refinements to the tree of life, which now resembles an overgrown bush more than a tree. We are gradually realizing that all life-forms, including ourselves, are interconnected within a vast biosphere.

Communities of cells (bacteria, archaea, and protists with their cellulose, silica, and calcium envelopes) form layers of microorganisms called *biofilms*.

Within biofilms, microorganisms exchange proteins and genes and secrete molecules that aid in adhesion, protection, and communication. Over time, biofilms transform and organic matter is recycled. Under favorable conditions, biofilms of bacteria and archaea adhering to rocks have mineralized, becoming fossils such as stromatolites, the oldest traces of microbial life on Earth—dating back more than 3 billion years.

The theory of evolution and the first trees of life

Understanding how living beings are linked to one another on the basis of shared common traits is a favorite pastime of biologists. In the 18th century, the Swedish naturalist Carl Linnaeus revolutionized this tradition. He began systematically classifying and naming plants and animals in his *Systema Naturae* (1735).

A century later, Charles Darwin returned to England after a five-year expedition to South America aboard the *Beagle*. The expedition was dedicated to finding evidence of God's existence in a world thought to be less than 6,000 years old. Darwin discovered fossils on mountaintops and witnessed earthquakes, which led him to understand the role of geological upheavals and extinctions. He became convinced that the world was much older, and intuited a branching evolution of all living things from a common ancestor.

After the *Beagle*'s return to England in 1836, Darwin drew his first picture of the tree of life in one of his notebooks. He reflected on his observations and experiments and laid the foundation for his revolutionary theory of evolution through natural selection. In 1859, Darwin and his contemporary, Alfred Wallace, made a groundbreaking proposal: The forces of selection, be it environmental or through breeding, would inevitably favor the fittest individuals in a population. This applies to multicellular organisms such as animals and plants, but also to populations of single cells. So, when a population of bacteria is treated with antibiotics, only a few individual bacteria that develop molecular mechanisms of resistance to antibiotics survive. These mutant bacteria are thus naturally selected from the population, and their descendants form a new, resistant population.

At the end of the 19th century, many naturalists, and Ernst Haeckel in particular, became obsessed with depicting evolution by drawing trees of life and the anatomical characteristics of organisms. Haeckel published over a hundred different representations of trees of life until his death in 1905. Fascinated by biodiversity, Haeckel drew countless protists and animals. The plates published in *Kunstformen der Natur* (*Art Forms in Nature*, 1904) are elegant masterpieces celebrating the fusion of art and science.

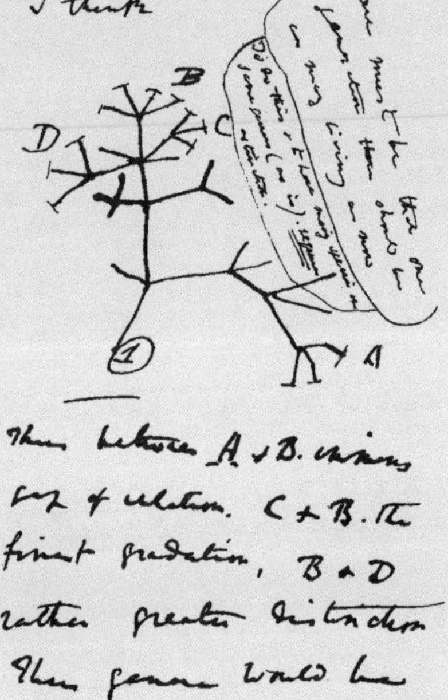

**CHARLES DARWIN DRAWS
THE FIRST TREE OF LIFE**

In 1837, Charles Darwin, author of *On the Origin of Species* (1852), sketched the famous "I think" in his notebook on the transmutation of species. This first conceptual drawing of the tree of life postulates that all life descends from a common ancestor. Life then diversified into many branches. Some of these forms are still around today, while others have become extinct (*the branches ending in bars labeled A, B, C, D represent the extinction of various groups*). The accompanying text, written by Darwin, states, "Case must be that one generation then should be as many living as now. To do this and to have many species in same genus (as is) requires extinction. Thus between A and B immense gap of relation. C and B the finest gradation, B and D rather greater distinction. Thus genera would be formed."

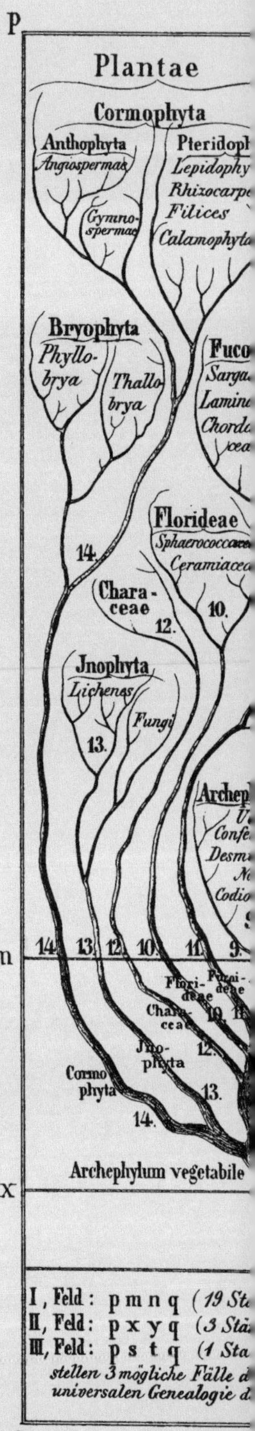

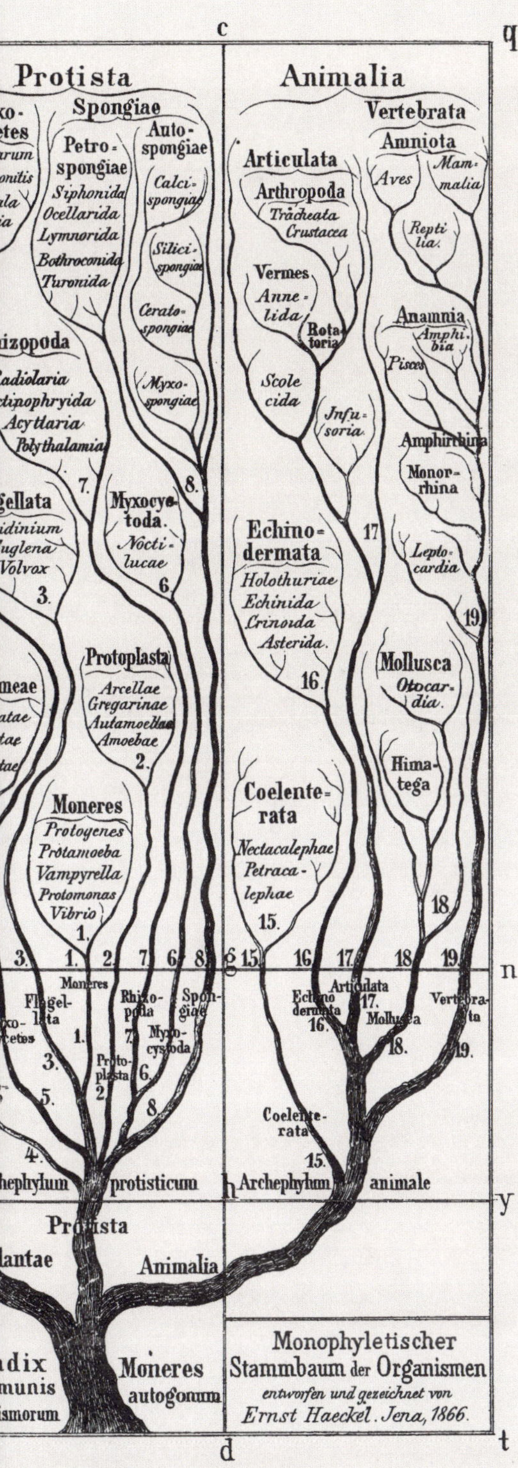

TREES OF LIFE IN THE 19th CENTURY, THE OBSESSION OF ERNST HAECKEL

Haeckel's drawings (this one dated 1866) illustrate the diversity of the animal and plant kingdoms and of single-celled beings—the protists—a name invented by Haeckel (inspired by the Greek *protos* meaning "very first"). These three kingdoms of life derive from a common ancestor at the base of the tree—the monera—a kind of primordial amoeba-like organism.

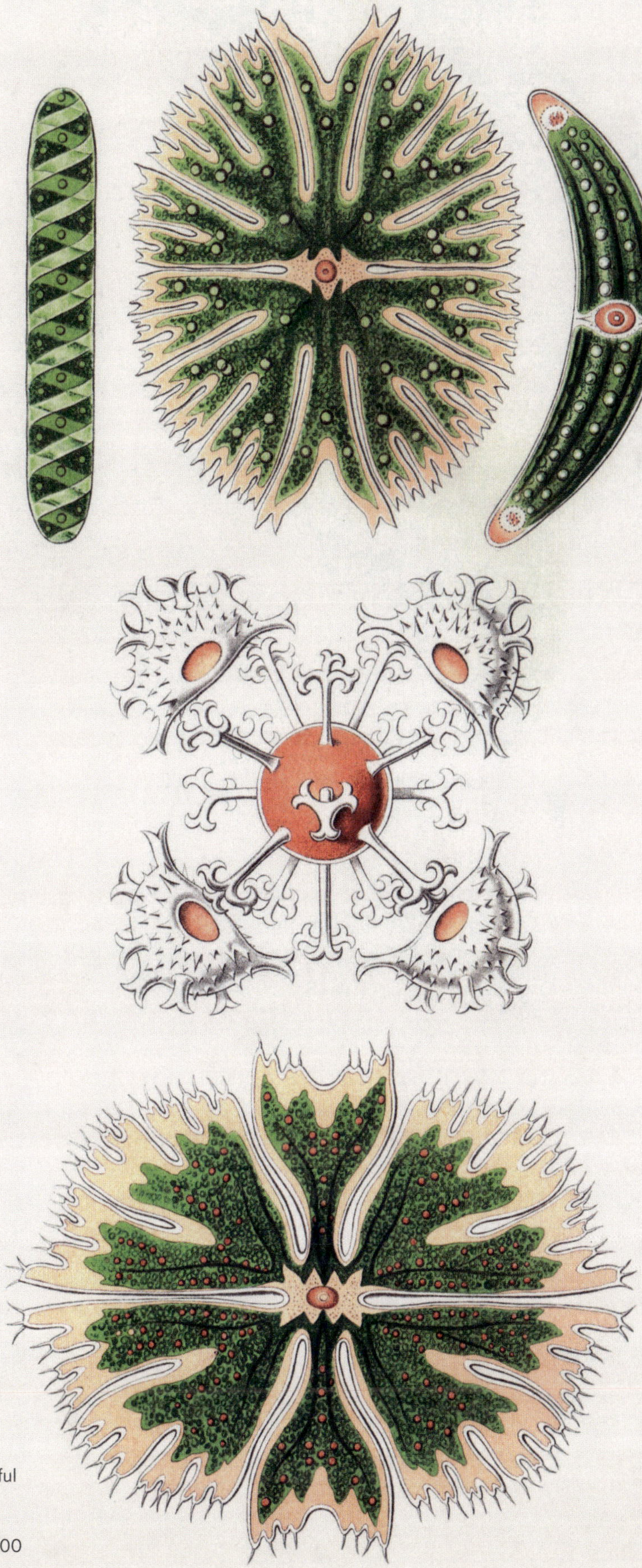

ERNST HAECKEL— THE INTERSECTION OF ART AND SCIENCE

Ernst Haeckel popularized the world of protists through his artful illustrations. Shown here are various *Desmidiaceae*, rich in greenish chloroplasts. Some 5,000 species of these microalgae are known to live in acidic freshwater.

From trees to webs— rethinking evolutionary connections

The tree of life representation of evolution and biodiversity was enriched with new branches and connections between the kingdoms of life from the 1980s onward. An influential book, *Five Kingdoms*, written by Lynn Margulis at this time, depicts five kingdoms of life:

1. prokaryotes (PROKAROTAE or MONERA)
2. protists (PROTOCISTA)
3. animals (ANIMALIA)
4. plants (PLANTAE)
5. mushrooms (FUNGI)

These trees of life include new proposals for relationships between the kingdoms of life. The main proposal is that eukaryotic organisms (protists, animals, plants, algae, and fungi) have inherited their energy-supplying organelles (mitochondria and chloroplasts) from bacteria. This integration process, known as *endosymbiosis*, was proposed in the late 1960s by Lynn Margulis, who drew inspiration from early Russian and French work that had fallen into oblivion. The theory of endosymbiosis struggled to gain recognition at first, but was gradually accepted thanks to evidence provided by microscopic and molecular analyses. This theory is the main explanation for the evolution of prokaryotic cells (bacteria and archaea) into unicellular eukaryotes (protists), and then into pluricellular ones (animals, plants, algae, and fungi).

It should be noted that archaea (ARCHAEBACTERIA in the diagram), a new type of prokaryotic microorganism discovered in 1977, were added to the base of the tree alongside bacteria (EUBACTERIA) in the latest editions of *Five Kingdoms*. The discoverers, Carl Woese and George Fox, showed that archaea constitute a third, new domain of life, distinct from bacteria and eukaryotes, by analyzing the gene sequences of microorganisms living in extreme environments. Over the last ten years, most researchers have adopted the view that archaea, by endosymbiosis, integrated and domesticated bacteria, which became the mitochondria of eukaryotic cells *(see chap. IV)*.

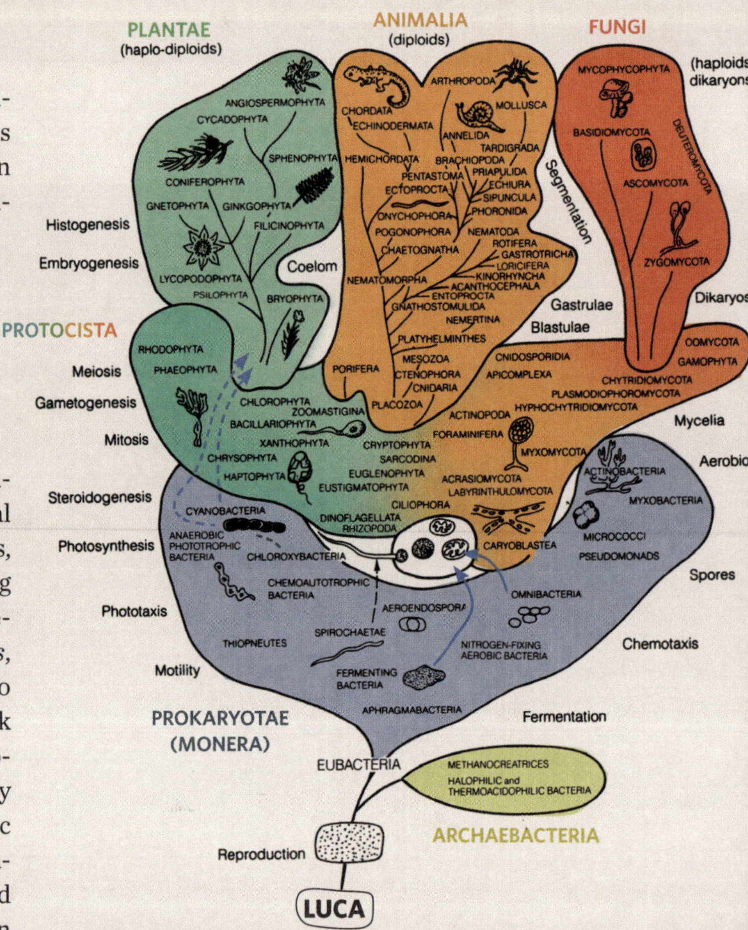

THE FIVE KINGDOMS OF THE TREE OF LIFE, WITH ARCHAEA AND ENDOSYMBIOSES

At the end of the 20th century, archaea (ARCHEABACTERIA) took their place at the base of the tree of life (where we have added the ancestral cell, LUCA). The contributions of bacteria (APHRAGMABACTERIA) and cyanobacteria (CYANOBACTERIA) are represented by dotted arrows, which we've amplified by adding blue arrows. These indicate that bacteria and cyanobacteria are the ancestors of the mitochondria and chloroplasts of eukaryotic cells (PROTOCISTA, PLANTAE, ANIMALIA, FUNGI). As early as 1967, the author of the book *Five Kingdoms*, American biologist Lynn Margulis (under the name Lynn Sagan; she was then married to the famous astrobiologist Carl Sagan), proposed that the mitochondria and chloroplasts of eukaryotic cells derive from endosymbiotic relationships with bacteria.

COLORIZED DIAGRAM

Based on the phylogenetic tree in the latest edition of *Five Kingdoms: An Illustrated Guide to the Phyla of Life on Earth* (1982 to 1988)

Trees of life in the 21st century are reshaped by molecular analysis

Since the late 20th century, analyses of gene and protein sequences have illuminated the evolutionary relationships among various phyla, genera, and species. Gene and protein sequences are subject to mutation and reorganization over time, which result in differences we can detect when two organisms' genomes are aligned and compared. These differences reveal the more- or less-distant kinship between those organisms. Sequence alignment comparisons allow us to place the analyzed organisms in relation to each other, in a tree structure with branches, nodes, and endings. Branch lengths represent the genetic distance between organisms, and nodes represent common ancestors. The root at the base of each phylum indicates the oldest common ancestor.

These sequence manipulations are based on in-depth knowledge of organism genomes. The first bacterial genome was sequenced in 1995, followed by that of yeast, a unicellular eukaryote, in 1997. Complete sequencing of the animal genome began with that of a small nematode worm, *Caenorhabditis elegans*, in 1998. Since then, the genomes of thousands of animals have been sequenced, but this represents less than 0.5 percent of living animal species. As for extinct species, just over a hundred have been analyzed at the genomic level. Two ambitious projects, the Darwin Tree of Life Project and the Earth BioGenome Project, plan to sequence hundreds of thousands of eukaryotic organisms, among the 1.8 million described species of protists, plants, algae, animals, and fungi. These projects focus on complex organisms. Other ongoing projects involve the mass sequencing of genes from all the organisms making up an ecosystem, known as *metagenomics*. These projects are building gigantic genetic databases of aquatic and terrestrial microorganisms, as well as those making up the microbiota of animals, plants, algae, and fungi. Together, these initiatives are producing extraordinary collective resources for understanding and preserving biodiversity in the context of climate and environmental disruption.

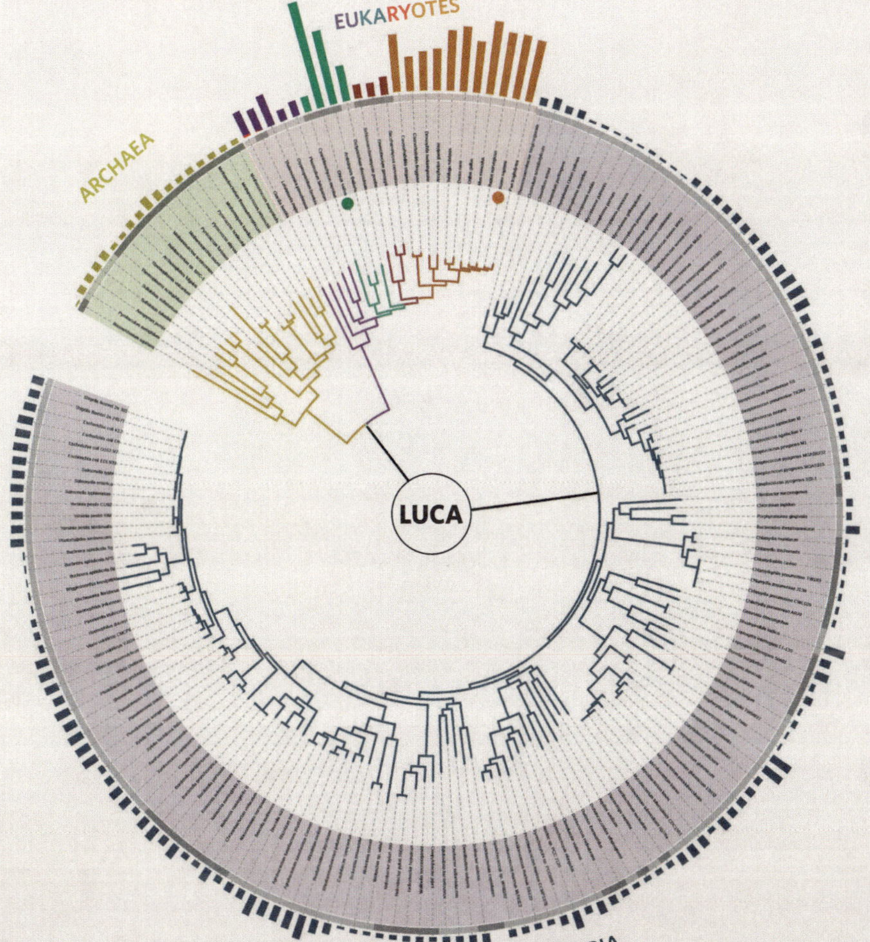

MOLECULAR PHYLOGENETIC TREE
Since 2005, the European Molecular Biology Laboratory (EMBL) has been offering interactive graphic representation tools (*iToL interactive Tree of Life*: see https://itol.embl.de). The phylogenetic tree in this graph is based on the sequenced genomes of 240 prokaryotic and eukaryotic organisms. We have added the presumed position of the ancestral cell LUCA in the center, and colored the graph so that the organisms' colors correspond to those on pages 30 and 33: From purple to green to ocher, here are about twenty eukaryotes (plants, animals, fungi, and protists) whose relative genome sizes are represented by the heights of the bars. The ocher dot indicates the location of the human species (*Homo sapiens*) and the green dot, that of rice (*Oryzia sativa*). In grey are 200 species of bacteria. In yellow, about twenty species of archaea.

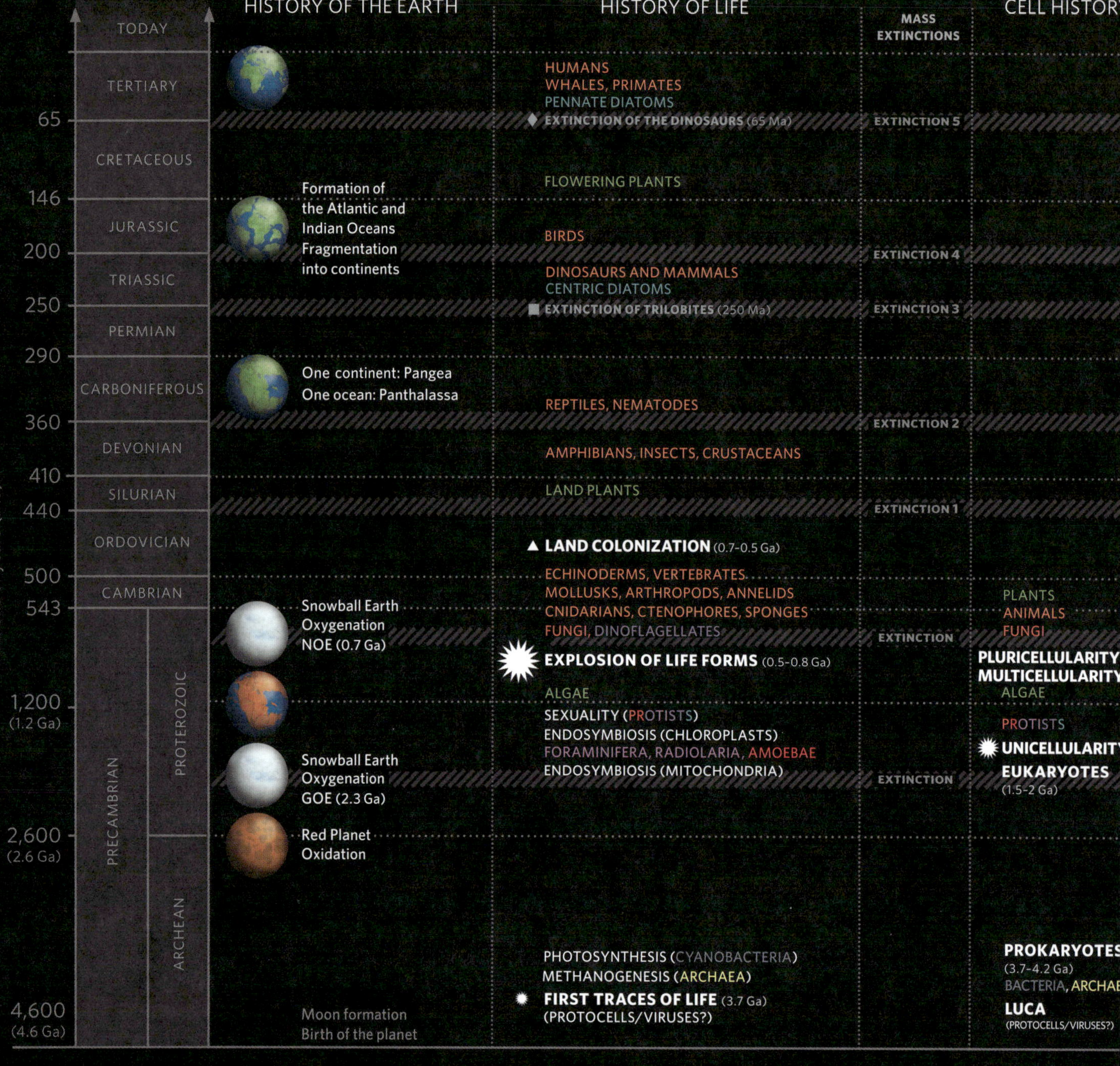

The evolution of the planet and of life

The first traces of life date back at least 3.7 billion years (Ga). The ancestral cell LUCA (last universal common ancestor) is at the origin of all known forms of life. Two domains of life (bacteria and archaea, the prokaryotes) appeared first, and viruses were probably already present. One type of bacteria (cyanobacteria) invented photosynthesis, making it possible to derive energy from the sun. These cyanobacteria generated considerable quantities of oxygen, causing a first massive oxygenation event (GOE) and subsequent oxidation of iron (Red Planet), followed by cooling and glaciation (Snowball Earth). This deep freeze wiped out many original life-forms. As the planet warmed up, endosymbiosis involving bacteria and archaea led to the emergence of more complex cell types (eukaryotes), endowed with nuclei and other organelles such as mitochondria and chloroplasts (for plants).

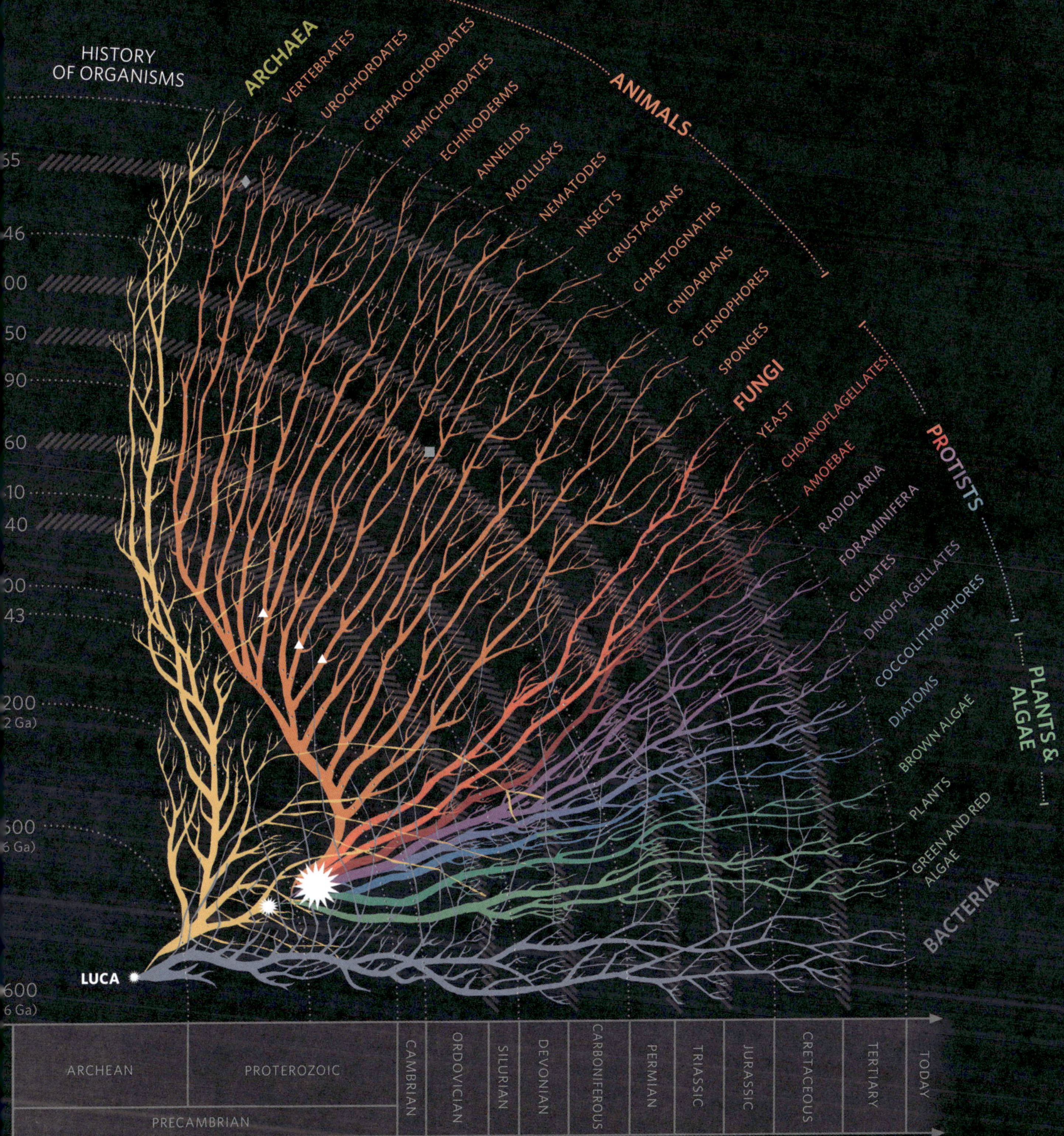

The first eukaryotic organisms were unicellular; they are the ancestors of animal protists (radiolaria, amoebas, yeasts, choanoflagellates, ciliates, etc.) and plant protists (dinoflagellates, diatoms, coccolithophores, etc.). Then, multicellular protists appeared, evolving into pluricellular organisms with different cell types, which are the ancestors of plants, animals, and fungi. Explosions of life (✹), probably due to a second great oxygenation episode (NOE) and the formation of a protective ozone layer (600 to 700 million years ago), gave rise to macroscopic pluricellular life forms. Many types of plants, fungi, and animals colonized the earth (▲). In the 500 million years since the Cambrian, five major mass extinctions have occurred (⁄⁄⁄). These extinctions led to the demise of numerous life-forms, including trilobites (250 million years ago ■) and dinosaurs (65 million years ago ◆). While some life-forms disappear, others continue to diversify after each extinction.

THE MATERIALS OF LIFE

DNA, RNA, proteins & nanomachines

H
H O
H
O
H
H
O
H
H
O
H
H
O
H
H
O
H
H
O
H
H
O
H
H
O
H
H
O
H
H
O
H
H
O
H
O
H
Cl
Ca
Ca
Na

ALL ELEMENTS ORIGINATE FROM STARDUST

The molecules and minerals that constitute living matter are made up of atomic elements born since the Big Bang, 13.7 billion years ago. It all began in the minutes following the Big Bang with the genesis of hydrogen, the primordial and most abundant element in the cosmos. Along with helium, hydrogen is the source of all other elements, from carbon to uranium, oxygen, nitrogen, and gold.

The elements we know today were born and still continue to be born from the transmutation of hydrogen and helium within stars, which are cauldrons of thermonuclear reaction. Projected into space, these elements ("stardust") clumped together into particles, meteorites, asteroids, and planets. Hydrogen (H) combined with oxygen (O), giving rise to water (H_2O), the essential solvent for minerals, molecules, macromolecules, and cells.

All living beings are composed of stardust, as the fundamental molecules of life are intricate assemblies of atoms, forged in the depths of the cosmos. Out of the hundred or so elements in existence, only a quarter are essential for life. The molecules of life, made from these elements, coalesced into ancestral cells, either somewhere in the universe or on Earth itself. And, over the past 4 billion years, these self-organizing molecular structures have evolved, giving rise to the immense diversity of organisms that figure in our planet's vast ecosystems.

Out of all the existing elements, six—carbon, hydrogen, nitrogen, oxygen, phosphorus, and sulfur, collectively known as the CHNOPS—are the main elements whose atoms combine into molecules of life. Among them, carbon occupies a special place. It constitutes the backbone of amino acids, nucleotides, lipids, sugars, and metabolites, the elementary building blocks of all organic matter.

Amino acids link together and combine to form proteins, and nucleotides pair up and wind into DNA and RNA helices. Lipids, which are fatty molecules, arrange themselves into hydrophobic (water-repelling) chains, spontaneously assembling into membranes that enclose and safeguard cells and their organelles. These molecular components do not function in isolation. Instead, they interact, transform, and collaborate to sustain the intricate processes of life.

All atomic elements originate from hydrogen (H) and helium (He), generated immediately after the Big Bang 13.7 billion years ago. They remain the most abundant elements in the cosmos.

Thermonuclear reactions in stars transform hydrogen and helium into about a hundred other elements, a quarter of which are essential for life.

The atmosphere is composed of 78 percent nitrogen (N) and 21 percent oxygen (O). It also contains a small amount of carbon (C) in the form of CO_2.

Living organisms are particularly rich in C, H, N and O, depicted here in fish, humans, and trees.

The continental crust is rich in oxygen (O), silica (Si), and aluminum (Al). The oceans are essentially composed of water (H_2O) molecules in which minerals are dissolved.

BiGBANG
He

The elements of life— from stars to the ocean

Since the origin of the universe, about one hundred elements have been forged in the furnace of dying stars. Atoms, the building blocks of matter, are dynamic assemblies of even smaller building blocks: protons, neutrons, and electrons. Smallest are the elementary particles with their strange names: bosons, fermions, leptons, quarks, etc.

Living matter consists of only a quarter of the existing elements, combined into an infinite variety of molecules of life, from simple ones like glucose, made of a few dozen atoms, to macromolecules such as proteins, containing hundreds to thousands of atoms. Some elements, such as hydrogen, oxygen, and carbon, are abundant, while others, like zinc and selenium, are present only in trace amounts.

We can wonder why the molecules of life use just over twenty of the hundred-plus available elements figuring on the famous periodic table, and not the others. The answer probably lies in the fact that these elements were abundant in the primitive ocean, where life developed over 3 billion years ago.

Hydrogen—fuel of the universe and of life

The name *hydrogen* comes from the Greek for "water generator." Antoine Lavoisier coined the term in 1783, because the explosive mixture of oxygen (O_2) and hydrogen (H_2) generated water (H_2O), wetting the walls of his container. Lavoisier's mixture revealed the nature of a chemical reaction for the first time.

Hydrogen, with a single proton and one electron, is the lightest and most elementary of all the elements. Hydrogen was born immediately after the Big Bang and remains the most abundant element (73 percent) in the cosmos, the one from which all other elements are derived inside the stars. Hydrogen is an inexhaustible source of energy. Every second, the sun burns 600 million tons of hydrogen into helium by thermonuclear fusion. The sun emits energy across the electromagnetic spectrum, including high-energy gamma rays carried by photons. Photons of visible light are captured by photosynthetic cells, enabling them to produce energy and organic matter, and nourish other organisms in turn. Hydrogen is truly the fuel of life, because in its ionic form (protons, denoted H^+), it is at the heart of energy metabolism. We will come back to this important topic in chapter VII *(see p. 126)*.

THE DEATH OF A STAR GENERATES ATOMIC ELEMENTS

The year-long evolution of the Monoceros nebula, photographed by the Hubble Space Telescope. In the center, the red giant star V838 Monocerotis, 20,000 light-years from Earth, is in the process of exploding, sending some of its matter into space, loaded with atoms of carbon, oxygen, nitrogen, and heavier elements.

THE ELEMENTS OF LIFE AND SEAWATER

The relative abundance of each element in seawater is represented by the height of the bars (on a logarithmic scale) from the lightest element, hydrogen (H) at top left, to the heaviest elements at the bottom, such as gold (Au) and platinum (Pt). The most abundant elements in cells and living organisms (the CHNOPS and the minerals Na, K, Ca, Mg, and Cl) are shown in yellow. Trace elements considered essential to life are in green (V, Mo, Fe, Co, Cu, Zn, B, Si, Se, and I). A few additional elements, shown in pink, play a role in the survival of particular organisms.

Published in *Philosophical Transactions of the Royal Society* A 373: 20140188 (2015) © Courtesy Rosalind Rickaby, Oxford University

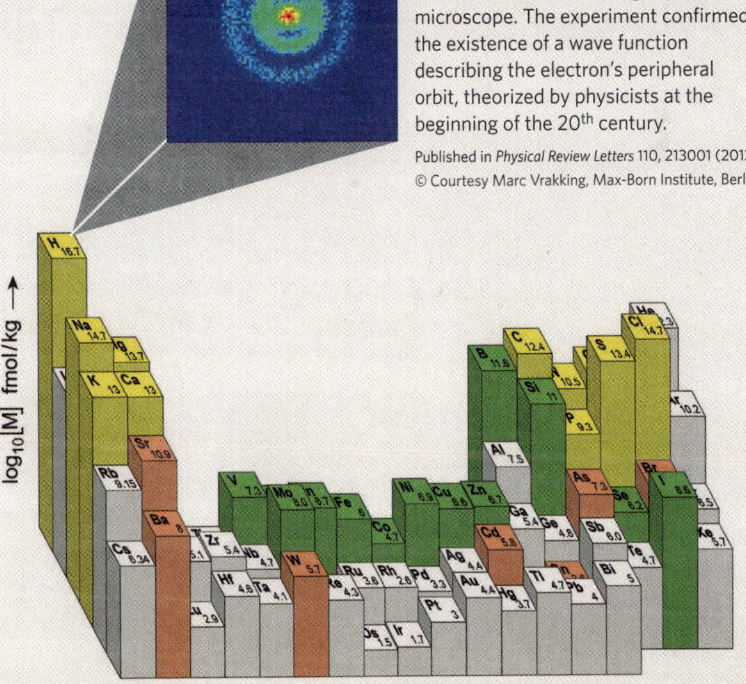

ANATOMY OF THE HYDROGEN ATOM

The hydrogen atom was observed for the first time in 2017 using a quantum microscope. The experiment confirmed the existence of a wave function describing the electron's peripheral orbit, theorized by physicists at the beginning of the 20th century.

Published in *Physical Review Letters* 110, 213001 (2013)
© Courtesy Marc Vrakking, Max-Born Institute, Berlin

Carbon—the all-purpose element

Carbon is the fourth most abundant element in the universe, after hydrogen, helium, and oxygen. Along with oxygen and hydrogen, carbon is a primary element in the molecules of life. The carbon atom is exceptional in that it establishes stable bonds—called *covalent* bonds—with a large number of other elements. Thus, among the 10 million known molecules, 95 percent are carbon-containing molecules.

Carbon atoms can associate with each other in many different formations. In coal, soot, and ashes, they attach to each other in a disorderly fashion; whereas in diamonds, they are arranged in crystals.

Thanks to its chemical versatility, the carbon atom is at the center of all life. Not only do carbon atoms constitute the backbone of most molecules of life by forming innumerable combinations with other atoms, but carbon dioxide (CO_2) provides the carbon atoms for living matter. Although representing only a tiny fraction of atmospheric gases (less than 1 percent), CO_2 is the elementary building block from which living organisms manufacture sugars, fats, amino acids, nucleotides, and all organic matter. Notably, carbon is fixed in organic matter via oxygenic photosynthesis—a metabolic process that produces oxygen. Photosynthesis uses light energy to drive a reaction by which plants convert CO_2 into glucose. Water serves as the electron donor molecule in this process, releasing oxygen as a by-product.

The carbon cycle—the recycling of carbon atoms—is at the heart of the climate system. Most of this cycle takes place between the atmosphere, soil, oceans, and living organisms, which exchange carbon via natural processes such as respiration, photosynthesis, and decomposition.

Oxygen—combustion and respiration

Oxygen is the gas of combustion. Thanks to oxygen, hydrocarbon molecules such as methane (CH_4) burn, heating stoves and boilers. This combustion produces carbon dioxide (CO_2) and water molecules (H_2O) in the form of gas and vapor.

Oxygen is associated with respiration. Animal and plant cells alike need to breathe oxygen to oxidize nutrients (sugars, amino acids, or fats) in order to obtain energy, while exhaling carbon dioxide and water.

A vital gas, oxygen can also be lethal. For bacteria that flee oxygen—called *anaerobic* bacteria, oxygen is toxic. Immune cells in our body exploit this toxicity by secreting molecules called super-oxidants, such as hydrogen peroxide (H_2O_2) and other ROS (reactive oxygen species). This oxidizing chemical arsenal annihilates pathogenic microbes, which are then ingested and eliminated by specialized cells of the immune system. Our cells protect themselves from ROS-type oxidizing molecules using enzymes and antioxidant molecules, such as vitamins C and E and melatonin, the sleep hormone.

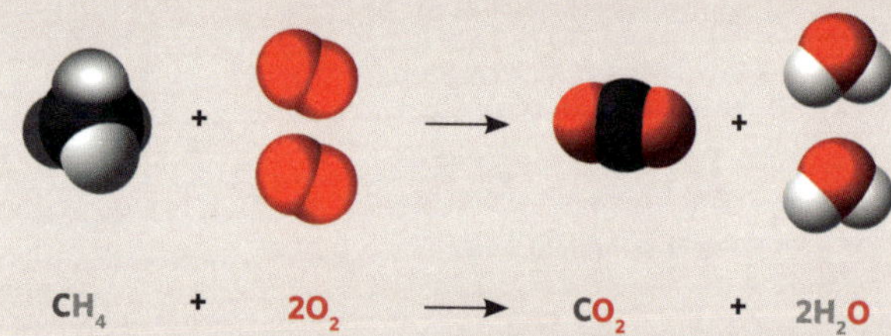

CHEMICAL COMBUSTION OF METHANE BY OXYGEN

Combustion of methane (CH_4) with oxygen (O_2) produces carbon dioxide (CO_2) and water (H_2O).

PHOTOSYNTHESIS FIXES CARBON FROM CO$_2$ AND PRODUCES GLUCOSE AND OXYGEN

From six molecules of carbon dioxide (CO_2) and six molecules of water (H_2O), photosynthesis produces one molecule of glucose ($C_6H_{12}O_6$) and six molecules of oxygen (O_2).

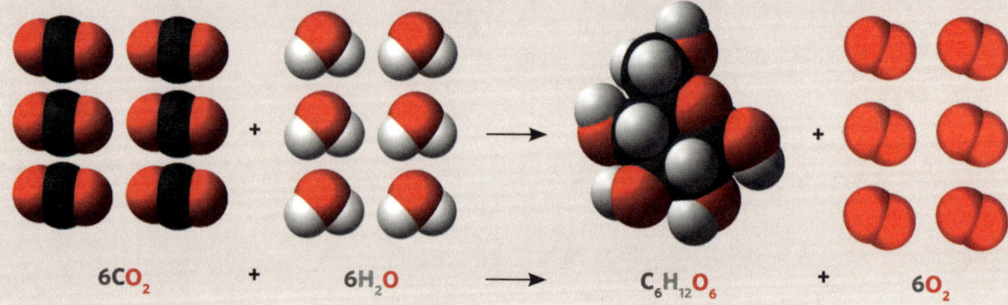

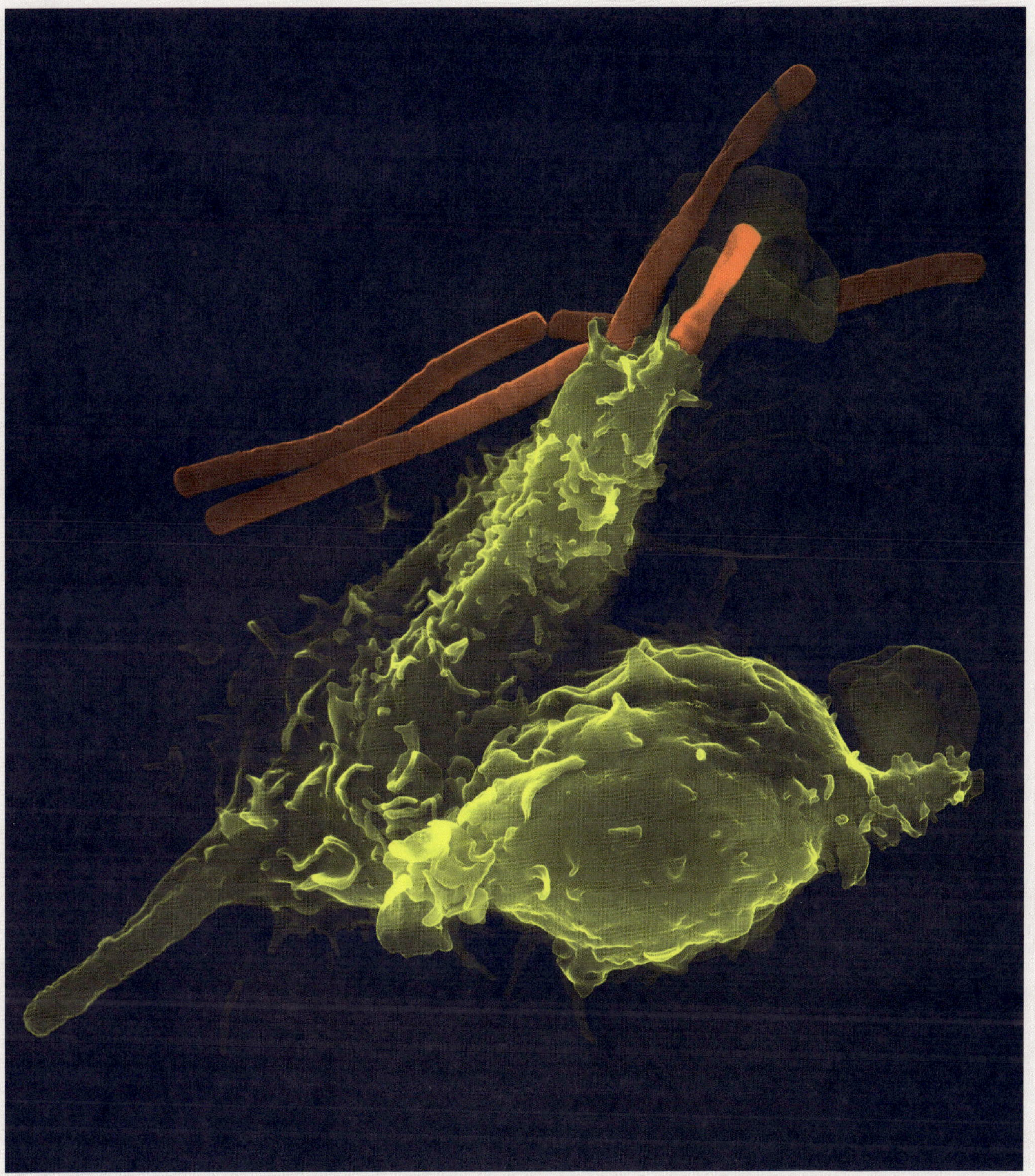

SUPER-OXIDATION BY IMMUNE CELLS
A neutrophilic blood cell (*yellow*) ingests *Bacillus anthracis* (anthrax, *brown rods*) in a process known as *phagocytosis* and kills them by secreting super-oxidants such as hydrogen peroxide or other ROS.

SCANNING ELECTRON MICROSCOPY
© Courtesy Volker Brinkmann, Max Planck
Institute for Infection Biology, Berlin

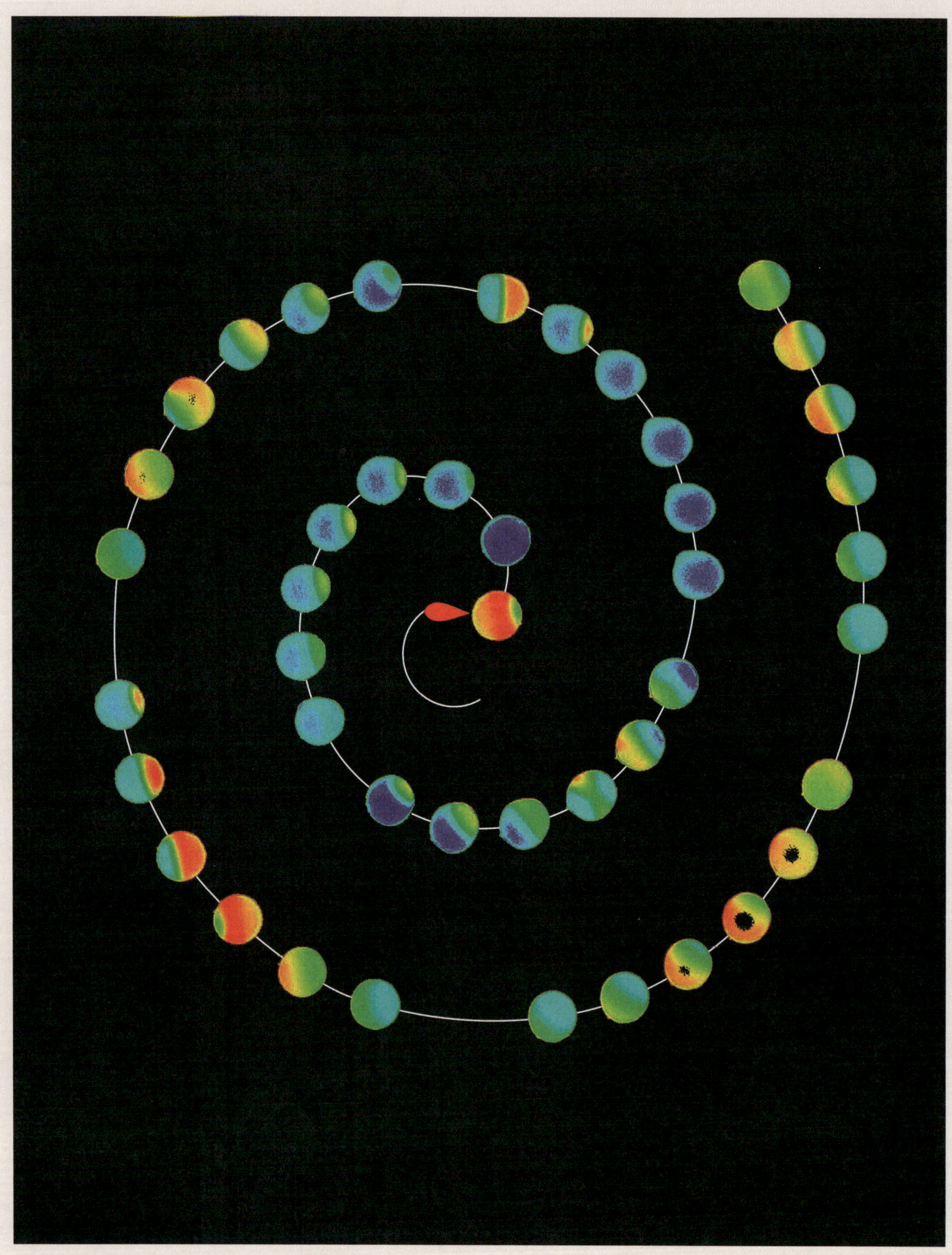

Vital minerals and metals

The interior of cells is a dense soup of molecules, macromolecules, organelles, biomolecular condensates, and constituents of the cytoskeleton. All bathe in an aqueous medium loaded with millions of small molecules and mineral elements in the form of ions of different sizes and charges.

Cells, thanks to their membranes, maintain an internal environment very different from the exterior. The interior of cells is richer in potassium ions (K^+), lower in sodium ions (Na^+) and chlorine (Cl^-), and extremely low in calcium ions (Ca^{2+}). Differences in ion concentration between inside and outside are maintained by pump proteins and channel proteins in the cell membrane, and also by the network of internal membranes and membranous organelles ubiquitous in eukaryotic cells. The different ion concentrations generate an electrochemical potential difference of the order of tens of millivolts across the membrane delimiting the cell. This electrochemical gradient is particularly important for generating energy *(see chap. VII)*.

Some minerals, such as magnesium (Mg^{2+}), manganese (Mn^{2+}), iron (Fe^{2+}), and iodine (I^-) are essential partners for macromolecules and enzymes. Hemoglobin—the protein that gives red blood cells their color—is a good example. One hemoglobin protein integrates four ferrous ions (Fe^{2+}), necessary for capturing and releasing oxygen into blood and tissues. As for thyroid cells, they pump and fix another mineral—iodine—essential for the functioning of thyroid hormones that regulate cellular metabolism in animals.

The most influential of all minerals for life is undoubtedly calcium. Cells need calcium—in the form of carbonates and phosphates—to build skeletons, shells, carapaces, and organs that regulate balance. Most importantly, calcium ions (Ca^{2+}) are used as a universal cellular activator. Cells maintain an extremely low calcium concentration in their cytoplasm. However, when certain hormones target a cell, or a spermatozoon fertilizes an oocyte, they trigger a flood of calcium ions. The calcium flood spreads rapidly from the stimulated area and invades the cell in one or more successive waves *(see p. 144)*. Such calcium tsunamis activate countless proteins and enzymes, awakening resting cells and regulating their metabolism, gene expression, secretions, communications, and mobility. Calcium waves also propagate through adjacent tissue cells via special openings known as *cell junctions (see p. 204)*. Calcium waves and detonations also underlie brain waves, heart contractions, root and shoot growth, and many other processes that involve stimulating a population of cells.

CALCIUM WAVES PROPAGATE THROUGH CELLS AND TISSUES

Waves of increasing calcium ion concentration (the highest concentrations of calcium ions are in *yellow* and *red*) propagate from the point of entry of a spermatozoon (*shown in center*) fertilizing an oocyte. From the center to the periphery, seven successive waves of calcium increase traverse the fertilized oocyte. The oocyte has been injected before fertilization with a molecule that changes its fluorescence according to the intracellular concentration of calcium ions.

CONFOCAL FLUORESCENCE MICROSCOPY
© Christian Sardet, Alex McDougall, Rémi Dumollard, Zoological Station, CNRS, Sorbonne University, Villefranche-sur-Mer

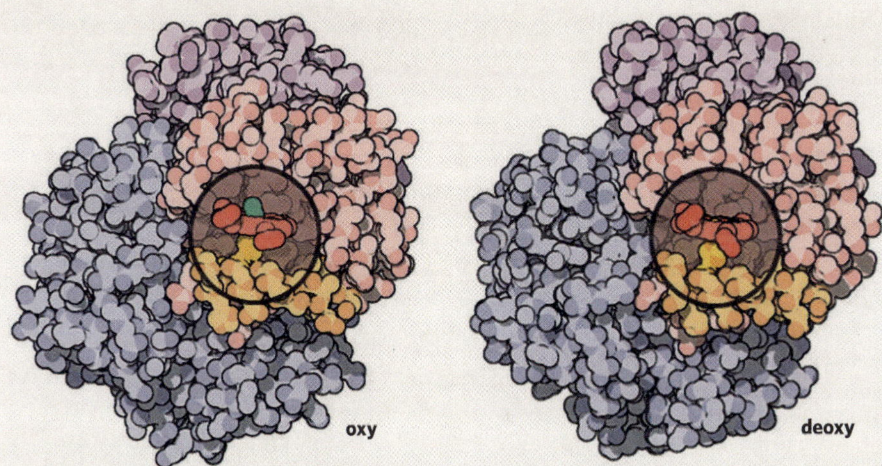

A HEMOGLOBIN PROTEIN NEEDS IRON

Hemoglobin is made up of four protein chains (*represented by different colors*). Each protein chain surrounds a small organic molecule (a heme molecule, *in red* inside the circle) incorporating an iron atom necessary for the capture of an oxygen molecule (*in blue*, oxy form *on the left*). By changing their shape, the protein chains release this oxygen into the blood and tissues (deoxy form *on the right*). Hemoglobin thus takes up oxygen in the lungs (oxy form) and releases oxygen in the tissues (deoxy form).

© Artwork by David Goodsell, for the RCSB PDB "Molecule of the Month"

WATER— UNIVERSAL SOLVENT AND ELIXIR OF LIFE

The universe is full of hydrogen gas (73 percent) and helium (25 percent), and also contains immense quantities of vaporized water molecules. Water vapor coats everything from cosmic dust to icy meteorites and frozen planets. Water is abundant in space and likely reached early Earth through torrential rains and impacts from icy meteorites carrying various molecules synthesized in the cosmos. Perhaps this ice even brought prototypes of cells (protocells) that could have seeded the early Earth.

Water (H_2O) is a simple molecule made up of two hydrogen atoms linked to one oxygen atom by strong (covalent) bonds. Water molecules are organized relative to each other in a compact, orderly fashion in ice, and in a fluid fashion in liquid water. In the vapor state, water molecules are spaced farther apart. Liquid water is an extraordinary solvent, a fluid within which a host of minerals, molecules, and macromolecules interact. In water, dissolved molecules attract, repel, and sometimes transform each other through chemical reactions facilitated by the solvent or other dissolved matter.

Since the beginning, the development of life on Earth has been closely linked to the presence of liquid water. Earth's ability to retain a surface layer of liquid water is due to its fortunate combination of a temperate climate and low atmospheric pressure. In addition, the circulation of water between oceans, seas, and rivers has favored the emergence, evolution, and dispersion of life in all its biodiversity.

The presence of liquid water and the sustenance of life are delicately maintained by Earth's stable climate. Will this change? It has previously—2 to 2.5 billion years ago, and 600 to 700 million years ago—when the oceans and the Earth froze en masse, exterminating the great majority of living beings (during the GOE and NOE periods, *see pp. 22-23*). Since then, other glaciations and cataclysms have contributed to less-drastic and more-localized extinctions. Each time, when favorable conditions were reestablished, life took off again.

With increasing efficiency and sophisticated technologies, space explorations look for liquid water on other planets. Sometimes we find credible traces of water, as on Mars—traces that encourage our fantasies of possible extraterrestrial life.

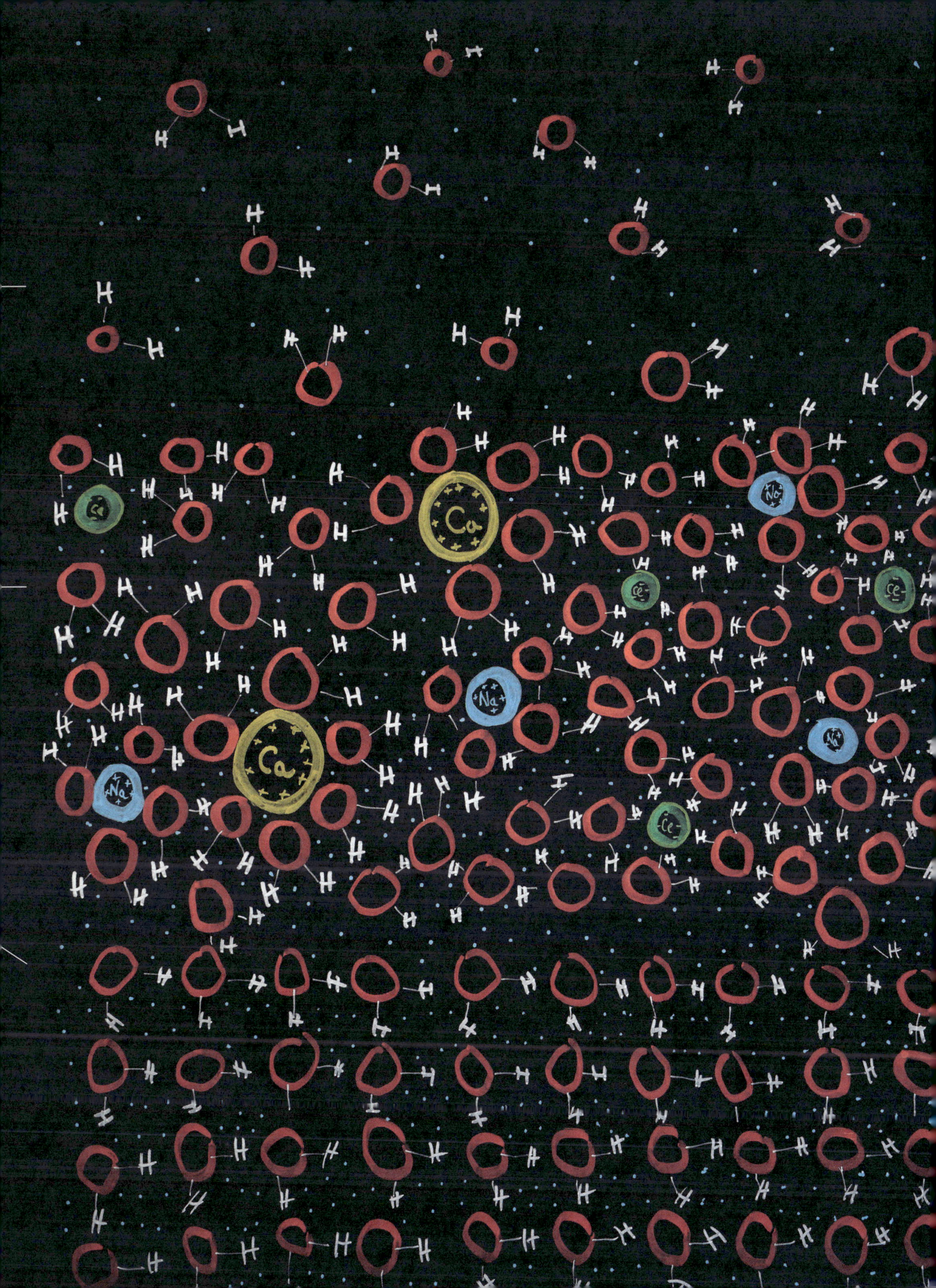

The molecular Legos of life

The universe abounds in reactive molecules and the building blocks of life—a kind of molecular Lego set.

Many molecules of life were assembled in the universe and on our planet from simpler, chemically reactive molecules—hydrogen, oxygen, water, carbon monoxide and dioxide, ammonia, methane, hydrogen sulfides and cyanides, etc. A variety of terrestrial environments—at the bottom of the oceans, and within geysers or hydrothermal springs—would have been ideal settings for countless chemical reactions to take place. These reactions were facilitated by the adsorption of molecules on the metallic and semicrystalline surfaces of pyrite rocks, clays, and micas *(see chap. III)*.

carbon monoxide
CO

water
H_2O

carbon dioxide
CO_2

hydrogen sulfide
H_2S

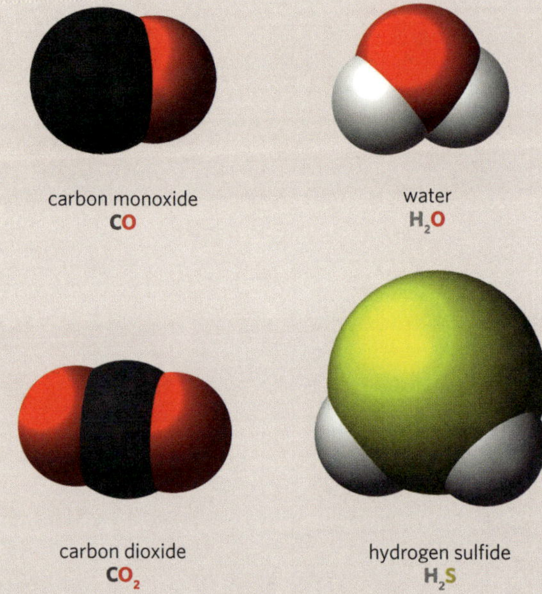

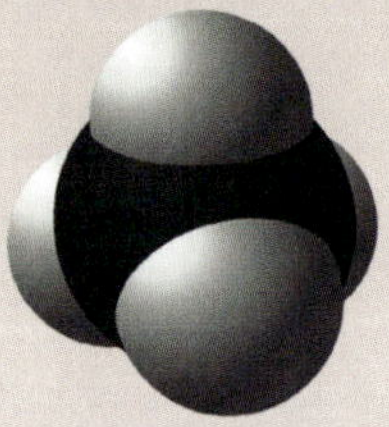

methane
CH_4

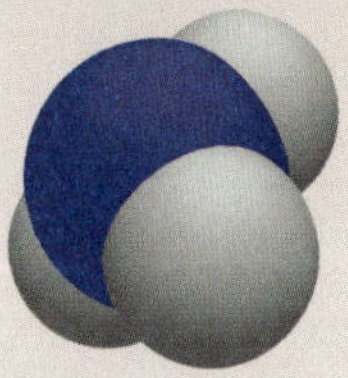

ammonia
NH_3

BASIC MOLECULES

Carbon monoxide (CO) and dioxide (CO_2), water (H_2O), hydrogen sulfide (H_2S), methane (CH_4), and ammonia (NH_3) are represented here in compact form (as space-filling models), taking into account the relative sizes of the atoms that compose them and their conventional colors (*white for hydrogen, red for oxygen, blue for nitrogen, black for carbon, yellow for sulfur*).

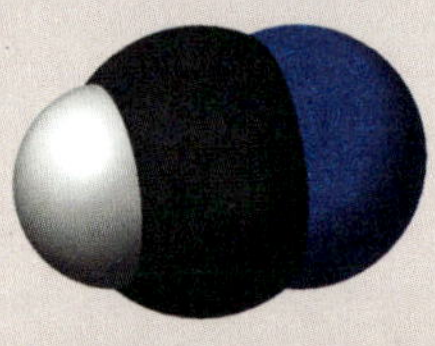

hydrogen cyanide
HCN

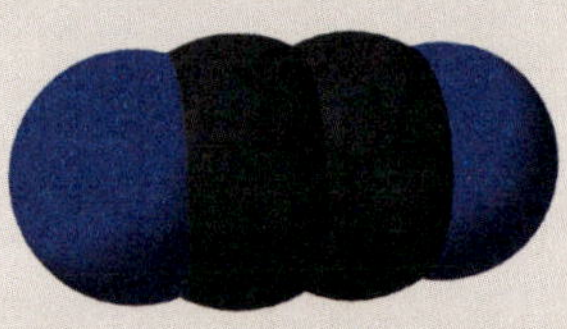

cyanogen
C_2N_2

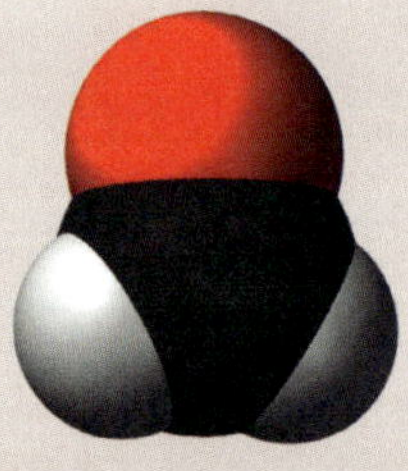

formaldehyde
CH_2O

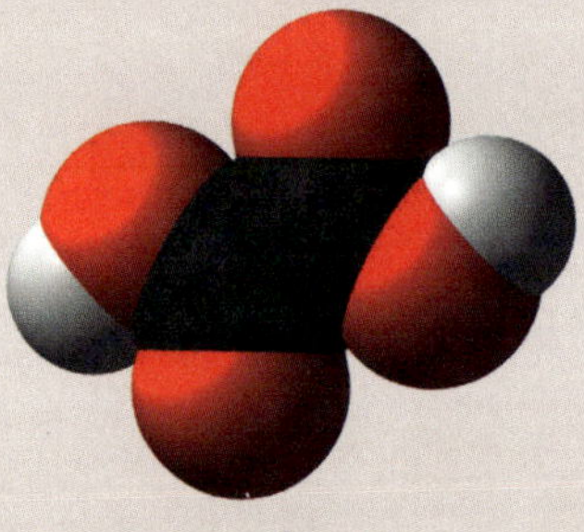

oxalic acid
$C_2H_2O_4$

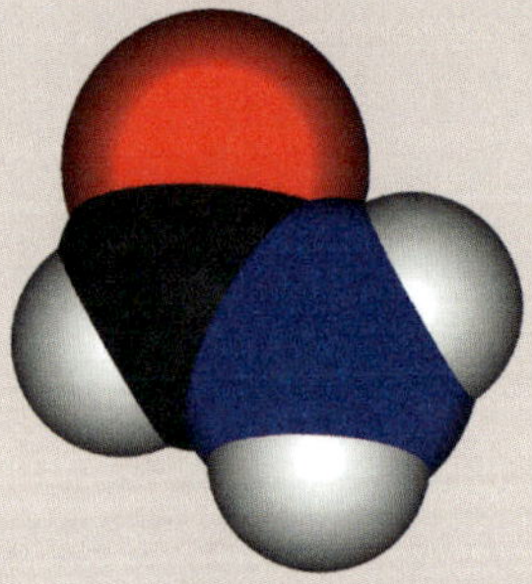

formamide
CH_3NO

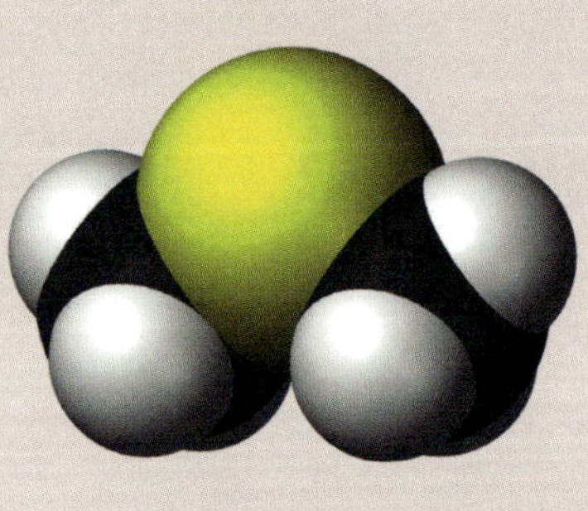

dimethyl sulfide
$(CH_3)_2S$

REACTIVE MOLECULES

The reactive molecules represented are hydrogen cyanide (HCN), cyanogen (C_2N_2), formaldehyde (CH_2O) and formamide (CH_3NO), oxalic acid ($C_2H_2O_4$), and dimethyl sulfide $(CH_3)_2S$.

1 nm

Elementary molecular building blocks

The "molecules of life" we've been discussing—the building blocks of cells and organisms—are sugars, amino acids, nucleotides, lipids, and metabolites. Some were already present in the cosmos and on our planet at the time of the origins, more than 4 billion years ago.

Certain of these molecules (twenty amino acids, five nucleotides, and a few sugars) became the building blocks of polymers that constitute the *macro*molecules of life: peptides, proteins, nucleic acids, and polysaccharides. As for the fatty molecules (lipids), they have the ability to spontaneously associate, to form a variety of micelles and membranes that coat and protect other molecular and cellular entities *(see next page and pp. 66–67)*.

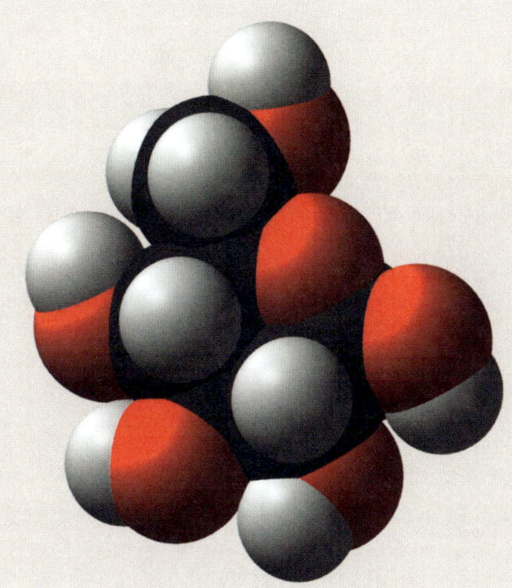

glucose
$C_6H_{12}O_6$

MOLECULES OF LIFE
A sugar (glucose) rich in oxygen (*in red*); two nucleotides (ATP or <u>a</u>denosine <u>t</u>riphosphate, and adenine), both rich in nitrogen (*in blue*) and with ATP also rich in phosphates (*orange*); an amino acid (cysteine) containing a sulfur atom (*yellow*); and a fatty acid (linoleic acid), a lipid consisting of a long hydrophobic (water-repelling) hydrocarbon chain (*carbon in black*) and a hydrophilic head containing oxygen atoms (*in red*).

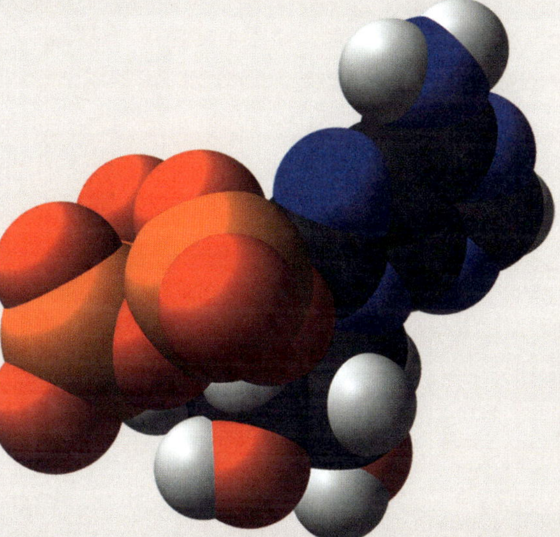

adenosine triphosphate (ATP)
$C_{10}H_{16}N_5O_{13}P_3$

cysteine
$C_3H_7NO_2S$

adenine
$C_5H_5N_5$

linoleic acid
$C_{18}H_{32}O_2$

Membranes— bubbles of life made of fatty molecules

Fat molecules (lipids) are very rich in carbon and hydrogen atoms that combine into hydrocarbon chains ten to thirty units long ($-CH_2-CH_2-CH_2-CH_2-$). The natural tendency of these nonpolar hydrocarbon chains is to band together closely to escape the polar molecules of water. The presence of water, therefore, forces the lipid molecules to organize themselves collectively into balls (micelles) or bubbles (membrane vesicles, *see pp. 66–67*), with the water-fearing hydrocarbon chains on the inside, and the water-loving ends of the lipids—the polar heads—on the outside. Inside the lipid membrane of a vesicle, the hydrocarbon chains face each other in a thin, double layer that acts as a semipermeable barrier to ions and hydrophilic molecules. This lipid bilayer is 3 to 4 nm (nanometers) thick, making the membrane 50,000 times thinner than a hair. This incredibly thin, insulating layer creates a selective barrier that serves to define the internal environment of a membrane vesicle or cell.

Innumerable proteins are embedded in the membrane delimiting a cell (the plasma membrane). Some of these membrane proteins penetrate and traverse the lipid bilayer. These proteins include pumps, channels, and receptors. They control exchanges and communications between the outside and inside of the cell by pumping, filtering, or receiving ions and molecules, such as hormones *(see chap. VIII)*. In addition, many proteins and lipids exposed on the outer surface are extended by linked sugar molecules known as *polysaccharides*. This forest of flexible sugar molecules on the surface plays a major role in cell adhesion and molecular recognition.

Finally, on both sides of the cell membrane, a dense population of proteins receives and transmits messages and signals *(see pp. 54–55)*. The interior of a eukaryotic cell hosts several compartments bounded by their own lipid membranes—the nucleus, mitochondria, chloroplasts, and other organelles. These membrane-bound organelles communicate with each other and with a vast, tubular membrane network—the endoplasmic reticulum—that reaches every nook and cranny of the cell *(see p. 91)*. All these cellular membranes form a continuous, dynamic surface dotted with distinct domains known as *membrane rafts*, in which particular lipids are closely associated with specific protein partners. Some proteins move laterally in the plane of the membrane, like corks on water. Other membrane proteins stay in a fixed position, like pillars, anchored by a network of proteins covering one or both sides of the membrane. The macromolecular structures on the inside and outside of the cell's plasma membrane constitute a complex entity called the *cortex*, a very dynamic peripheral region of the cell.

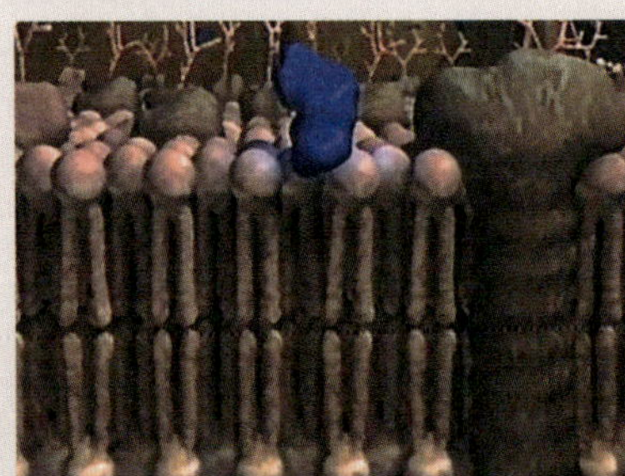
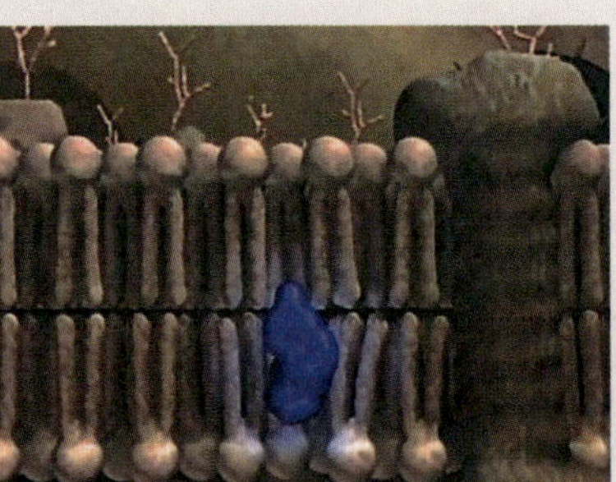

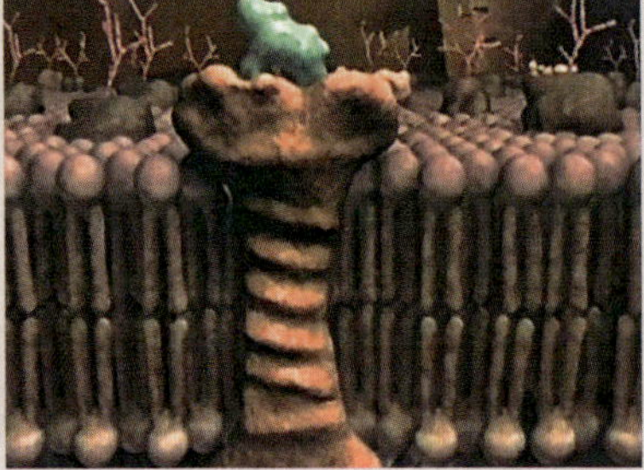

CELL MEMBRANES FILTER MOLECULES

Cell membranes are selective barriers to molecules.
Top: A small hydrophobic corticoid hormone *(in blue)* crosses the lipid bilayer.
Bottom: Insulin *(in green)*, a larger hydrophilic hormone received by its receptor protein on the cell surface. Reception of the hormone triggers the production of messenger molecules on the inner face of the membrane. (See a molecular representation of insulin and its receptor protein on page 54.)

View *Voyage Inside the Cell*, 3D digital creation © by Andreas Koch, Laurent Larsonneur, Christian Sardet, Digital Studio, 1999

Fats and sugars— stored energy

Cells have a constant need for energy, and for this, they mainly burn sugars and fats (in fact, they oxidize them). Stocks of sugars and fats are generally chained sugar molecules (polysaccharides) and lipid molecules rich in hydrocarbon chains, such as triglycerides.

In plants that produce glucose and energy through photosynthesis, individual glucose molecules (monomers) are linked together by enzyme proteins (which "run" chemical reactions) to form polysaccharides (polymers) with chains of varying lengths and branching patterns. These polysaccharides include starch, the primary energy reserve, stored in granules; and cellulose, which is synthesized by cells and secreted externally to construct protective cell walls. In fact, plants inherited cellulose through gene transfer from bacteria and protists that were already making cellulose envelopes, long before vegetation appeared. The cellulose fibers secreted en masse by plant cells combine with other polymers, such as pectins and lignins, to form walls between cells. An armada of enzyme proteins on the cell surface enables this process. As the main structural support for trees, plants, and algae, cellulose is the most abundant biopolymer on the planet.

In animals and fungi, glucose molecules are stored as glycogen—a polysaccharide polymer of glucose with multiple branches. Glycogen is compacted into granules for storage inside animal cells, mainly of the liver and muscles—tissues that require a lot of energy. When a cell needs energy, it breaks the glycogen down into glucose again, using enzymes such as the amylases in saliva and pancreatic juices.

Animals also store lipids in large membrane sacs that fill fat cells. In mammals, there are several types of adipose tissue—white, beige, or brown—in varying proportions. Brown adipose tissue is particularly abundant in hibernating marine mammals. The accumulated lipids (fatty acids) are used for heat production by a dense population of iron-rich mitochondria, giving a characteristic brown color to some adipose cells and tissues.

FROM DNA AND RNA TO PROTEINS AND NANOMACHINES

The cell is tightly packed with various molecules: thousands of different kinds of macromolecules (proteins and the nucleic acids DNA and RNA) working in concert with each other and with sugars, lipids, and metabolites. Proteins and nucleic acids are polymers made up of strings of monomers—amino acids for proteins, and nucleotides for nucleic acids. Each strand of protein or RNA has its own particular shape, dynamic, and function. Each one also has a specific lifespan and place in the cell.

Proteins are the most numerous. There are a few million per bacterium and billions in human cells. Proteins associate and work together with their partners like pieces of machinery. These molecular machines made of proteins, sometimes combined with RNA, are called *nanomachines* because of their minute size, ranging from 5 to 50 nm (1 nanometer = 1 billionth of a meter). Some nanomachines act as tiny motors, while others carry out essential functions such as cutting, grinding, pumping, or filtering. All these tasks rely on the catalytic properties of enzyme proteins and RNA molecules which facilitate chemical reactions.

A substantial percentage of proteins serve as main architectural components of a cell's intricate internal structures: organelles such as mitochondria, as well as biomolecular condensates and aggregates like nucleoli and chromosomes. Additionally, certain proteins dynamically assemble and disassemble to form and dismantle networks of microfilaments and microtubules—the primary constituents of the cytoskeleton, essential for cellular movement and intracellular transport.

Within the cell, proteins, DNA, RNA, and the nanomachines into which they assemble are confined within compartments defined by membranes of varying permeability. These membranes consist of lipid bilayers that function as selective filters and platforms for numerous proteins and nanomachines, including pumps, channels, receptors, and signal transducers. Proteins that regulate ion flow and are coupled with energy production span and line these membranes. The symphony of life depends upon the regulation of ions and mineral elements—such as hydrogen, sodium, potassium, calcium, and iron—and the pivotal role of ancestral metal complexes involving iron that continue to govern and facilitate most chemical reactions. Moreover, enzymes and nanomachines require tiny amounts of copper, nickel, vanadium, zinc, and numerous other trace elements to function optimally. These intricate molecular interactions are the result of billions of years of macromolecular evolution in the presence of minerals and ions, all within the aqueous environment that constitutes 70 to 90 percent of the cellular mass in all living organisms.

PROTEIN
DNA
RNA

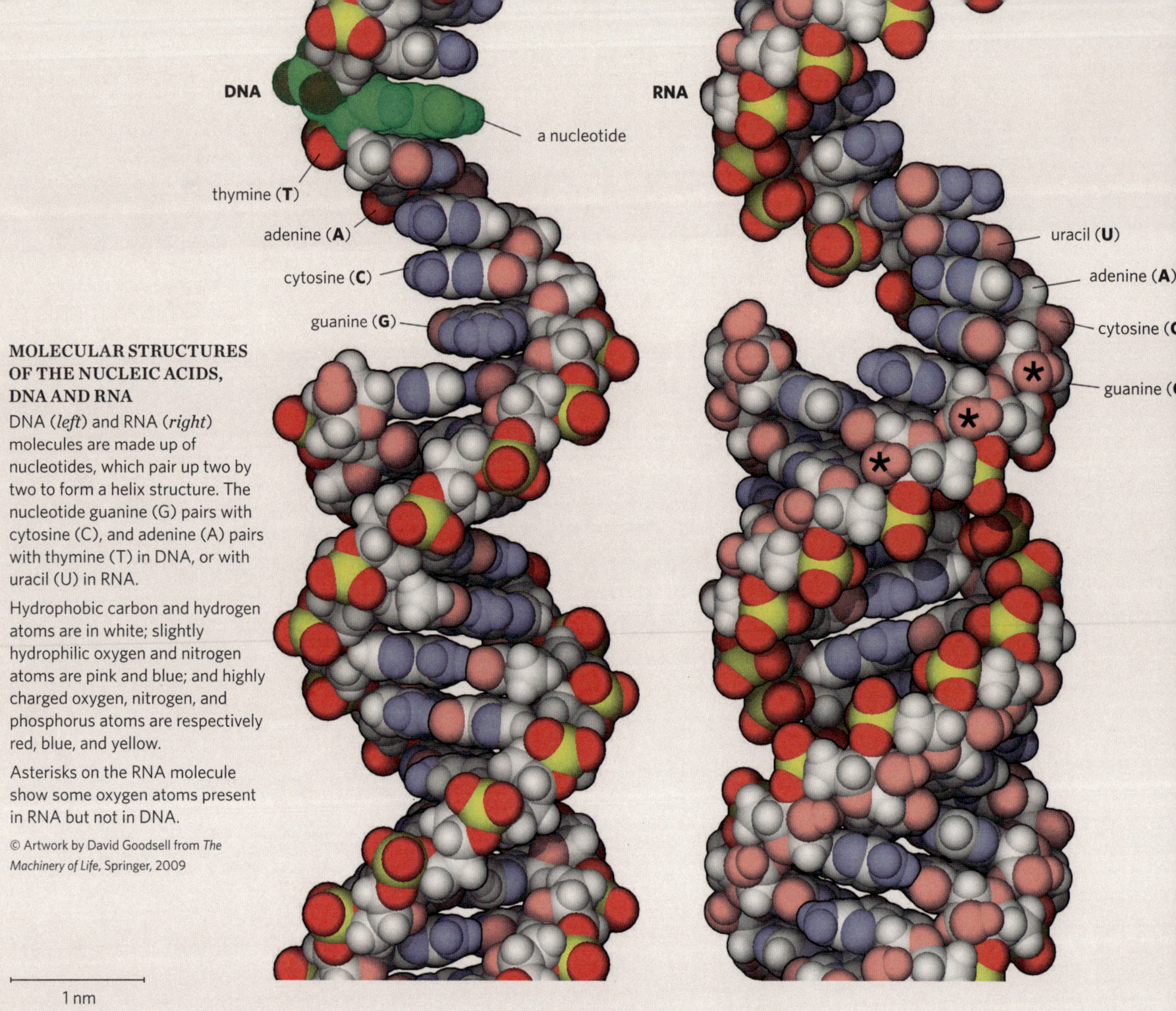

MOLECULAR STRUCTURES OF THE NUCLEIC ACIDS, DNA AND RNA

DNA (*left*) and RNA (*right*) molecules are made up of nucleotides, which pair up two by two to form a helix structure. The nucleotide guanine (G) pairs with cytosine (C), and adenine (A) pairs with thymine (T) in DNA, or with uracil (U) in RNA.

Hydrophobic carbon and hydrogen atoms are in white; slightly hydrophilic oxygen and nitrogen atoms are pink and blue; and highly charged oxygen, nitrogen, and phosphorus atoms are respectively red, blue, and yellow.

Asterisks on the RNA molecule show some oxygen atoms present in RNA but not in DNA.

© Artwork by David Goodsell from *The Machinery of Life*, Springer, 2009

From molecules of life to polymers—DNA, RNA, and proteins

Proteins and nucleic acids (DNA and RNA) are macromolecules, composed of specific monomers linked sequentially, much like beads threaded onto a necklace. In proteins, these monomers are twenty distinct amino acids, each one contributing unique chemical properties that influence the protein's overall function.

For nucleic acids, the monomers are nucleotides—five in total: adenine (A), guanine (G), cytosine (C), and thymine (T) in DNA, or uracil (U) in RNA. These nucleotide "beads" connect to form the long, linear polymers of DNA and RNA, which carry genetic information essential for life.

The twenty types of amino acid that make up proteins, and the five types of nucleotide that make up DNA and RNA, attract and repel each other to such an extent that they cause these long polymer chains to fold up in space. In proteins, the folding process is influenced by the hydrophobic or hydrophilic nature and the electric charges of the constituent amino acids. The amino acid chains form helices, sheets, flexible or rigid regions, and hydrophilic or hydrophobic pockets.

Each protein is made up of one or more distinct chains, a few dozen to several thousand amino acids in length. These different regions and foldings give each protein its characteristic shape and personality. Some proteins curl into balls, others stretch into rods. All proteins have

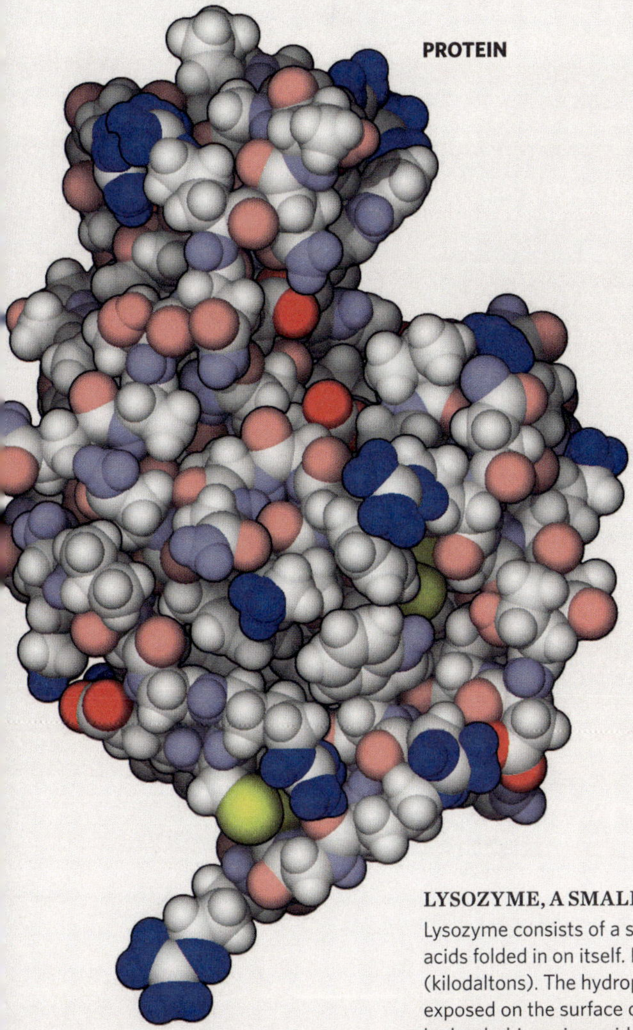

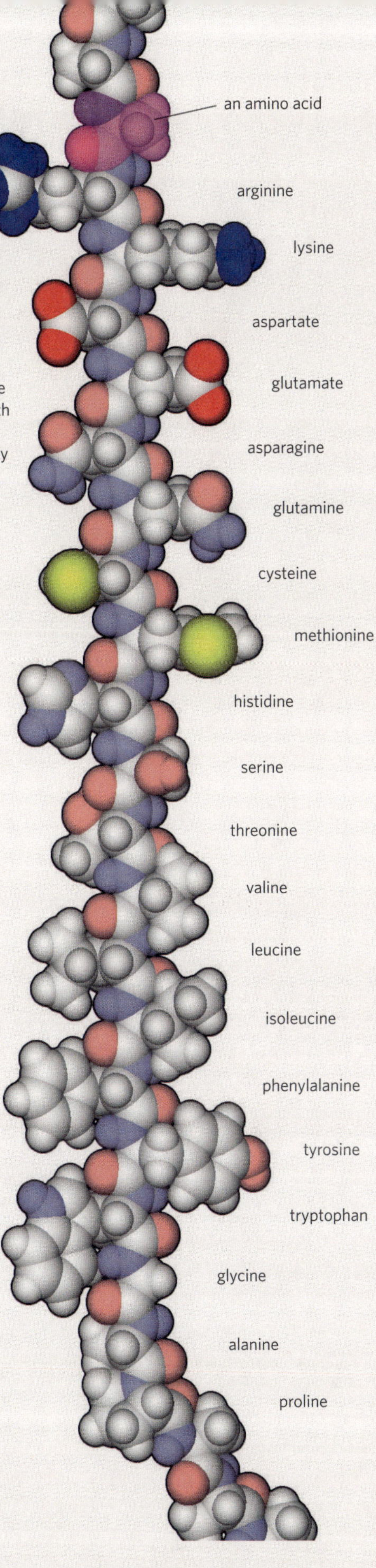

CHAINED AMINO ACIDS

The twenty amino acids of life are shown chained together. Carbon and hydrogen atoms are in white, oxygen in red and pink, sulfur in yellow, and nitrogen in shades of blue. At the top are the four hydrophilic amino acids, with strong positive or negative charges. In the middle are slightly charged amino acids, and at the bottom are the hydrophobic amino acids.

© Artwork by David Goodsell from *The Machinery of Life*, Springer, 2009

LYSOZYME, A SMALL PROTEIN

Lysozyme consists of a single chain of 130 amino acids folded in on itself. Its molecular mass is 14 kDa (kilodaltons). The hydrophilic amino acids are exposed on the surface of the protein, and the hydrophobic amino acids are on the inside. Cells secrete lysozyme as an antibacterial agent because this enzyme protein breaks down the polysaccharide chains which make up the cell walls of bacteria.

© Artwork by David Goodsell from *The Machinery of Life*, Springer, 2009

multiple pockets or indentations. These pockets constitute micro-spaces that recognize and bind other molecules, just as a lock takes a key.

As for nucleic acids, they form different kinds of helices, like the famous DNA double helix, thanks to the affinities of their component nucleotides, which form complementary pairs: Adenine (A) pairs with thymine (T), and guanine (G) with cytosine (C). RNA features a greater variety of helix, loop, and ladder structures due to a single molecular substitution—uracil (U), replacing the thymine (T) molecule present in DNA.

Proteins— macromolecular workhorses

Proteins are the versatile workhorses of the cell, orchestrating countless biological functions. They catalyze chemical reactions, serve as signal receptors and transmitters, regulate the exchange of molecules across membranes, and participate in the encoding and decoding of genetic information. Other proteins are building blocks of cellular structures—microtubules and microfilaments of the cytoskeleton, organelles, biomolecular condensates, and chromosomes. Many are associated with membranes. For instance, hormone receptor proteins, like those for glucagon or insulin, span the bilayer lipid membrane, recognizing and binding their respective hormones outside the cell. These receptors, composed of long chains of hundreds of amino acids, relay hormonal messages into the cell, triggering specific responses. The hormones themselves are also proteins—glucagon consists of a single chain of 29 amino acids, while insulin is composed of two chains (29 and 30 amino acids long) linked by disulfide (S-S) bonds between molecules of the amino acid cysteine. Disulfide bridges also play a crucial role in stabilizing the connections between the four chains that form immunoglobulins—the Y-shaped antibodies of the immune system. The different classes of immunoglobulins (IgG, IgM, IgD, IgA, and IgE) share 90 percent of their amino acid sequence, yet possess variable regions of extraordinary diversity. As a result, over 100,000 antibody variants circulate in our blood, primed to recognize and bind to specific antigens, providing a robust defense against pathogens.

All proteins are encoded by genes. The DNA corresponding to a protein is first transcribed into RNA, which is then translated into a protein by entities called *ribosomes* (see chap. IX). These ribosomes, measuring tens of nanometers, are nanomachines made up of RNA and dozens of proteins *(see following pages)*. Proteins are very dynamic entities. They modulate their functions by changing shape on contact with partner molecules—ions, minerals, hormones, metabolites. In addition, some proteins react to environmental signals such as light, acidity, heat, or pressure. Cells large and small are teeming with thousands of kinds of proteins, all constantly being produced, degraded, and renewed. Together, they weigh much more than all other cellular components, and account for 80 percent of Earth's biomass.

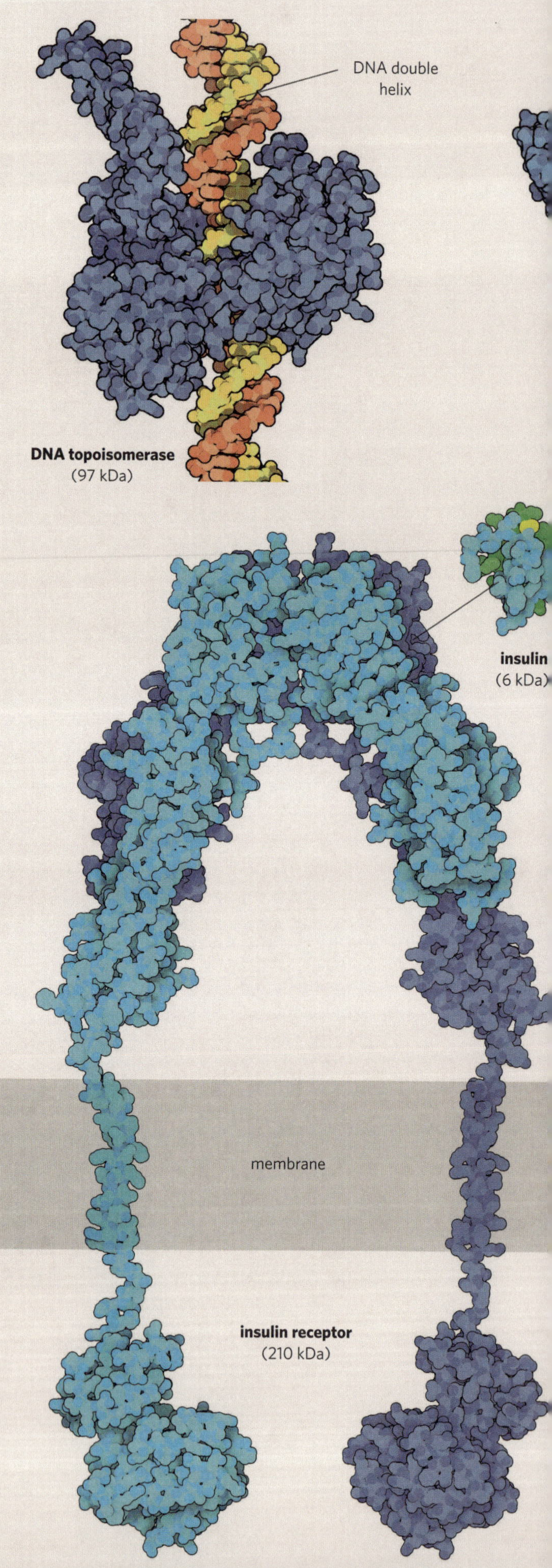

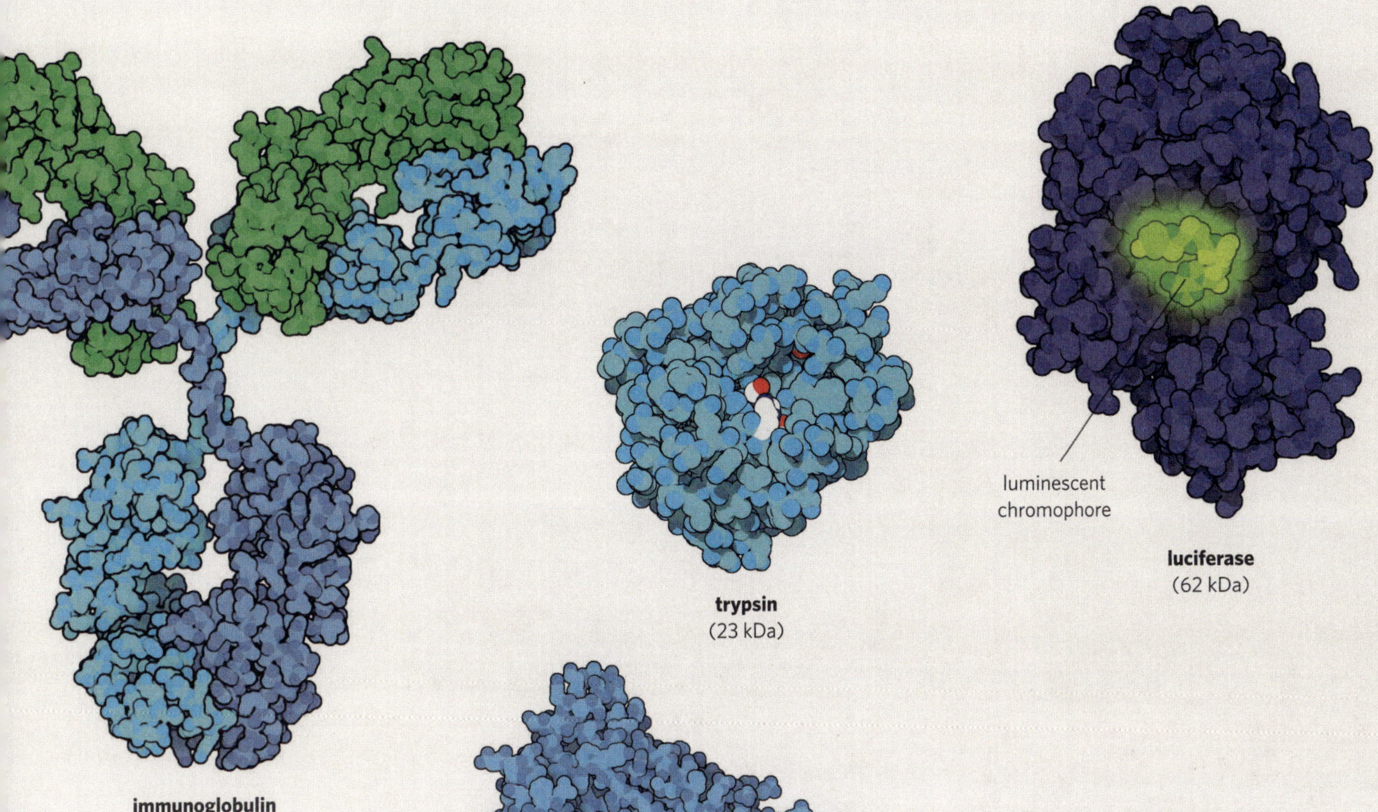

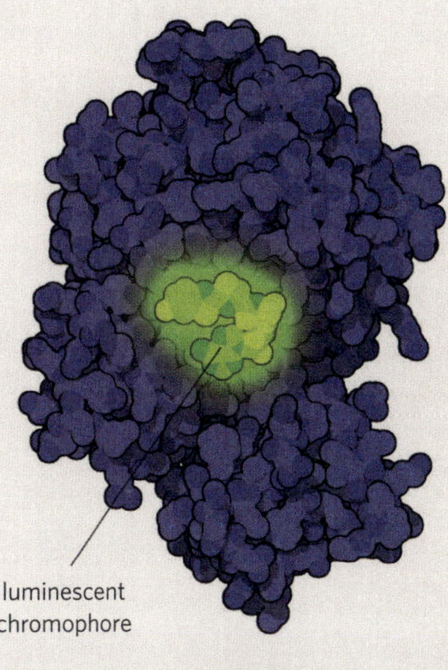

immunoglobulin
antibody
(150 kDa)

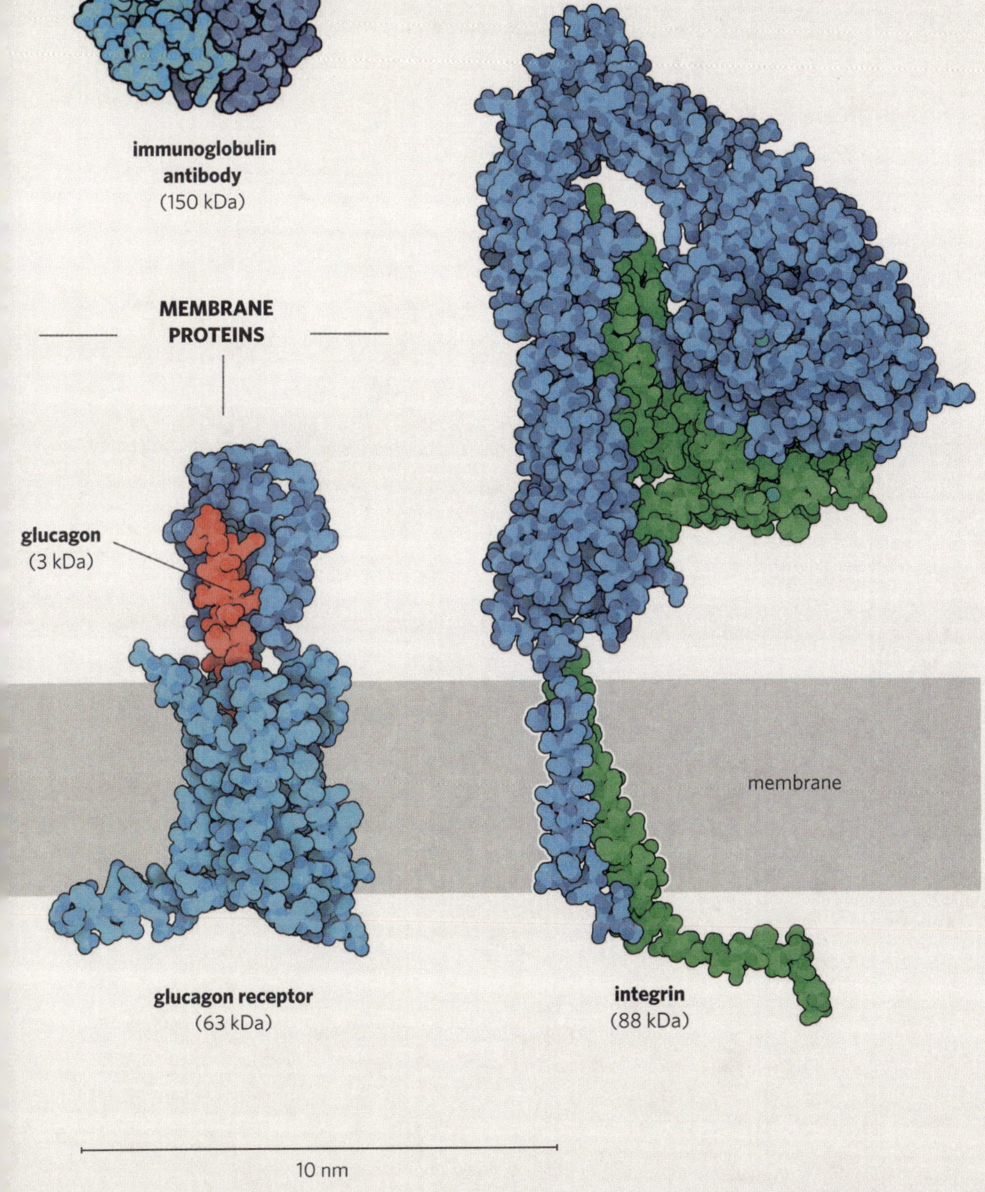

trypsin
(23 kDa)

luminescent
chromophore

luciferase
(62 kDa)

**MEMBRANE
PROTEINS**

glucagon
(3 kDa)

membrane

glucagon receptor
(63 kDa)

integrin
(88 kDa)

10 nm

A GALLERY OF PROTEINS—
FROM SMALLEST TO LARGEST

Proteins come in different shapes and sizes (their molecular masses are indicated in kilodaltons). The smallest proteins are made up of a single amino acid chain; larger proteins consist of several chains. Here, each chain is given its own color, so the proteins in one, uniform color each consist of just one chain. Shown are the small hormone proteins glucagon and insulin; the enzyme protein trypsin (which cuts protein chains into pieces), DNA topoisomerase (which copies the DNA helix, shown in *yellow* and *orange*), and luciferase (which emits light); and an antibody protein, an immunoglobulin composed of four amino acid chains. The lower part of the image shows proteins that span the lipid bilayer of the membrane (shaded): two hormone receptor proteins (insulin and glucagon, which regulate sugar metabolism) and integrin, a membrane protein that maintains contact between cells and with extracellular structures (such as collagen fibers) and intracellular structures (like microfilaments and microtubules).

© Artwork by David Goodsell for the RCSB
PDB "Molecule of the Month"

See more molecules on the RCSB
PDB "Molecule of the Month" site

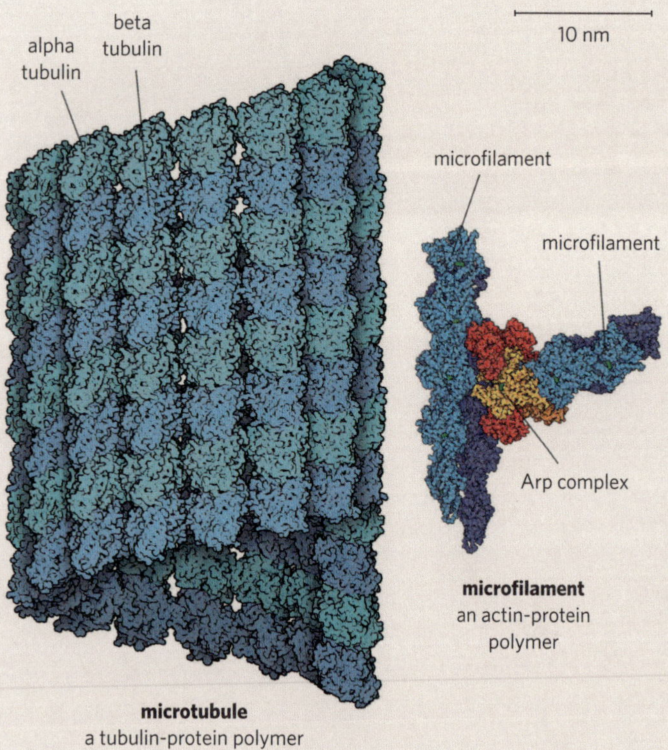

microfilament
an actin-protein
polymer

microtubule
a tubulin-protein polymer

MICROFILAMENTS AND MICROTUBULES
Microtubules are tubes, 20 nm in diameter, that polymerize from the assembly of tubulin proteins (alpha and beta, with molecular masses of 55 kDa). Microtubules grow or shrink as the tubulin proteins attach at one end and detach from the other at different speeds. Microfilaments are made up of two helices of actin proteins in single file. Branches can be created along the microfilaments by partner proteins such as Arp2/3, a nanomachine made up of seven protein subunits. This creates dense networks of contractile microfilaments within the cell.

© Artwork by David Goodsell for the RCSB PDB "Molecule of the Month"

Actins and tubulins— proteins that polymerize

The cellular cytoskeleton is a dynamic network of microtubules and microfilaments forming an articulated, branching structure. Microtubules and microfilaments appear and disappear, plugging in as needed. Their dynamic properties are due to the tubulin proteins (alpha and beta) that make up microtubules and the actin proteins that make up microfilaments, both of which easily polymerize (arrange themselves into larger superstructures). A host of partner proteins control the polymerization and depolymerization rates of microtubules and speed the assembly and disassembly of microfilaments, depending on a cell's needs.

When an animal cell is about to divide, two nucleation sites called *centrosomes* organize countless microtubules into two star-shaped structures known as *asters*. These microtubules capture chromosomes and organelles from the mother cell and maneuver them into the two daughter cells *(see pp. 148–49)*.

Microfilaments play a major role in the way a mother cell divides and splits into two daughter cells, as thousands of microfilaments contract at once, creating a massive constriction of the cell cortex overall. Microfilament movements are the work of motor proteins called *myosins*. On a larger scale, these same myosin proteins enable the microfilaments in our muscles to contract. Microtubules also bind specific motor proteins (dyneins and kinesins) which can slide lengthwise along the microtubules. Dynein proteins and kinesin proteins move macromolecules, nanomachines, and organelles in different directions along the microtubules, as if on rails.

It must be noted that the dynamics of microfilaments and microtubules differ significantly. Microfilaments, being thinner (7 nm across) and more flexible than microtubules (20 nm), do not grow exclusively at their ends, as microtubules do. Growth, shrinkage, and branching rely on partner proteins such as Arp2/3 (actin-related protein 2/3), which help establish branching networks of microfilaments.

Finally, the dynamics of microfilaments and microtubules depend on their environment, and in particular on local concentrations of metabolites such as ATP and calcium and magnesium ions (Ca^{2+} and Mg^{2+}).

Nanomachines— molecular machines

DNA serves as the repository of genetic information, guiding the synthesis of RNA and proteins. This process begins with transcription, where enzyme proteins known as *RNA polymerases*—a type of nanomachine—encase the DNA strand to translate its genetic code into messenger RNA (mRNA). The translation of mRNA into protein chains is then carried out by ribosomes, intricate nanomachines composed of around sixty proteins, assembled around ribosomal RNA. Working in chains or clusters, ribosomes envelop the mRNA strand, systematically reading and translating its genetic instructions into a

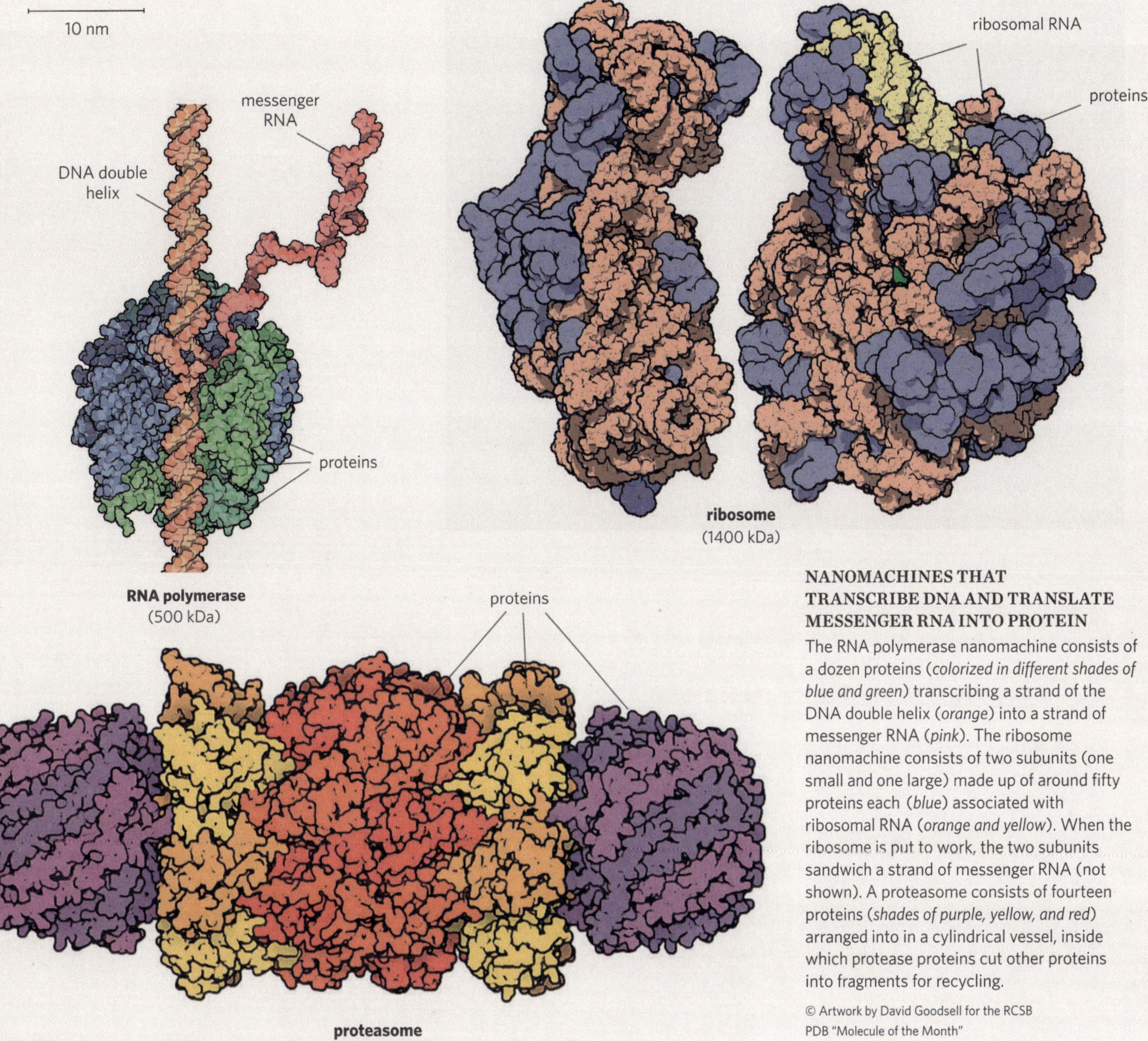

NANOMACHINES THAT TRANSCRIBE DNA AND TRANSLATE MESSENGER RNA INTO PROTEIN

The RNA polymerase nanomachine consists of a dozen proteins (*colorized in different shades of blue and green*) transcribing a strand of the DNA double helix (*orange*) into a strand of messenger RNA (*pink*). The ribosome nanomachine consists of two subunits (one small and one large) made up of around fifty proteins each (*blue*) associated with ribosomal RNA (*orange and yellow*). When the ribosome is put to work, the two subunits sandwich a strand of messenger RNA (not shown). A proteasome consists of fourteen proteins (*shades of purple, yellow, and red*) arranged into in a cylindrical vessel, inside which protease proteins cut other proteins into fragments for recycling.

© Artwork by David Goodsell for the RCSB PDB "Molecule of the Month"

precise sequence of amino acids, the building blocks of proteins. The complex process of RNA translation into a protein is orchestrated with the assistance of numerous protein and RNA partners (*see p. 167*).

RNA polymerases and ribosomes are prime examples of nanomachines. These intricate molecular assemblies are composed of specialized proteins that provide structural support and/or exhibit enzymatic activity. Often, the proteins within nanomachines are linked to sugars, lipids, RNA, or DNA, forming dynamic complexes essential for cellular function.

A host of other nanomachines like ribosomes operate within the cell. Proteasomes, for example, are "garbage cans" equipped with enzyme proteins called *proteases*, which chop up proteins and recycle their fragments.

Thanks to electron microscopy imaging techniques using flash-frozen macromolecules, three-dimensional (3D) representations of nanomachines, sometimes even "in action," are being published daily. Recently, artificial intelligence has helped crack the protein-folding code. It allows us to predict the 3D structure of any protein from its amino acid sequence, and therefore from its DNA sequence. And designing artificial proteins with novel functions is now a possibility.

A TALE OF ORIGINS

From LUCA
to bacteria & archaea

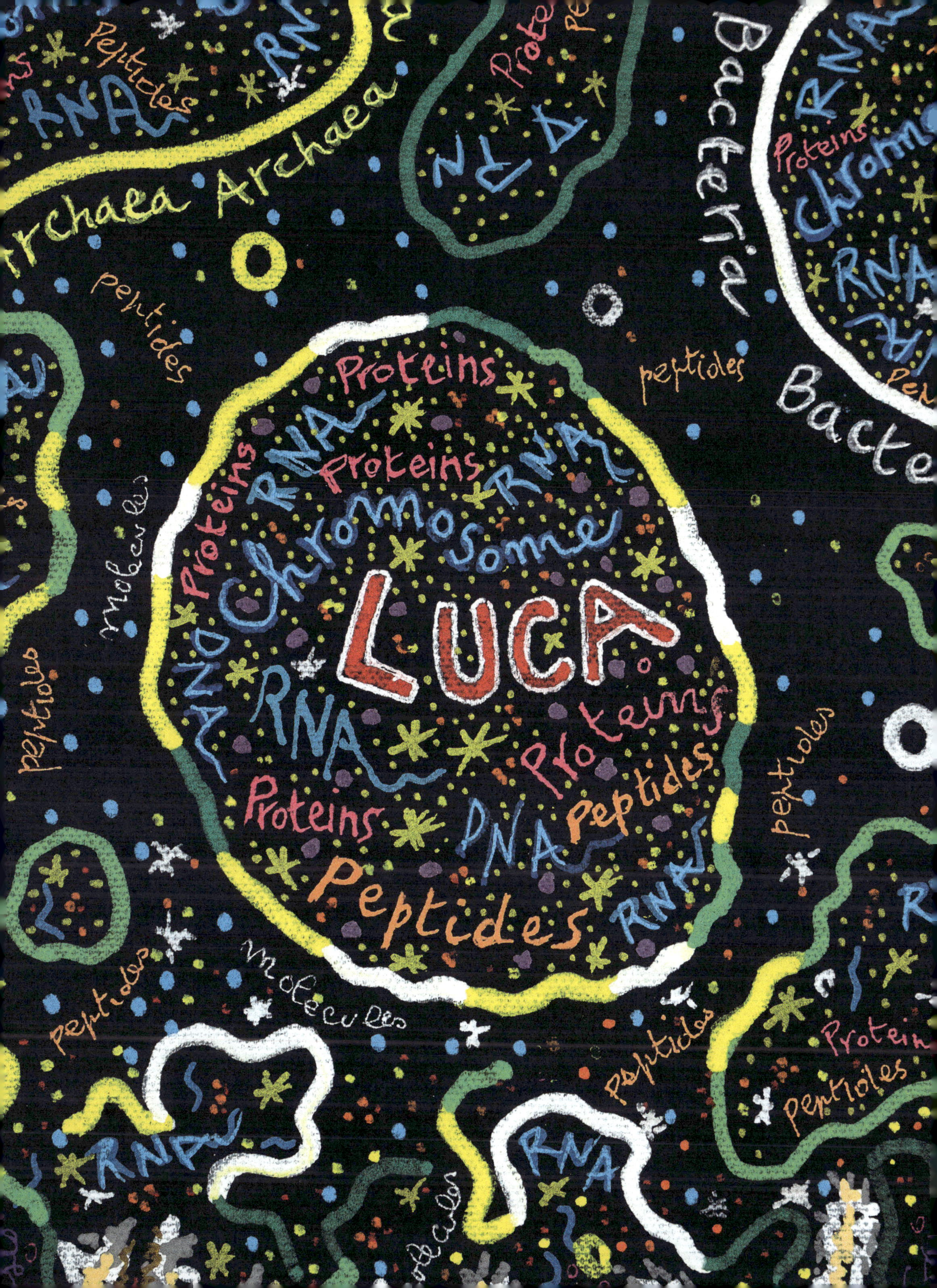

LUCA

Proteins

RNA

Chromosome

DNA

Peptides

Molecules

Archaea

Bacteria

Peptides

FROM STARDUST TO LUCA— THE LAST UNIVERSAL COMMON ANCESTOR

Atomic elements have been forged in stars since the Big Bang, 13.7 billion years ago *(see pp. 36-40)*. The origin of life—when, how, and where it first emerged on Earth—is a far more complex and uncertain story. It begins nearly 4 billion years ago, in a world teeming with basic molecules—life's essential building blocks, composed of six main elements (the CHNOPS)—and an abundance of minerals and energy sources.

It is likely that, over hundreds of millions of years, molecular assemblies capable of increasing complexity and self-replication evolved through natural selection in favorable environments. From these dynamic aggregates or macromolecular condensates, primitive cells—protocells—emerged. Among them, the ancestral cell LUCA (the last universal common ancestor) is believed to have given rise to the first autonomous microorganisms enclosed within membranes. From LUCA's lineage, bacteria and archaea appeared, two microbial domains that have endured and evolved for over 3 billion years. Other primordial cells may have existed, but, unlike LUCA, they left no surviving descendants.

The central mystery of life's origin is: How did the inanimate world of stardust, and the molecules and macromolecules of life, self-organize into the first living cells? To answer this question, astrophysicists, geologists, chemists, and biologists share and combine their knowledge of planetary conditions, energy sources, and the self-organizing and evolutionary properties of molecular assemblies, including the precursors of proteins and nucleic acids.

Some researchers propose that the first protocells arose in alkaline, hydrothermal vents on the ocean floor, where mineral-rich fluids provided a stable energy source. Others argue that intermittent geysers near uranium deposits, a source of ionizing radiation, or surface thermal springs, may have offered equally favorable conditions. All these environments were characterized by powerful, continuous energy flows, driving the countless out-of-equilibrium chemical and energetic reactions essential for life's emergence. Among many plausible hypotheses, one particularly compelling scenario envisions life originating in the depths of the ocean—a journey from molecular chaos to the dawn of biological organization.

Six main elements—the CHNOPS: carbon (C), hydrogen (H), nitrogen (N), oxygen (O), phosphorus (P), and sulfur (S). Here they are shown making up the basic molecules: hydrogen (H_2), oxygen (O_2), hydrogen sulfide (H_2S), carbon dioxide (CO_2), methane (CH_4), carbon monoxide (CO), ammonia (NH_3), phosphoric acid (H_3PO_4), and water (H_2O).

From these basic molecules, reactive molecules such as formaldehyde (CH_2O), dimethyl sulfide ((CH_3)$_2$S), and others were formed that chemically reacted to produce the molecules of life with carbon skeletons—such as glucose ($C_6H_{12}O_6$) or the amino acid cysteine ($C_3H_7NO_2S$), which are the building blocks of polysaccharides and proteins *(see molecular models on pp. 46–47)*.

A SCENARIO FOR LIFE'S EMERGENCE IN THE DEPTHS OF THE OCEAN

Let's journey back 4 billion years. The planet is cooling, and deep beneath the ocean's surface, volcanoes are releasing clouds of gas—hydrogen, methane, nitrogen, ammonia, and sulfides—which seep through intricate networks of rocky microchannels. Immersed in seawater saturated with CO_2, colossal mineral chimneys emerge from the abyss. These towering rock formations are riddled with countless microchambers, their crystalline surfaces lined with metal complexes rich in iron, sulfides, and nickel, creating ideal conditions for electrochemical reactions to take place.

This scenario of life's emergence from alkaline hydrothermal vents at the bottom of the primordial ocean hinges on a powerful energetic force that may have sparked the origin of life. According to some researchers, this force arose from a difference in proton concentration (H^+, the ionized form of hydrogen) across the porous walls of abyssal vents. This proton gradient drove the reduction of CO_2 by hydrogen, giving rise to the simplest carbon and sulfur molecules: methane (CH_4), formaldehyde (CH_2O), cyanides (HCN), and sulfides (H_2S). Gradually, these basic compounds combined to form the molecules of life—a vast array of amino acids, sugars, nucleotides, lipids, and more. This abundant molecular diversity was further enriched by numerous organic molecules delivered to Earth and its ocean by meteorites, adding to the primordial chemical repertoire.

Tens of millions of years later, within the countless microchambers of the abyss, increasingly complex chemical reactions and molecular structures emerged. Thus began the "world of metabolism," as reactions continuously generated the molecules of life, which combined into the earliest macromolecules. Short chains of amino acids (peptides) formed, sketching the first outlines of proteins. Nucleic acids, with RNA possibly among the earliest, would have appeared, alongside lipids that spontaneously coalesced into micelles and membranes. On rare occasions, these macromolecular assemblies would have naturally concentrated and lingered together, laying the groundwork for the next evolutionary step. This critical phase led to the refinement and evolution of molecules capable of storing and transmitting information. RNA, in association with peptides, would have played a key role, forming molecular entities capable of self-replication with variations. This heralded a form of chemical evolution driven by natural selection of the most successful molecular groups. An "RNA–peptide world" would have emerged, as molecules acquired new enzymatic properties and the ability to store information. Gradually, this nascent world of information and chemical transformation would have stabilized and evolved further, integrating DNA molecules to carry instructions and enzyme proteins to carry them out, paving the way for the first living entities.

Sea water
PROTOCELLS CO2 PROTOCELLS
lipid membrane
Ribosomes
Proteins
DNA
nanomachine
RNA RNA
Proteins
Ribosomes
RNA
lipid membrane
Proteins
RNA
Ribosomes
nanomachine
RNA
DNA
Ribosomes
Proteins
nanomachine lipids
Proteins Ribosomes
nanomachine
RNA RNA
lipids
Proteins
RNA
lipids
nanomachine
lipids
RNA
Proteins
RNA
lipids
nucleotides
Proton Motive Force
Primitive Metabolism
amino acids
NH3
H2S
amino acids
NH3
Proton Motive Force
amino acids
Primitive Metabolism
H2
amino acids
H2
amino acids
H2
nucleotides
NH3
CO2
H2S
H2
NH3
H2
Proton Motive Force
Sea water
CO2
amino acids
Primitive Metabolism
NH3
H2
H2
Alkaline hydrothermal vent
NH3

FROM LUCA TO BACTERIA AND ARCHAEA

The emergence of protocells implies that after hundreds of millions of years of encounters, within billions and billions of microchambers in the abyss, the "world of metabolism" and the "RNA–peptide world" came together. Perhaps these two primordial worlds initially organized themselves into aggregates or molecular condensates—veritable crucibles of chemical reaction and replication. We can imagine that nascent forms of life—the first protocells and possibly even LUCA, the ancestral cell—remained dependent on the electrochemical forces permeating the mineral walls of their reaction microchambers. Over millions upon millions of years, every conceivable combination of primitive proteins, RNA, and the nanomachines they formed would have been tested and embraced or discarded, ultimately giving rise to a robust, established genetic code.

To complete the picture, we must envision that on countless occasions the "world of membranes," composed of diverse lipids associated with minerals and proteins, encapsulated the functional, growing worlds of metabolism, RNA–peptides, and nanomachines. These encapsulations by a variety of membranes emerging from the mineral matrix would have produced the first autonomous protocells, including LUCA. From these primordial life-forms, bacteria and archaea emerged, each endowed with membranes featuring distinct lipid compositions that define their respective lineages.

In the depths of the abyss, ancestral cells thrived without the need for light, drawing their energy from the inorganic molecules, minerals, and gases abundant in their surroundings. To this day, their microbial descendants—certain archaea and bacteria—continue to harness mineral and gas flows on the ocean floor, sustaining oases of deep-sea life, including rich animal ecosystems.

Imagining such distant events has its limits, but the field attracts and unites philosophers, chemists, biologists, and astrophysicists around a common passion. Some researchers probe the universe in search of possible traces of life on other planets. Others use chemical reaction chambers to re-create molecular assemblies in the environmental conditions that presumably prevailed at the time of the origins. Though recreating life in a test tube may seem illusory, perhaps one day we will converge on a consensus around the fundamental principles that sparked the emergence of life—much like our understanding of the Big Bang.

View our cartoon *O as Origin*, a collaboration with Y. Mahé, G. Macagno, and M. C. Maurel, including J. Iwasa's computer animations and images, which are presented on the following pages

Archaea Archaea Archaea
Archaea
Archaea Archaea
Archaea Archaea Archaea
Peptides
RNA
RNA
Proteins
Peptides
Chromosome
Proteins Peptides
RNA
RNA
DNA
Proteins Peptides Proteins
DNA
Bacteria Bacteria
Bacteria
Bacteria
Bacteria Bacteria
Peptides
RNA RNA
Peptides
Proteins Chromosome
RNA Proteins
DNA Proteins
Peptides
peptides
peptides
DNA RNA
RNA
Peptides
Proteins
DNA
Proteins
RNA
Proteins
RNA Proteins
Chromosome
DNA
LUCA
RNA
Proteins
Proteins
DNA
Peptides
RNA
Proteins
Peptides
molecules
molecules
molecules
molecules
molecules
peptides
Peptides
peptides
RNA
Proteins
Peptides
Peptides
RNA
RNA
RNA
molecules molecules
molecules molecules molecules molecules molecules molecules molecules molecules molecules molecules molecules molecules

Imagining and imaging protocells

We imagine that the very first cells capable of reproducing and evolving—the protocells—were made up of (at least) a membrane vesicle protecting a nucleic acid macromolecule.

With computer imaging, we can represent the organization of molecules into structures (such as protocells) both realistically and artistically, taking rigorous account of the known sizes, shapes, and dynamic behaviors of molecules and macromolecules.

These images are taken from computer animations created by Janet Iwasa, a pioneer in the field *(see QR code on p. 64, and below)*. They show lipid molecules spontaneously self-organizing into micelle-like spheroid aggregates and a membrane vesicle consisting of a double layer of lipids. Nucleotides and nucleic acids (here, RNA in the form of a helix) are represented at proper scale.

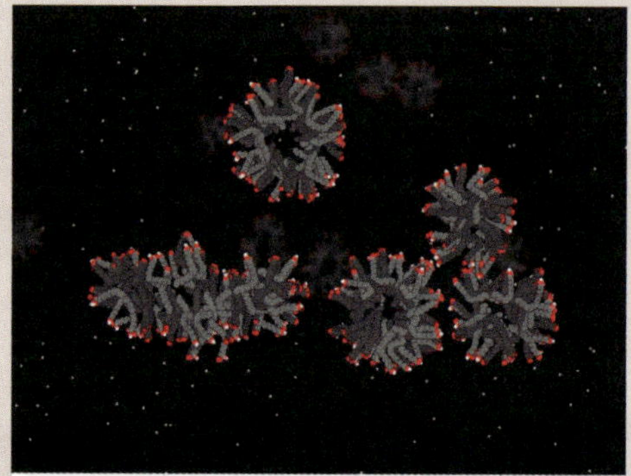

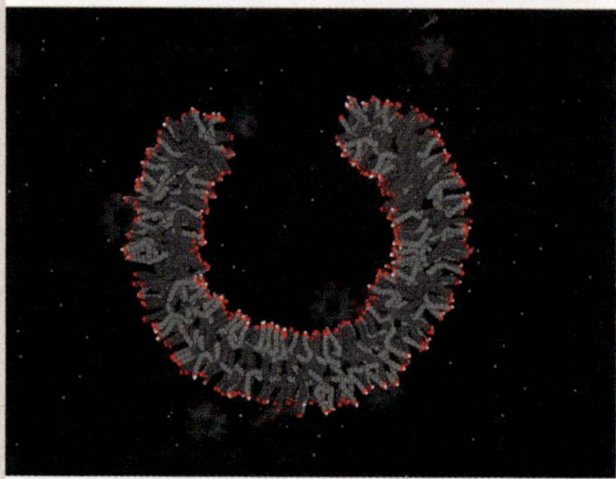

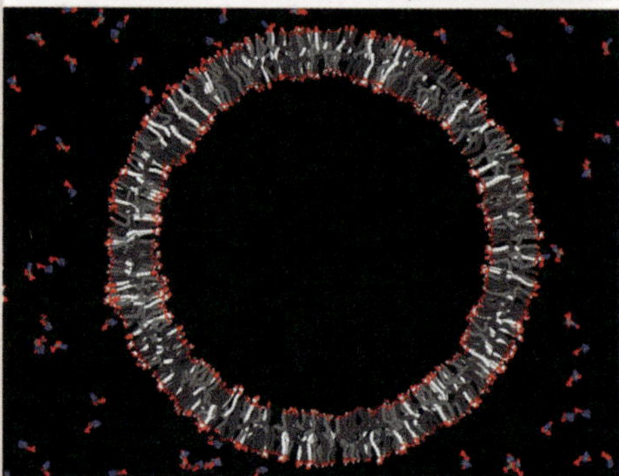

This page

MEMBRANOUS MICELLES AND VESICLES

Lipid molecules—in this case fatty acids (linoleic acid, *see p. 47*)—are so-called *amphiphilic* molecules. This means they have a hydrophilic part (represented by red dots) and a hydrophobic part (the hydrocarbon chains, represented by grey rods). In water, they form micelles with the hydrophilic part facing outward and the hydrocarbon chains facing inward. When micelles fuse, their lipid molecules spontaneously combine in two layers—a membrane—which closes in on itself to form a vesicle. The double-layered lipid membrane separates the inside of the vesicle from the outside, forming a selective barrier that's permeable to certain molecules and ions, and impermeable to others. Nucleotides (*in red and blue*) cross the membrane from outside the vesicle, to position themselves inside.

SEQUENCE FROM AN ANIMATED FILM
© Digital creation by Janet Iwasa, laboratory of Jack Szostak,
Massachusetts General Hospital, Harvard Medical School

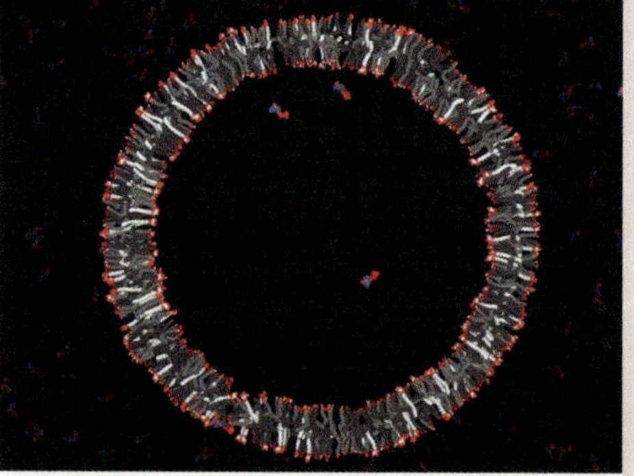

Opposite page

**A MEMBRANE VESICLE
CONTAINING NUCLEOTIDES AND RNA**

A vesicle made of a double layer of lipids (fatty acids) encloses small RNA-type nucleic acids (the blue-and-white helix-shaped molecules) and nucleotides (the small blue-and-white molecules). Some nucleotides are shown crossing the membrane. The hydrophilic parts of the lipids are acid molecules (*red dots);* the hydrophobic parts are long carbon chains (*grey rods).*

© Digital creation by Janet Iwasa, laboratory of Jack Szostak,
Massachusetts General Hospital, Harvard Medical School

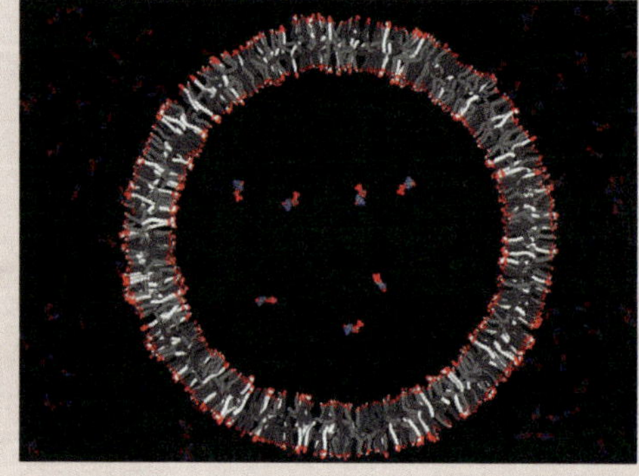

View computer animations of molecular processes by Janet Iwasa

Life in extreme environments

The earliest forms of microbial life emerged on Earth in the absence of oxygen and under extreme conditions of temperature, pressure, salinity, and acidity. Although such harsh environments are now rare, they still persist in certain isolated pockets, reminiscent of the conditions that likely dominated the surface and depths of the primordial oceans. Today, some of these extreme habitats are colonized by bacteria and archaea species that have evolved over billions of years. These so-called *extremophile* microorganisms—including not only bacteria and archaea but also some single-celled protists and fungi—have continually adapted to hostile conditions to ensure their survival. Among them are hyperthermophilic organisms that withstand boiling water, as well as acidophiles and halophiles that thrive in acidic or salt-saturated waters.

Their remarkable resilience is due to chemical adaptations that stabilize the structures and functions of their constituent molecules, which include nucleic acids, lipids, and proteins. One such bacterium, *Thermus aquaticus,* discovered in a hot spring in Yellowstone Park, gained fame in the 1970s for its ability to produce a thermostable DNA polymerase. This enzyme made it possible to copy and amplify DNA using PCR (polymerase chain reaction) techniques, opening the way to gene analysis and manipulation.

Enzyme proteins derived from extremophilic organisms are highly sought after due to their remarkable stability, making them invaluable for industrial applications. With the guidance of artificial intelligence, knowledge of their three-dimensional structures can be harnessed to introduce favorable mutations into other proteins through genetic manipulation.

One promising avenue of research is the optimization of enzyme proteins capable of breaking down plastics—petroleum-derived polymers—at relatively high temperatures (70°C–100°C), offering the potential for recycling or even composting plastic pollutants.

THE WHITE SMOKERS OF THE ABYSS

The ocean floor is punctuated by countless hydrothermal vents scattered along the vast network of undersea volcanic ridges. Known as *white smokers* and *black smokers*, these vents emit diverse mixtures of gases, minerals, and metals, providing the chemical energy essential to bacteria and archaea that thrive on inorganic compounds such as sulfur or iron. This remarkable ability to harness chemical compounds as an energy source is known as *chemosynthesis*. Among these hydrothermal sites, one of the most remarkable is the Lost City hydrothermal field, discovered in the middle of the Atlantic Ocean in 2000. There, alkaline hydrothermal vents—with pH values ranging from 9 to 11—release gas mixtures rich in hydrogen and methane at moderate temperatures (40°C–90°C), which are compatible with life. These towering, white carbonate chimneys can reach up to 60 m in height, forming a striking landscape on the ocean floor.

© Courtesy Debbie Kelley and Mitch Elend, University of Washington, Seattle

EXTREME SURFACE ENVIRONMENTS

The hydrothermal pools and vents of the Danakil Depression in Ethiopia—where the oceanic crust rises to the surface—present a striking mosaic of corrosive environments. These waters, saturated with potash salts and sulfides, display extraordinary hues of yellow, green, and red, their colors reflecting varying copper and iron concentrations. In these extreme settings, researchers investigate the potential for life among archaea and bacterial populations, enduring harsh conditions of temperature, salinity, and acidity. Such analyses are particularly challenging, as they demand meticulous precautions to prevent any external microbial contamination that could compromise the results.

Bottom

LIFE IN THE ABYSS

At many hydrothermal sites, microorganisms that thrive on gases and minerals through chemosynthesis form the foundation of vibrant ecosystems, supporting a wealth of marine invertebrates. Among them are colonies of giant annelid worms, like *Riftia pachyptila*, which can grow up to 3 m long. These remarkable creatures flourish in the abyss through a unique symbiosis with bacteria that extract energy from the oxidation of sulfur and nitrogen gases released by nearby black smokers. The existence of these giant worms was discovered by chance in 1977 at the bottom of the Pacific Ocean, during an expedition to study black smokers with the American bathyscaphe (deep-sea submersible) *Alvin*. The revelation that life could thrive through chemosynthesis, without any dependence on light, was groundbreaking. Since then, numerous abyssal ecosystems teeming with crustaceans, annelids, and mollusks have been explored by underwater expeditions aboard bathyscaphes, reshaping our understanding of the potential for life in extreme environments.

THE WORLD OF BACTERIA AND ARCHAEA

Since the dawn of time, two types of microorganisms have colonized the planet: bacteria—familiar to all—and another, far less known to the public: archaea. Discovered only in the 1970s in volcanoes and hot springs, archaea resemble bacteria in their microscopic appearance and size, typically just a few thousandths of a millimeter. Yet, despite their outward similarities, archaea are fundamentally different from bacteria in the molecular makeup of their membranes, walls, appendages, and other structures.

Bacteria and archaea are surrounded by a cell membrane protected by one or more walls. Their genetic material is carried on a single circular chromosome, which is not compartmentalized within any internal membrane. This is why both bacteria and archaea are classified as prokaryotes—cells without a nucleus. Traditionally, the absence of any internal compartments (organelles) surrounded by membranes in the cytoplasm was considered a defining feature of prokaryotic cells. However, this notion has evolved with the discovery that many bacteria and archaea do, in fact, contain compartments, both membrane-bound and membrane-free. For example, cyanobacteria (the most numerous cells on the planet) contain internal membranes performing photosynthesis, as well as biomolecular condensates called *carboxysomes* that fix CO_2.

While most bacterial and archaeal species live independently, some form colonies, either temporarily or permanently, and can even develop two or three distinct cell types within a single colony. For 3 billion years, bacteria and archaea have colonized every conceivable habitat, including the most extreme environments. They often coexist, evolving side by side and exchanging genes directly or through viruses—a phenomenon known as *horizontal gene transfer (see p. 198)*. Different species frequently associate with and complement one another, forming consortia or biofilms where cooperation and competition are the rule. Their metabolic capacities are both immense and astonishingly varied, allowing them to adapt to countless ecological niches. Many bacterial and archaea species live in or on eukaryotic organisms—as symbiotes, commensals, or parasites—forming complex microbiota with which they coevolve. This intricate interplay between microbes and their hosts is a fundamental aspect of life on Earth.

When it comes to the planet's biomass, bacteria and archaea weigh in heavily—at 77 Gt (gigatons) in total. This represents 35 times more biomass than all animals combined (2 gigatons) and 5 times more than that of fungi (12 gigatons). Only trees and plants surpass them, with a total biomass of over 450 gigatons. In short, bacteria and archaea are not only ancient and resilient life forms but also major pillars of Earth's living world.

Bacteria ○ are microorganisms that live alone or in colonies. Bacterial cells are delimited by a membrane ▌ protected by a wall ▌. Bacteria have a single circular chromosome ▌ and appendages for movement. Some bacteria (cyanobacteria) have photosynthetic membranes ○.

Archaea ○, like bacteria ○, have various shapes, appendages, and a single circular chromosome ▌. The lipid molecules that make up the membranes ▌, walls ▌, and appendages of archaea ○ are different from those of bacteria. Some species live in colonies, but there are no photosynthetic archaea.

The different species of bacteria ○ and archaea ○ each have their own cohorts of viruses, called *bacteriophages* and *archaeal viruses*, represented in the background.

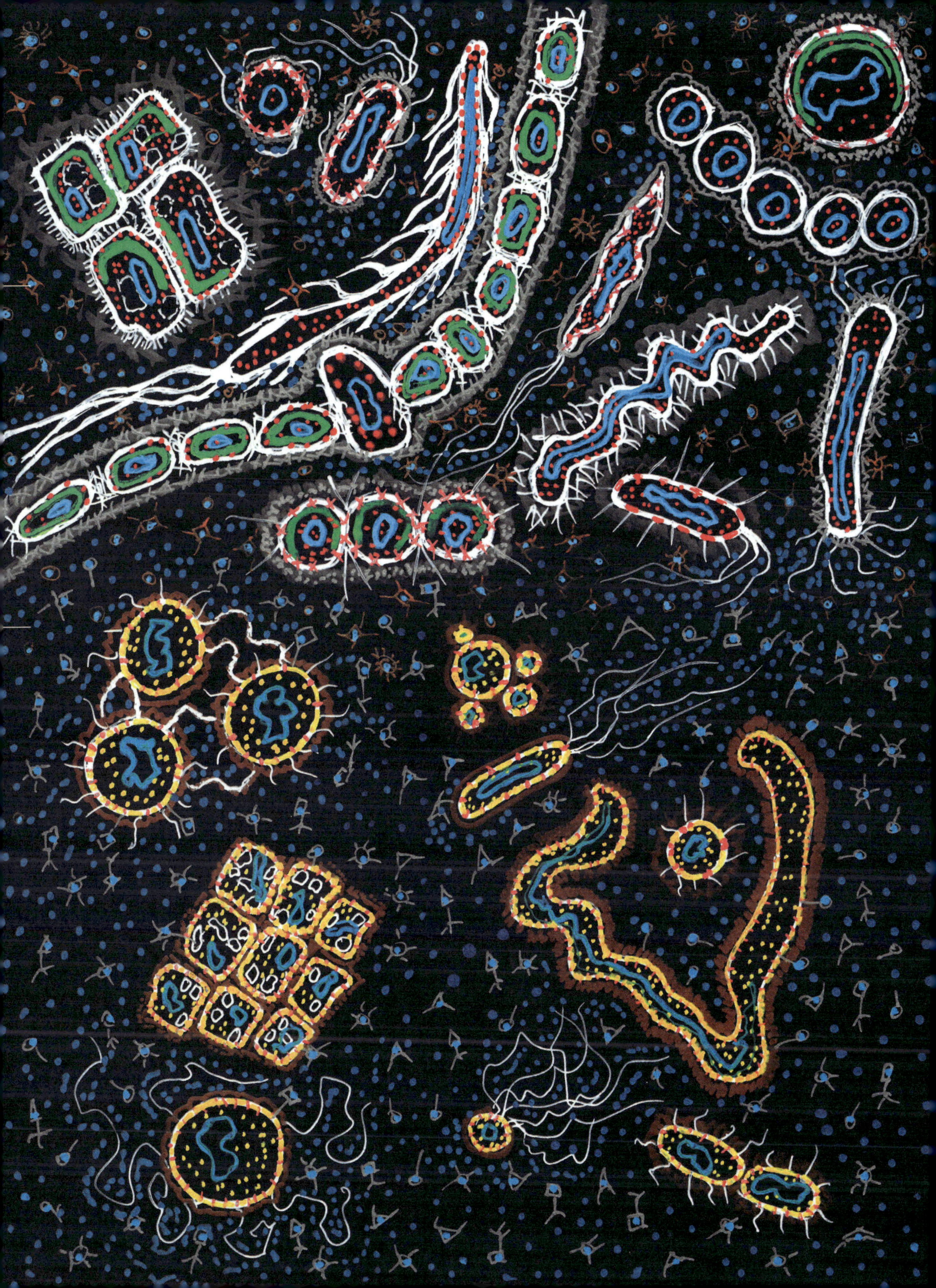

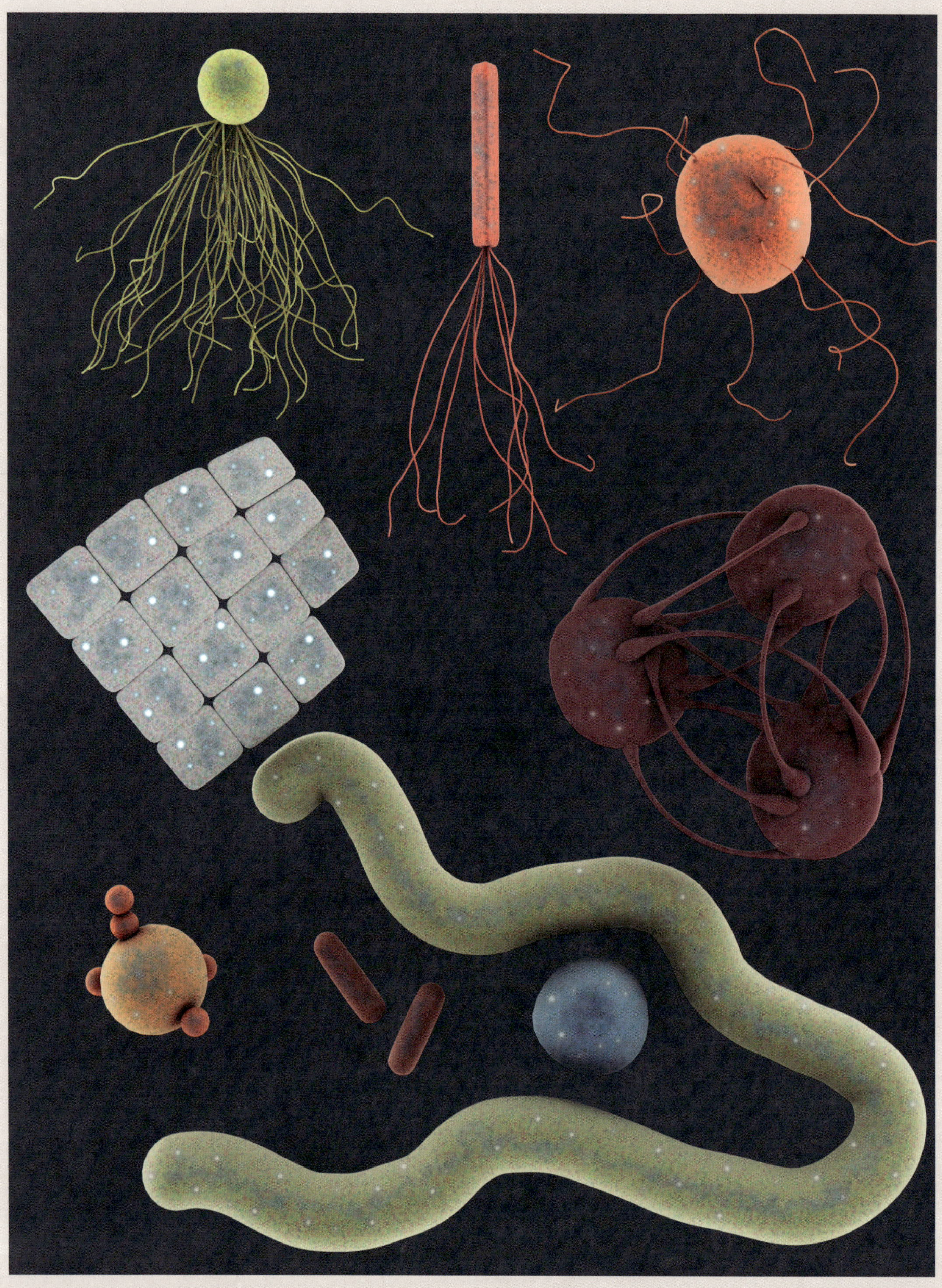

Archaea—a surprise discovery fifty years ago

Archaea were discovered in the 1970s, when molecular biologists analyzed the genes of microorganisms living in volcanoes and hot springs *(see pp. 30 and 68)*. To their surprise, the nucleotide sequences of their ribosomal RNA—the nanomachines responsible for protein synthesis—revealed the existence of not just one type of prokaryote (microorganism without a nucleus), but two: bacteria and archaea. This discovery fundamentally changed our understanding of prokaryotes: In fact, a diverse menagerie of bacterial and archaeal species have evolved and interacted side by side for the past 3 billion years. However, despite their shared ancient history, bacteria outnumber archaea on the planet by a factor of ten.

Although bacteria and archaea are comparable in size and appearance, they exhibit notable differences at the molecular level. These differences include the lipid molecules that make up their membranes, the proteins and sugars within their cell walls, and the mechanisms by which they replicate and transcribe DNA and translate genetic information into proteins. Archaea also possess their own viruses. One unique metabolic characteristic of many archaea is their ability to produce methane (a process called *methanogenesis*), which bacteria cannot do. This trait is considered a hallmark of ancestral microorganisms *(see p. 62)*. Due to these distinctive features, archaea were initially named *archaebacteria*, from *archae-* meaning "ancient" or "primitive," although it soon became clear that they are no more ancestral than bacteria.

Contrary to early assumptions, archaea are not the only extremophilic microorganisms, as many bacterial species are also adapted to inhospitable conditions, including extreme temperature, salinity, or acidity. Both kinds of prokaryote have demonstrated remarkable ecological versatility and are found in virtually all ecosystems, from deep-sea hydrothermal vents to animal and plant microbiomes, including our own.

Within the human body, anaerobic archaea constitute around 10 percent of the microorganisms in our intestinal microbiota, playing a role in digestion and appearing to have coevolved with their hosts. Interestingly, while bacteria include numerous pathogenic strains, not a single clearly identified archaea has yet been shown to cause disease.

In 2015, an important discovery revealed the existence of a group of archaea known as *Asgard archaea*, found in deep marine sediments off the coast of Norway. Strikingly, these archaea proved to be closer to eukaryotic cells (cells with nuclei) than to bacteria. Asgard archaea contain genes that were previously thought to exist only in eukaryotes. This finding has led to the hypothesis that the ancestors of Asgard archaea, through interactions with other prokaryotes—possibly including an ancient symbiosis with a bacterium—played a central role in the emergence of eukaryotic cells. This process, known as *eukaryogenesis*, is explored in the next chapter *(see p. 86)*. The discovery of archaea has not only expanded our understanding of microbial diversity but also deepened our knowledge of life's evolutionary history.

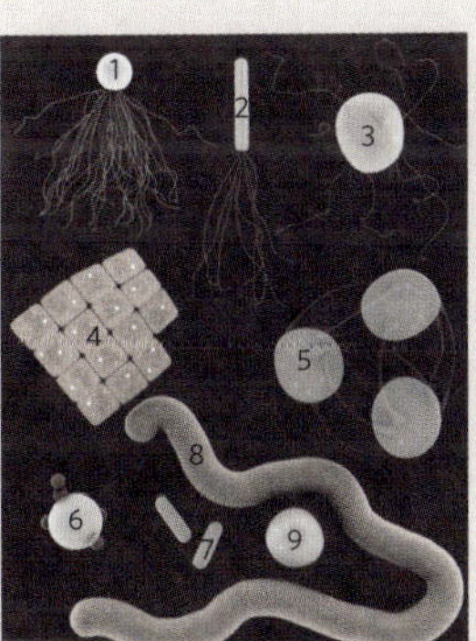

1. *Methanococcus jannaschii*
2. *Pyrobaculum aerophilum*
3. *Sulfolobus acidocaldarius*
4. *Haloquadratum walsbyi*
5. *Pyrodictium abyssi*
6. *Ignicoccus hospitalis*
7. *Nanoarchaeum equitans*
8. *Methanothermus fervidus*
9. *Methanogenium frigidum*

THE DIVERSITY OF ARCHAEA

These computer-generated images, based on observations of archaea with electron microscopes, provide insight into the varied shapes and sizes of different species of archaea, some of which associate together in colonies. The genus names given to archaea generally come from the environments in which they live, or from their widely varying metabolic capacities (*pyro*: hot, *halo*: salty, *sulfo*: sulfurous, *methano*: producing methane).

© Digital creation Janet Iwasa, University of Utah, Salt Lake City

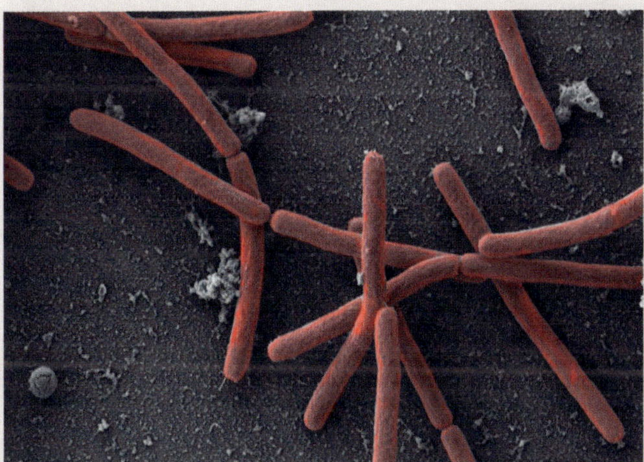

METHANE-PRODUCING ARCHAEA

Archaea of the genus *Methanothermobacter* (here, the species *Methanothermobacter thermautotrophicus*) produce methane gas (CH_4) by reducing carbon dioxide (CO_2) with hydrogen gas (H_2). They generally live in warm environments between 50°C and 70°C.

COLORIZED SCANNING ELECTRON MICROSCOPY

Cyanobacteria— ancient bacteria still abundant today

Cyanobacteria are a remarkable group of bacteria that harness sunlight to generate energy through photosynthesis. Over 3 billion years ago, ancient cyanobacteria began producing oxygen, profoundly transforming the composition of Earth's primitive atmosphere—and culminating in an atmospheric upheaval more than 2 billion years ago, known as the Great Oxygenation Event. This GOE marked a turning point in Earth's history, cooling the planet into a "snowball Earth" and dramatically altering the trajectory of life (see pp. 20–22).

Beyond their impact on the atmosphere, cyanobacteria played a fundamental role in the evolution of eukaryotic plant cells, over a billion years ago. Absorbed by other microorganisms, ancient cyanobacteria became integrated within their host cells—a remarkable occurrence known as *endosymbiosis*—and eventually gave rise to chloroplasts, the membrane-bound organelles responsible for photosynthesis in all vegetal eukaryotic cells, including protists, algae, and plants. This remarkable event was not a singular occurrence but rather a repeated process, giving rise to primary, secondary, and tertiary endosymbioses that shaped the evolution of many, diverse types of plant cell (see p. 88).

These successive endosymbioses enabled plant cells to acquire specialized organelles called *plastids*, each with distinct functions:

- Chloroplasts contain chlorophyll and drive photosynthesis by capturing solar energy.
- Chromoplasts, rich in pigments, give fruits their vibrant colors.
- Amyloplasts are specialized in storing starch.

Today, cyanobacteria continue to thrive in illuminated aquatic environments around the world. Marine species from the genera *Prochlorococcus* and *Synechococcus* are among the most abundant cells on the planet. However, cyanobacteria can also become a source of pollution and toxicity when their populations explode, fueled by nitrate and phosphate pollution in coastal waters, lakes, and ponds. Often referred to as "blue-green algae"—a misleading term—these microorganisms can form dense blooms that disrupt ecosystems and pose health risks. Some genera of cyanobacteria, such as *Anabaena*, form long filaments in which cells are linked together, sharing a common outer wall. To adapt to environmental changes, including oxygen deprivation, filamentous cyanobacteria can rapidly differentiate into two or three complementary cell types. Most cells remain photosynthetic, fixing CO_2, while a minority of cells—called *heterocysts*—fix nitrogen. This enables cyanobacterial species to adapt, grow, and proliferate in a wide range of aquatic and wetland environments.

FILAMENTOUS CYANOBACTERIA IN PLANKTON

Filamentous cyanobacteria are ubiquitous in plankton nets dragged along the surface. These bacteria proliferate in the form of colonies made up of tangled filaments visible to the naked eye, floating on the surface. We collected these *Tricodesmium* cyanobacteria off the coast of Japan.

DARKFIELD MICROSCOPY
© Christian Sardet and the Macronauts, *Plankton Chronicles*, www.planktonchronicles.org

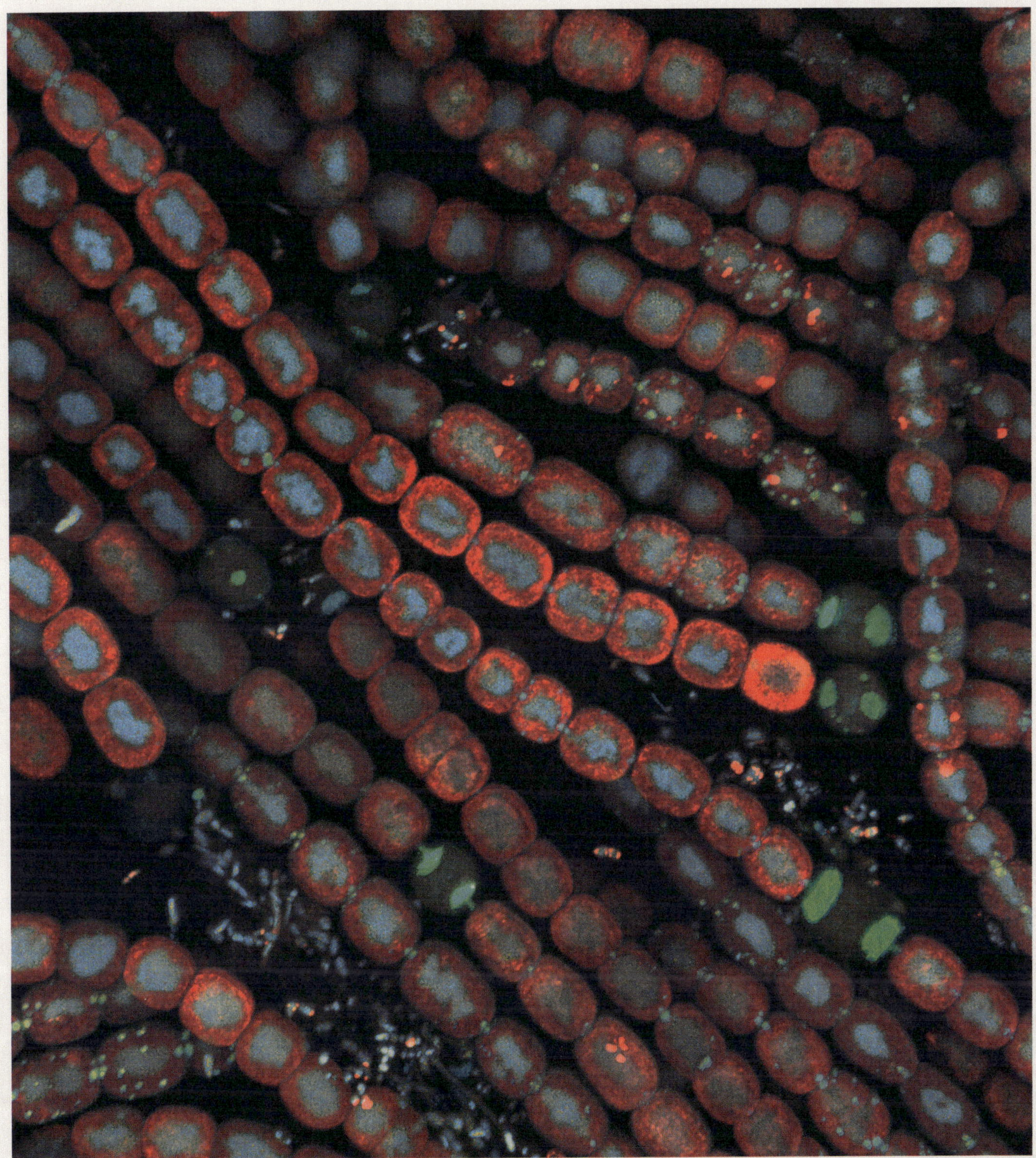

CHAINED CYANOBACTERIA ADAPT

These chains of cyanobacteria of the genus *Anabaena* contain two types of cells. They allow this colony to adapt to changing environmental conditions. If oxygen is lacking, some cells in the chain (about one in twenty) transform into heterocyst cells (*in green*) capable of fixing nitrogen. The genetic material (chromosomal DNA), which fluoresces (*in blue*), is visible in the center of most cyanobacteria cells. Their photosynthetic pigments also fluoresce (*in red*).

CONFOCAL FLUORESCENCE MICROSCOPY
© Courtesy Tagide deCarvalho, Keith R. Porter Imaging Facility, University of Maryland, Baltimore County

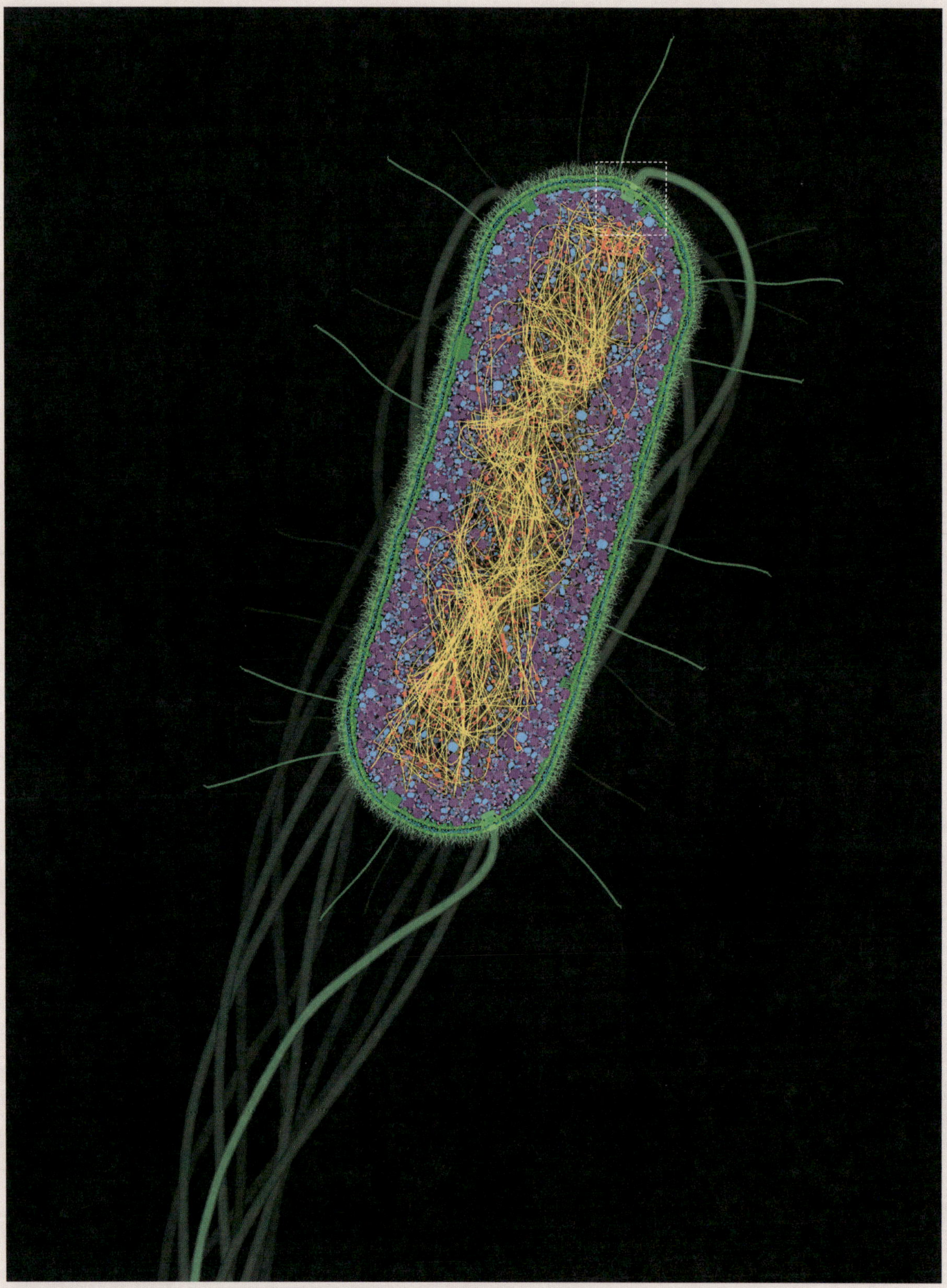

ART MEETS SCIENCE IN A PORTRAIT OF _E. COLI_—A MODEL BACTERIUM

An _E. coli_ bacterium is surrounded by a membrane topped by a hairy wall (_green_). The bacterium moves by means of a cohort of flagella, and its surface bristles with shorter filaments called _pili_. In the center of the bacterium, the cytoplasm is occupied by a chromosome (_yellow filaments_) containing the genes. The cytoplasm is filled with macromolecules and nanomachines (_blue and violet_). The dotted frame highlights the anchoring region of a flagellum, which is magnified ten times in the image at right.

Escherichia coli— a model bacterium

Escherichia coli, commonly known as *E. coli,* is a prolific bacterium inhabiting the intestines of warm-blooded animals. While most strains are harmless and play essential roles in the gut microbiota, a small number are pathogenic and can cause diarrhea. Discovered in 1885 by Theodor Escherich, *E. coli* has since become an invaluable research model. Its beneficial presence in the human digestive system helps produce vitamin K and prevents colonization by harmful microorganisms.

The bacterium itself is rod shaped, measuring between 1 and 4 μm (micrometers) in length. In optimal conditions—with ample oxygen and nutrients—it divides rapidly, roughly every twenty to thirty minutes. This impressive growth rate has made *E. coli* a preferred model organism in microbiological and genetic research. One particularly famous laboratory strain, known as K-12, was among the first organisms to have its complete genome sequenced in 1997, marking a milestone in molecular biology. An *E. coli* cell houses an impressive array of 3 to 4 million proteins, encoded by approximately 4,200 genes on a single circular chromosome, composed of a single DNA molecule.

There is great genetic diversity among different strains of *E. coli,* which share less than one third of their genes. The variable genes enable different strains to adapt to specific ecological niches, such as the intestines of different animal hosts. This adaptability underscores the bacterium's evolutionary success and its ability to occupy diverse environments. Its genetic diversity is not just a scientific curiosity—it has practical applications as well. By analyzing the genes of *E. coli* strains present in contaminated water, it is possible to trace the source of fecal contamination, whether human or animal. Such techniques are crucial for public health monitoring and environmental safety. Industrial, chemical, and pharmaceutical researchers wield their mastery of genetic manipulation and the robustness of *E. coli* to create valuable compounds and recombinant proteins. This model bacterium remains at the forefront of innovation in both pure research and practical applications.

This iconic cell is beautifully depicted in David Goodsell's *Molecular Landscapes,* watercolors that consider the sizes, shapes, and concentrations of cells' molecular constituents. To view his work, scan the QR code below.

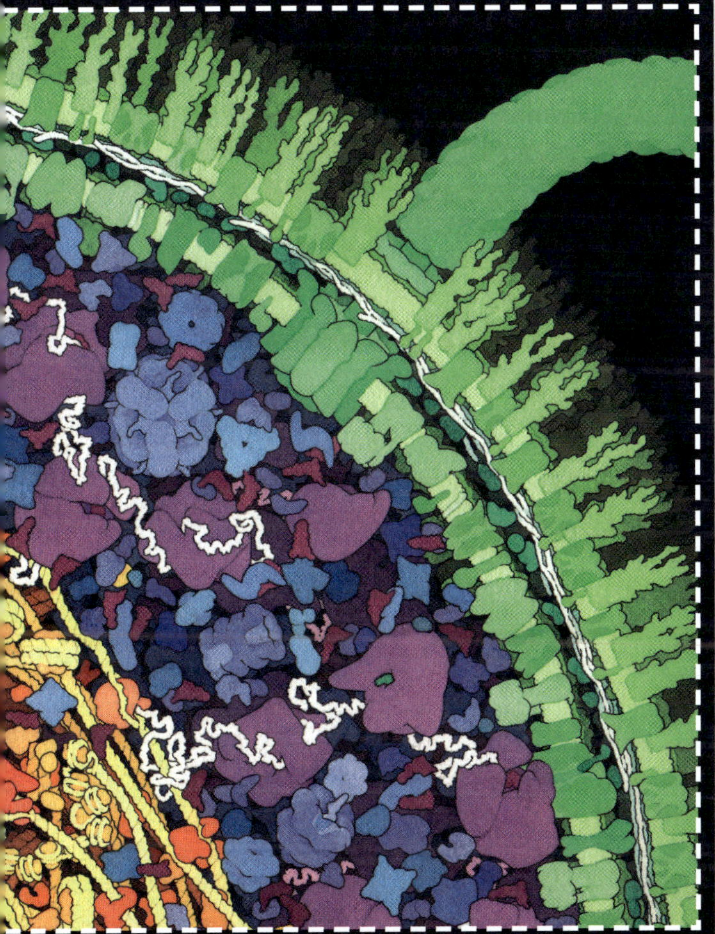

CLOSE-UP OF THE FLAGELLUM ANCHORING REGION OF AN *E. COLI* BACTERIUM

Periphery: The bacteria's lipid membrane (*pale green*) is traversed by proteins (*green*) and protected by an outer, polysaccharide-rich wall (*shades of green*).

Flagellum: The anchoring and rotation zone of the flagellum (*green*) is represented by a ring of proteins forming part of a rotary motor, inserted into the membrane.

Cytoplasm: This region between the chromosomes (*yellow*) and the surface (*green*) contains a dense population of proteins and nanomachines. Ribosomes (*violet*) are shown translating threadlike messenger RNA (*beige filaments*) into proteins of various sizes and shapes, which assemble into nanomachines (*blue*).

Chromosome: The chromosome is represented in the form of filaments (*yellow*) to which are attached nanomachines (*orange*) that copy, transcribe, or repair genes.

© Artwork: *Molecular Landscapes* by David Goodsell, Scripps Research Institute / Protein Data Bank

CHAPTER IV

LIFE BECOMES COMPLEX

The advent of protists

CHIMERIC CELLS FROM BACTERIA AND ARCHAEA

During the first 2 billion years of Earth's history, primitive bacteria and archaea evolved and proliferated on an oxygen-free planet. Then, oxygen-producing cyanobacteria appeared, fundamentally altering Earth's geochemical and climatic balances. Hundreds of millions of years later, more complex composite cells emerged from the fusion of archaea and bacteria. These chimeric cells, known as *eukaryotes*, diversified into a vast array of unicellular organisms, including protists. Despite being considered outdated, the term *protist*, coined by Ernst Haeckel from the Greek *protos*—meaning "very first"—is still appropriate to describe this remarkably diverse group of unicellular organisms equipped with nuclei, organelles, and a cytoskeleton.

Protists likely arose gradually from prokaryotic ancestors. It is well established that ancestral archaea and bacteria exchanged genetic material with one another, as well as through viruses, just as their modern descendants continue to do. To endure harsh conditions, certain species cooperated and complemented one another, forming resilient communities. Many of these close interactions occurred within biofilms, fossilized remnants of which date back over 3.5 billion years. Beyond exchanging molecules and genes, some bacteria and archaea merged—engulfing and coexisting within one another—and evolved to a state of "domestication." On rare occasions, these intimate partnerships gave rise to viable chimeric cells. Over millions of years and countless cell divisions, these chimeras transformed into complex cells (eukaryotes) endowed with energy-producing, membrane-bound organelles. This evolutionary journey led to the development of mitochondria and chloroplasts, descendants of bacteria and cyanobacteria that were incorporated within other cells more than a billion years ago.

The earliest eukaryotic cells were likely animal protists, possessing both mitochondria and nuclei. Subsequent chimeric events gave rise to plant protists, which also acquired chloroplasts. Since that time, protists have diversified alongside archaea and bacteria, interacting with them and with viruses throughout evolutionary history.

When and how did the protists' ancestor, the first eukaryotic cell, come into being? As with LUCA, the origin of LECA (the last eukaryotic cell ancestor) is the source of much debate. But, a plausible account of LECA's genesis is emerging and continues to develop.

4 – More than 1 billion years ago, photosynthetic bacteria (cyanobacteria ◯) were ingested and domesticated by protists ◯. These cyanobacteria ◯ evolved into chloroplasts ●, the organelles that allow plants to thrive on of sunlight using photosynthesis.

3 – More than 1.5 billion years ago, bacteria ◯ and archaea ◯ joined forces to create the first protists, ancestors of all eukaryotes. An archaeon ingested and domesticated a bacterium, which then gradually evolved into a mitochondrion ●, the organelle that provides energy to eukaryotic cells.

2 – LUCA would have evolved into autonomous cells, ancestors of the first bacteria ◯ and archaea ◯, which then colonized the primordial ocean over hundreds of millions of years.

1 – We imagine that, among other protocells, the ancestral cell LUCA appeared in the abyss nearly 4 billion years ago.

Bacteria
Eukaryotes Animal
Vegetal
Protists
Archaea
Bacteria
Bacteria
cyanobacteria
Cyanobacteria
Cyanobacteria
Bacteria
Protists
Bacteria
Bacteria
Bacteria
Bacteria
Archaea
Archaea
Bacteria
LUCA

THE EMERGENCE OF EUKARYOTIC CELLS

Eukaryogenesis (the genesis of LECA, the ancestral cell of all eukaryotes) is a complex process involving two fundamental mechanisms: syntrophy and endosymbiosis. Syntrophy occurs when one organism feeds on the metabolic products of another, while endosymbiosis involves the integration of one organism within another. The potential for a eukaryotic cell began when bacteria and archaea first established a syntrophic partnership. On rare occasions, this evolved into a more intimate relationship—endosymbiosis—which culminated in the formation of a chimeric cell: LECA, the ancestral eukaryote.

It's important to understand that this process took place over hundreds of millions of years, with viable chimeric cells arising only in exceptional cases. Indeed, just as LUCA represents the sole ancestor of all modern cells, LECA is likely the unique progenitor of all protists, animals, plants, fungi, and algae. Astonishing!

Numerous scenarios exist in which syntrophy and endosymbiosis might occur. Typically, they involve two or three types of bacteria and archaea that exchange and exploit various gases, notably hydrogen. Over time, a bacterium incorporated within a host cell gradually evolved into a mitochondrion, becoming highly specialized for energy production.

The implications of this were profound, reshaping life's evolutionary trajectory. Most researchers agree that the presumed host cell—the one that engulfed a bacterium—was an archaeon, specifically from the Asgard lineage. Discovered in deep-sea sediment samples *(see p. 86)*, these archaea possess genes characteristic of eukaryotic cells, suggesting that an Asgard archaeon engulfed a bacterium around 1.5 to 2 billion years ago, eventually giving rise to mitochondria. Nevertheless, competing hypotheses continue to emerge. One intriguing proposition is that the host cell might have been a large bacterium belonging to the *planctomycetes* genus, known for its remarkable ability to engulf other bacteria. Our knowledge of the countless species of archaea and bacteria remains limited, and new discoveries frequently reveal giant and unexpected forms that challenge our established understanding.

Archaea ○ and bacteria ○ live together through mutual exchange, a relationship called *syntrophy*. Each bacterium and archaeon has a circular chromosome ▮. This chromosome is not contained within a nucleus, which is why bacteria and archaea are called *prokaryotes*, meaning "before the nucleus" (the Greek *karyon* means "nut" or "kernel").

Protists ○ are unicellular eukaryotes (eu means "true"). The cell nucleus is an organelle built of intracellular membranes ▮ encircling several linear chromosomes ▮. Protists are characterized by the presence of other organelles, including mitochondria ●, which act as the energy powerhouses of eukaryotic cells. Mitochondria evolved from bacteria that were absorbed and domesticated by other microorganisms, a process known as *endosymbiosis*.

Prokaryote Archaea Prokaryote
Archaea
Prokaryote
Archaea
Bacteria Prokaryote Bacteria Prokaryote Archaea
Bacteria Prokaryote Bacteria
Prokaryote Bacteria
Chromosome
Chromosome
Chromo some
Prokaryote
Archaea
Prokaryote
Archaea
Prokaryote Archaea Prokaryote Archaea
Prokaryote Archaea Prokaryote Archaea Prokaryote
Protist Eukaryote Protist Eukaryote
Eukaryote
Protist
Mitochondria
Chromosome
Chromosome
Chromosome
Chromosome
Chromosome
Mitochondria
DNA
Protist
Eukaryote
Protist Eukaryote Protist Eukaryote Protist
Eukaryote Protist Eukaryote Protist Eukaryote

FROM LECA, THE LAST EUKARYOTIC COMMON ANCESTOR, TO PROTISTS

LECA emerged over 1.5 to 2 billion years ago, likely resembling an animal protist that fed on bacteria and archaea. This primordial protist possessed one or more mitochondria, gained through endosymbiosis with bacteria.

Over generations of LECA's descendants, numerous genes transferred from the domesticated bacterium's chromosome to that of the cell itself—until only a few genes of bacterial origin remained within the mitochondria, which became specialized for energy production.

The first eukaryotic cells capable of harnessing light energy appeared hundreds of millions of years later. These were the plant protists, which contained not only mitochondria but also at least one chloroplast—a photosynthetic organelle derived from a cyanobacterium domesticated by an animal protist. This initial endosymbiotic event was followed by secondary and tertiary endosymbiosis, during which plant protists absorbed other cyanobacteria. These complex chimeric cells gave rise to new evolutionary lineages in the plant kingdom, endowed with chloroplasts and various other plastids carrying pigments or nutrient reserves *(see p. 74)*.

The origins of membrane organelles other than mitochondria and chloroplasts remain enigmatic. Opinions differ as to how eukaryotic cells acquired the nucleus, which envelops the chromosomes within a double membrane, part of an extensive membrane network known as the *endoplasmic reticulum* *(see p. 90)*. One popular hypothesis suggests that the presence of mitochondria within the host cell dramatically amplified its energy and metabolic capacities, paving the way for increased complexity: the formation of intracellular membranes and additional organelles. Another hypothesis points to the possible influence of giant viruses, capable of inducing nucleus-like structures within infected cells. The enigma deepens as scientists continue to uncover giant bacteria harboring multiple chromosomes enclosed by membranes, which challenge the notion that bacteria categorically lack a nucleus—and add fuel to the ongoing debate about the origins of eukaryotic complexity.

ENDOSYMBIOSES AT THE ORIGIN OF ANIMAL AND PLANT PROTISTS

One or more bacteria ◯ were internalized and then domesticated by an archaeon ◯, a process known as *endosymbiosis*. This is thought to be the origin of mitochondria ● in the ancestor of all eukaryotic cells, LECA (the last eukaryotic common ancestor). Then, LECA evolved into protists.

A protist ◯, the ancestral prototype of an animal-like cell, contains mitochondria ● and a network of internal membranes ▮, forming a nucleus that envelops several linear chromosomes ▮. Protists also possess a cytoskeleton made up of microtubules ▮ and microfilaments ▮.

In later endosymbiotic events, protists internalized and then domesticated photosynthetic bacteria (cyanobacteria ◯). These cyanobacteria evolved into chloroplasts ●. These events were the origin of plants, all of which possess chloroplasts ● (in addition to mitochondria ●, other organelles, and a cytoskeleton).

Secondary and tertiary endosymbioses, involving protists ◯ ingesting cyanobacteria ◯ and plant protists, gave rise to new evolutionary lineages.

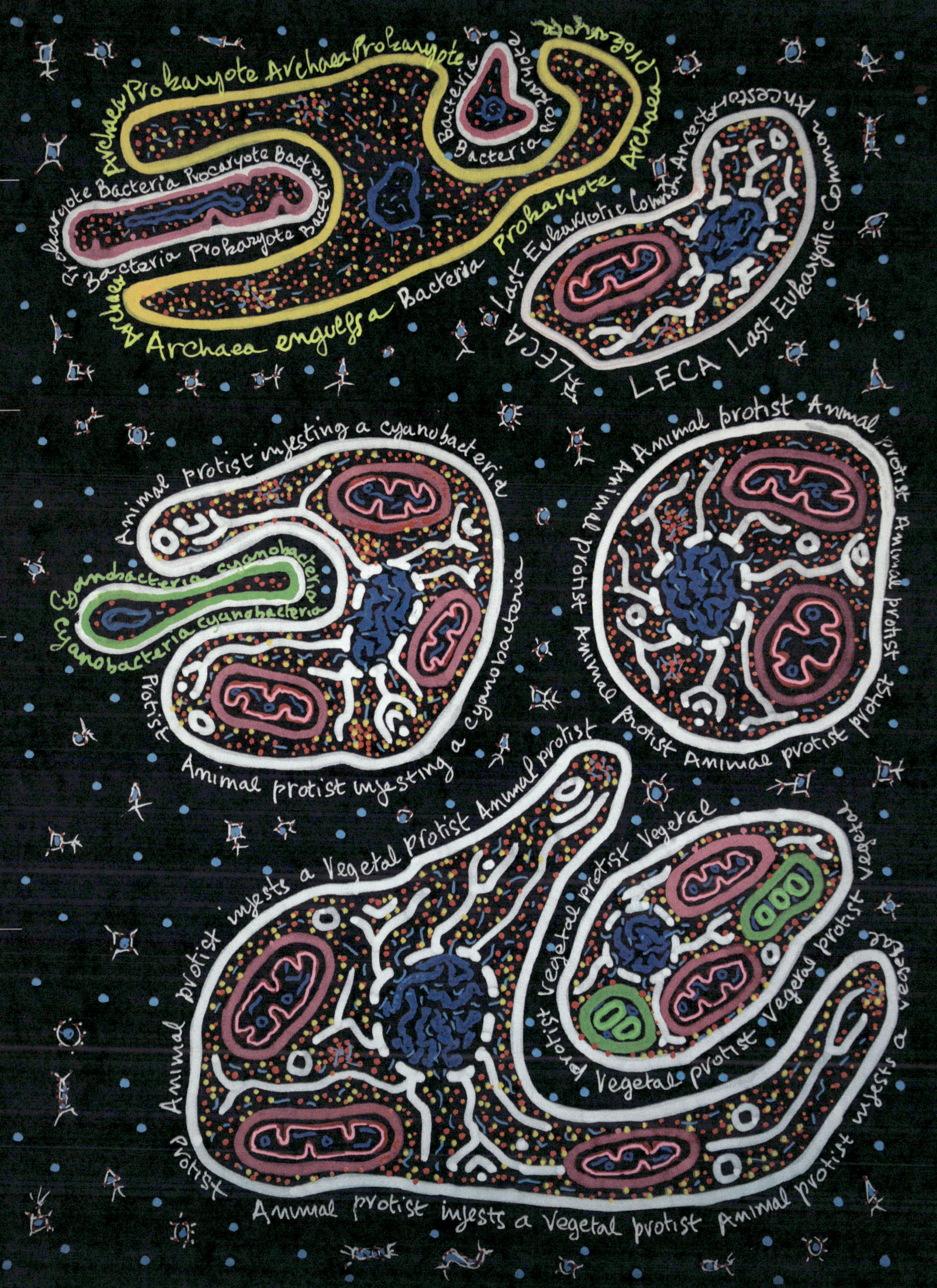

Archaea Prokaryote Archaea Prokaryote
Prokaryote
Bacteria
Bacteria
Archaea Prokaryote Bacteria Prokaryote Bacteria
Bacteria Prokaryote Bacteria
Archaea Bacteria Prokaryote Bacteria
Archaea engulfs a Bacteria Prokaryote Archaea
Last Eukaryotic Common Ancestor
Last Eukaryotic Common Ancestor
LECA
LECA Last Eukaryotic Common
Animal protist ingesting a cyanobacteria
Cyanobacteria cyanobacteria
Cyanobacteria cyanobacteria
Cyanobacteria cyanobacteria
Protist
Animal protist ingesting a cyanobacteria
Animal protist Animal protist Animal
protist Animal protist Animal protist Animal
protist Animal protist Animal protist protist
Animal protist protist Animal protist protist
Animal protist ingests a Vegetal protist Animal protist Vegetal
Vegetal protist Vegetal
Protist Vegetal protist Vegetal protist Vegetal
Vegetal protist Vegetal protist
Animal protist Vegetal protist a vegetal
Protist
Animal protist ingests a Vegetal protist Animal protist

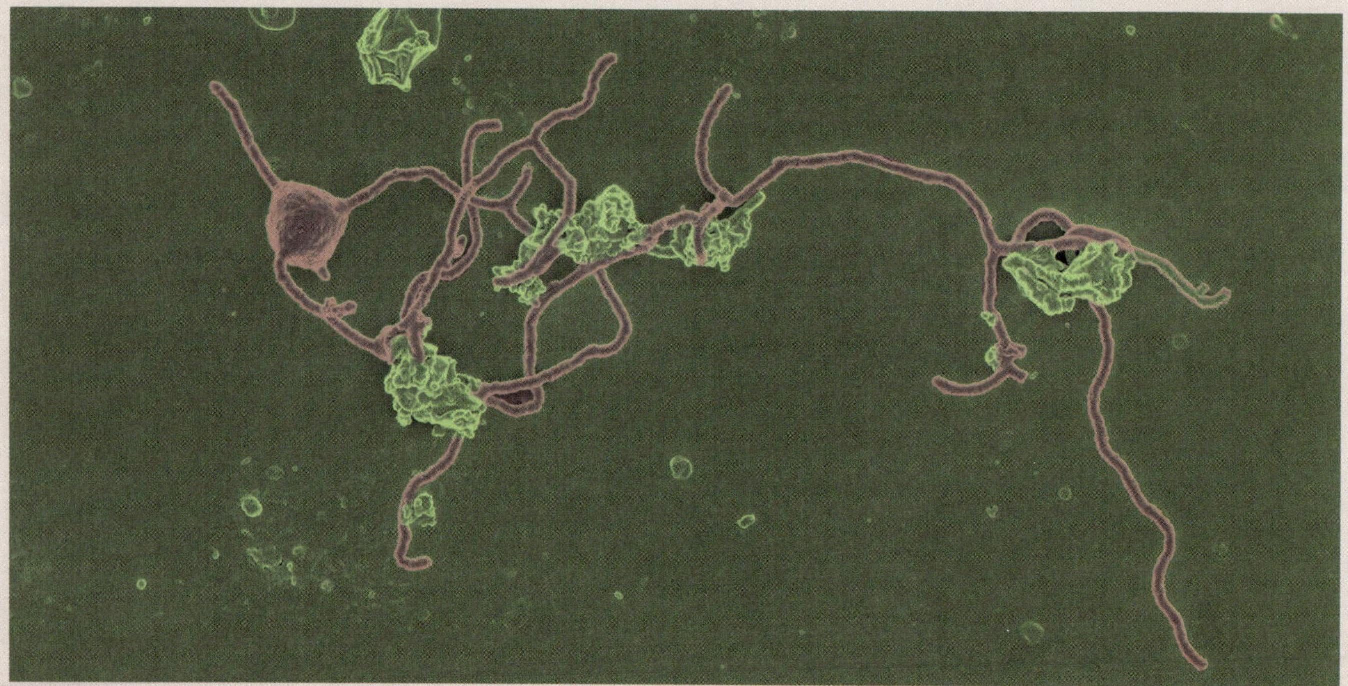

The mystery of eukaryotic origins

Within the past twenty years, genetic research has revealed that archaea are more closely related to eukaryotic cells than they are to bacteria. This hypothesis gained significant traction in 2010 with the discovery of new archaea, known as *Asgard archaea*, in the marine sediments of Greenland. These microorganisms were named for Asgard, the mythical realm in Norse mythology, as they were found near abyssal, hydrothermal springs called Loki's Castle—a nod to Loki, the trickster god who thrives on chaos and deception. Today, most researchers consider the Lokiarchaeota, a subgroup of Asgard archaea, to be closely related to eukaryotes. This assertion stems from the presence of genes encoding proteins involved in internal membrane transits and the actin cytoskeleton, both of which are hallmark features of eukaryotic cells. Despite these similarities, no archaea have yet been observed to internalize bacteria, a critical step toward potential endosymbiosis. Instead, cultured Asgard archaea are known to live in syntrophy with bacteria, sharing metabolic by-products rather than engulfing them.

An alternative hypothesis posits that eukaryotes originated from planctomycetes, a group of large marine bacteria. Unlike most other bacterial species, planctomycetes exhibit two remarkable characteristics in

ASGARD ARCHAEA—EUKARYOTES' NEAREST COUSIN?

After a decade of effort, a Japanese research team—Hiroyuki Imachi, Masaru K. Nobu, et al.—recently cultivated the first Asgard archaea (*Prometheoarcheum syntrophicum*) successfully grown in culture. This archaeon, whose cell body measures 0.5 μm (micrometers), is characterized by long, slender tendrils. *Promethearcheum syntrophicum* possesses many proteins characteristic of eukaryotic cells, but lacks an internal membrane network, and does not exhibit bacteria-engulfing activity. However, these slow-growing archaea depend on the presence of other bacteria and archaea species in their cultures. This necessary cooperation between microbes is called *syntrophy*.

COLORIZED SCANNING ELECTRON MICROSCOPY
© Courtesy Hiroyuki Imachi and Masaru K. Nobu, JAMSTEC, Japan

common with eukaryotic cells. First, they possess a network of internal membranes within their cytoplasm. Second, some planctomycete bacteria from the *Candidatum* genus have the ability to engulf smaller bacteria—a rare property in prokaryotes. However, it is important to note that *Candidatum* bacteria do not use the typical protein machinery of phagocytosis (the process by which eukaryotic cells, such as macrophages, engulf foreign particles or microorganisms). Eukaryotic cells use specialized protein families for this, which planctomycetes lack. Notably, these phagocytosis-related proteins are found in many protists, including amoebae, which excel at capturing and internalizing prey. The search to trace the evolutionary origins of the eukaryotic cell continues, drawing clues from both archaea and bacteria.

A giant bacterium shakes our certainties

A filament-like bacterium visible to the naked eye, discovered in the mangroves of Guadeloupe, has challenged conventional wisdom about the differences between prokaryotes (cells without nuclei) and eukaryotes (cells with nuclei). Indeed, the giant bacterium *Thiomargarita magnifica* possesses three characteristics generally associated with eukaryotic cells.

First, the size: Some *T. magnifica* bacteria measure more than 2 cm. They are 10,000 times larger than typical bacteria, such as *Escherichia coli*, which measures 1 to 4 μm. That said, at least four types of bacteria (including planctomycetes) were already known to be comparably large. Before the discovery of *T. magnifica*, the size record was held by bacteria of the same genus discovered in sediments off the Namibian coast: *Thiomargarita namibiensis*, nicknamed "sulfur pearls," the largest of which measure nearly 1 mm in diameter.

Second, *Thiomargarita* bacteria exhibit a remarkable cellular organization. Like other macro-bacteria that thrive in sulfur-rich environments, *Thiomargarita* possesses multiple types of organelles. At the periphery, thousands of sulfur-filled globules give the bacterium its lustrous appearance. At its center lies a voluminous sac that stores water and, in some species, nitrates. In *T. magnifica*, this large central vacuole occupies over 70 percent of the bacterium's volume. It is surrounded by a thin layer of cytoplasm, 3 to 4 μm thick, containing a network of internal membranes, which supply energy in the form of ATP, as well as hundreds of thousands of membrane-bound organelles, known as *pepins* (or pips), which contain DNA and ribosomes. They enable the bacterium's polyploid genome (containing multiple sets of chromosomes) to be distributed all along its length.

Finally, the genome of *T. magnifica* is exceptionally large, comprising 11,000 genes—that's around three times larger than the average bacterial or archaeal genome. It remains uncertain whether this giant bacterium harbors the genes and proteins of the cytoskeleton, or manifests the membrane "transit system" that characterizes eukaryotic cells.

The unique features of *Thiomargarita* raise intriguing questions: Are the pepins akin to nuclei? Could these giant bacteria represent a missing link between prokaryotes and eukaryotes? While it is too early to draw conclusions, the discovery challenges our preconceptions and highlights the potential of exploring the diversity of bacteria and archaea.

***THIOMARGARITA MAGNIFICA*, GIANT BACTERIA FROM THE GUADELOUPEAN MANGROVES**

In the early 2000s, biologist Olivier Gros discovered whitish filaments attached by one end to the surface of decomposing leaves in the mangroves of Guadeloupe. He suspected a bacterium. His collaborator Jean-Marie Volland continued and expanded this research in the United States. This giant cell of the genus *Thiomargarita* is one of several chemolithotrophic gammaproteobacteria that thrive in environments rich in hydrogen sulfide (H_2S) and organic matter. Unlike most microscopic bacteria, which divide by binary fission (splitting in two) within minutes or hours at most, *T. magnifica* take from several days to weeks to reproduce by budding daughter cells at one end of the filament. Little is known about their development cycle, as laboratory culture of these bacteria has not yet been mastered.

LIGHT MICROSCOPY
© Courtesy Tomas Tyml, The Regents of the University of California, Lawrence Berkeley National Laboratory

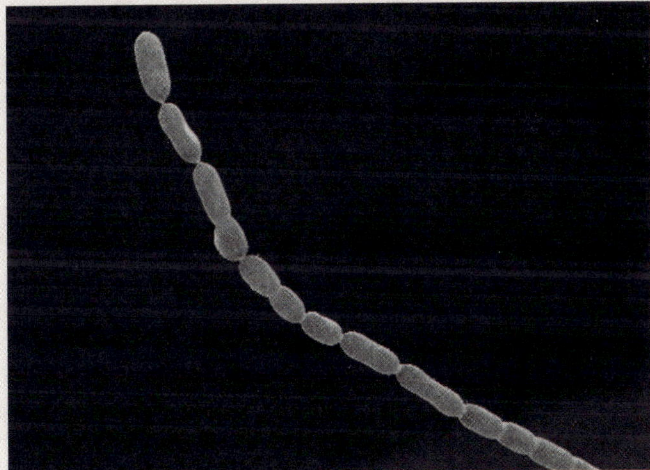

***THIOMARGARITA MAGNIFICA* BUDS DAUGHTER CELLS**

This giant bacterium is also atypical when it comes to reproduction. It doesn't need to double its volume, as most bacteria do before dividing, since it reproduces by budding a daughter cell at its free end. Only one daughter cell separates at a time from the rest of the bacterium, ensuring their dispersal.

SCANNING ELECTRON MICROSCOPY
© Courtesy Jean-Marie Volland, The Regents of the University of California, Lawrence Berkeley National Laboratory

Mitochondria and chloroplasts— energy powerhouses descended from bacteria

Mitochondria and chloroplasts are unique membrane-bound organelles with bacterial ancestry. Mitochondria originated from a bacterium that was integrated and domesticated via endosymbiosis by another microorganism, around 1.5 to 2 billion years ago *(see pp. 82–84)*. Chloroplasts, on the other hand, emerged from an ancient cyanobacterial endosymbiosis over a billion years ago.

Since these pivotal events, numerous genes of bacterial origin have transferred to the genome of the eukaryotic cell. So, most mitochondrial and chloroplast proteins are encoded by the eukaryotic cell's nuclear genes, produced in its cytoplasm, and finally imported into the mitochondria or chloroplasts. Nevertheless, mitochondria and chloroplasts have also retained a small set of their own bacterial genes that produce crucial proteins for their energy operations, giving these organelles a certain degree of autonomy.

Remarkably, both organelles continue to divide independently during mitosis, although they are attuned to the separation of the chromosomes and other cellular components. Mitochondria and chloroplasts engage in continuous communication with the nuclear genome and cell-cycle regulators, synchronizing their divisions with the cellular tempo *(see chap. VII)*.

Mitochondria and chloroplasts function as the powerhouses of the cell, playing vital roles in energy production and metabolic regulation. They also actively participate in signaling pathways and interact with other organelles, such as the endoplasmic reticulum. Beyond their role in cellular metabolism, mitochondria are central to programmed cell death (apoptosis) and are implicated in various pathologies *(see p. 184)*. Over time, mitochondria and chloroplasts have diversified into other membrane-bound organelles. Mitochondria have given rise to hydrogenosomes and mitosomes, which lack genetic material and energy-production capabilities but have specialized metabolic functions. Similarly, chloroplasts have evolved into various plastids with specialized roles: for example, chromoplasts, which store the pigments that give flowers and fruits their vivid colors, and amyloplasts, which warehouse starch granules. These specialized plastids reflect chloroplasts' evolution and adaptation to fill diverse metabolic and storage functions.

CHLOROPLASTS CARRY OUT PHOTOSYNTHESIS IN THE CELLS OF PLANTS AND ALGAE

These two chloroplasts in a plant cell contain stacks of thylakoid membranes (from the Greek *thylakos*, meaning "bag," and *oïdes*, "similar"). The stacks of bags (*yellow*) contain chlorophyll, which captures luminous photons and converts light energy into chemical energy. This energy and the capture of CO_2 enable the production of sugars (*see p. 40*), which are stored in the form of starch (*black grains*).

THIN SECTION OF A PLANT CELL OBSERVED UNDER AN ELECTRON MICROSCOPE AND COLORIZED

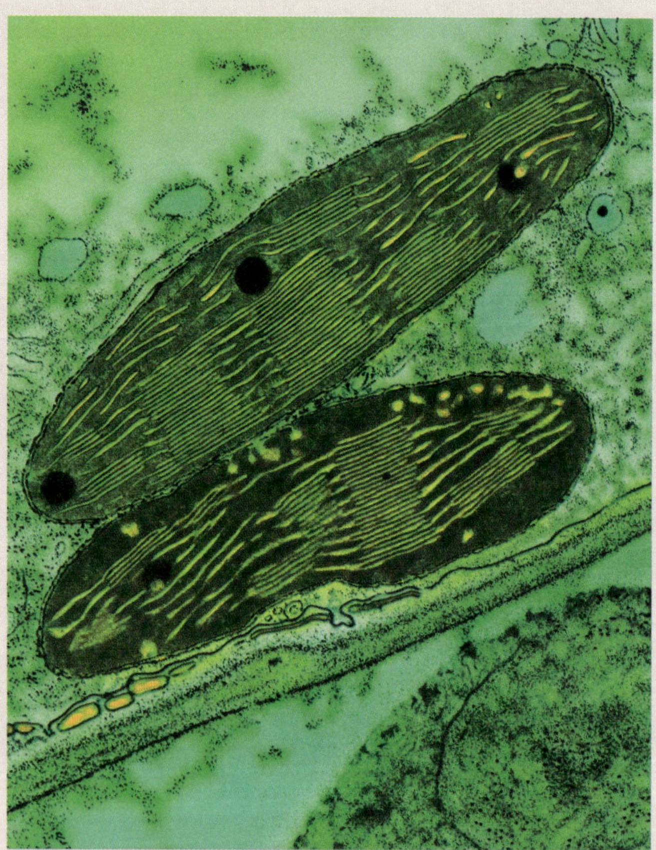

Opposite

MITOCHONDRIA SUPPLY EUKARYOTIC CELLS WITH ENERGY

Mitochondria are present in all eukaryotic cells. Here, the spaces between the inner and outer mitochondrial membranes are colored red. Mitochondria are particularly numerous in cells with high energy requirements, such as this muscle cell. Muscle cells contain contractile fibers made up of rows of actin filaments (*colored brown*). The filaments slide over each other during muscle contraction thanks to motor proteins called *myosins*, which use energy in the form of ATP, produced inside the inner membranes of mitochondria (*see pp. 132–33*).

THIN SECTION OF A MUSCLE CELL OBSERVED UNDER AN ELECTRON MICROSCOPE AND COLORIZED

© Christian and Dana Sardet

The nucleus— emblematic organelle of eukaryotic cells

The nucleus, a ubiquitous organelle in eukaryotic cells, houses chromosomes and nucleoli, as well as smaller biomolecular condensates. Like the mitochondria, the nucleus plays a central role in cellular function. While the bacterial ancestry of mitochondria is undeniable, the evolutionary pathway that led to the formation of the nucleus is still mysterious.

The nucleus is enveloped by a double membrane pierced by nuclear pores—openings that allow the passage of ions, molecules, and nanomachines from the nucleus to the cytoplasm and vice versa *(see p. 162)*. The double membrane delimiting the nucleus is not an isolated structure, but rather a specialized extension of the cell's internal membrane network, known as the *endoplasmic reticulum*. In most eukaryotic cell types, this membrane network disassembles during mitosis to allow chromosome separation, only to be reassembled at the end of division. Outside of mitosis, the nuclear membrane effectively isolates the chromosomes from the rest of the cell, safeguarding the genome.

Within the nucleus, chromosomal DNA is duplicated and transcribed into RNA. Ribosomes (the nanomachines responsible for protein synthesis) are assembled in the nucleoli, which are condensates within the nucleus.

However, the final step of protein production—the translation of RNA into proteins—occurs outside the nucleus, in the cytoplasm. To accomplish this, the ribosomes and messenger RNA, originally produced within the nucleus, must exit through the nuclear pores. This separation between transcription (DNA into RNA) and translation (RNA into proteins) fundamentally distinguishes eukaryotes from prokaryotes. In bacteria and archaea, which lack a nucleus, the processes of transcription and translation are physically coupled and take place in the cytoplasm *(see p. 167)*. However, some large bacteria (planctomycetes and *Thiomargarita* sp.) possess networks of internal membranes, reminiscent of nuclei, surrounding the chromosomes. These peculiar bacteria could represent an evolutionary bridge between prokaryotes and eukaryotes *(see pp. 86–87)*. An alternative hypothesis posits that the nucleus may have originated from membrane structures known as *virus factories* that form during viral infections *(see p. 117)*. The mystery of the nucleus's origin, like that of many other organelles, remains unsolved.

**ÉDOUARD CHATTON
DRAWS THE NUCLEUS**

At the center of a cell sits a nucleus containing chromosomes (*in blue*) and a nucleolus (*in red*). Beside it is a star-shaped structure of the cytoskeleton—an aster—made of microtubules (*blue lines*), which emanate from an organizing center—the centrosome—containing two centrioles (*blue dots*) in its center. These drawings are the work of Édouard Chatton, to whom we owe the two main categories of cells: eukaryotes (cells with a nucleus) and prokaryotes (cells without), which he distinguished and named in 1925. Chatton represents the nucleus in two states: At the beginning of cell division, the chromosomes (*in blue*) begin to form filaments, and the nucleolus (*in red*) begins to disappear, while the cellular skeleton (the cytoskeleton) forms an aster. Then, the centrosomes move away from each other to prepare for the division of the cell by mitosis *(see p. 146)*.

THE ENDOPLASMIC RETICULUM: AN EXTENSIVE NETWORK OF INTERNAL MEMBRANES

Right

Eukaryotic cells are characterized by a nucleus. The nuclear membrane is part of a continuous network of membranes—namely the endoplasmic reticulum, an organelle that performs a multitude of functions. Here, this network of tubular membranes, labeled with a red-fluorescing protein, fills a cell in culture.

PROJECTION OF CONFOCAL FLUORESCENCE
MICROSCOPY SECTIONS
© Courtesy Andy Moore and Jennifer Lippincott-Schwartz,
Advanced Imaging Center, Howard Hughes Medical
Institute, Janelia Research Campus, Ashburn, Virginia

Below

An enlarged view of the endoplasmic reticulum, made up of membrane tubes and sheets (*red*) and partly covered by ribosomes (*yellow*), the protein-synthesizing nanomachines. In the background, we can see the inner surface of the cell membrane (*brown*), covered with cytoskeletal microfilaments (*orange*) and endocytotic vesicles (*blue*).

ELECTRON MICROSCOPY, FAST FREEZING/DEEP ETCHING TECHNIQUE
© Christian Sardet

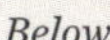

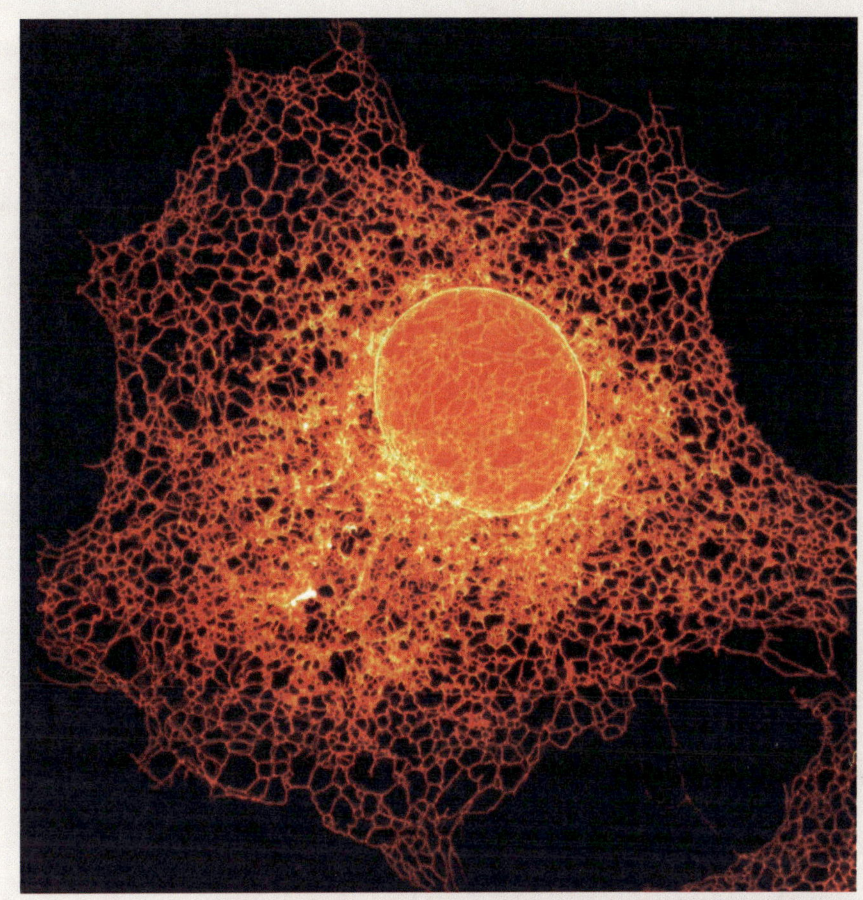

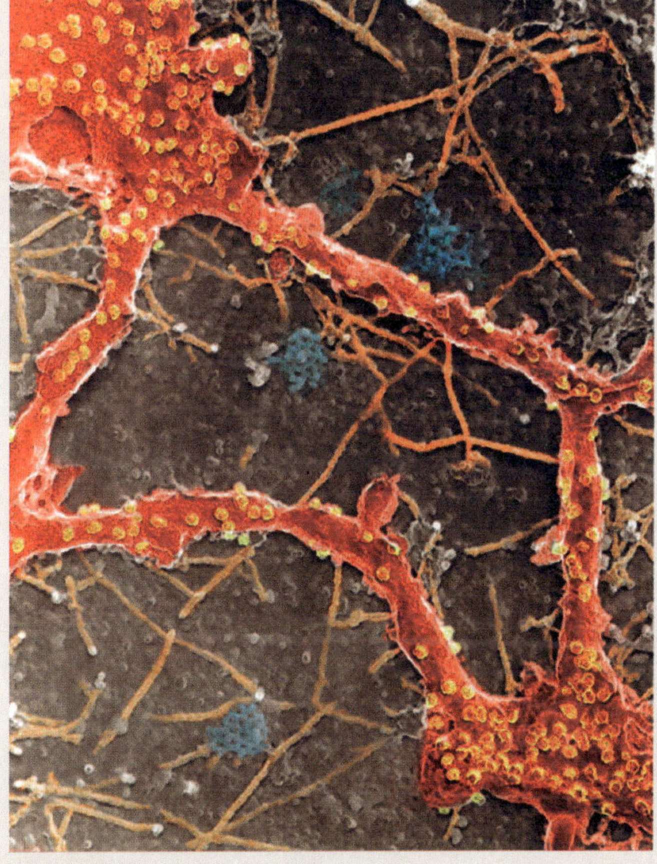

THE TEEMING WORLD OF PROTISTS

Life took a remarkable leap forward in complexity with the emergence of eukaryotic cells, which appeared as single-celled protists over 1.5 billion years ago. Since then, protists have continued to evolve, and, depending on their environment, they have specialized to live either as single cells or in colonies. Throughout the past 500 million years, countless protist lineages have risen and fallen, shaped by waves of extinction of varying magnitude. Photosynthetic organelles—chloroplasts—enabled certain protists, such as microalgae and diatoms, to harness light for energy, earning them the title *autotrophs*. By contrast, other protists—such as amoebae and paramecia—are *heterotrophs*, equipped with specialized organelles that ingest prey and organic matter. Some protists, like dinoflagellates and euglenoids, are *mixotrophs*, capable of thriving on both light and organic matter, either simultaneously or alternately.

Protists epitomize the astonishing evolution of biodiversity within the living world. Despite millions of known species classified into eight principal groups, the full inventory remains far from complete. These single-celled organisms, much like animals, exhibit an impressive array of functions: They move, feed, sense light, attack, and defend themselves—all thanks to their specialized organelles. Some protists even demonstrate signs of intelligent behavior.

A drop of pond or sea water, seen through a microscope, reveals a vibrant world teeming with life—paramecia, amoebas, diatoms—existing singly or in colonies. Among the most notorious protists are parasitic species that cause diseases like malaria, sleeping sickness, and toxoplasmosis.

Other protists, such as the ciliates and flagellates of plankton, are not much bigger than bacteria and archaea, with which they form unicellular communities. But there are also giant protists, such as *Stentor*, that can be seen even with the naked eye, and the famous "blob"—the social amoeba *Physarum*—endowed with learning, memory, and extraordinary regenerative capabilities.

Many protists have complex life cycles. They are also the originators of two major cellular innovations: sexual reproduction and multicellularity, which we explore in the following chapters. In the meantime, let's get to know a few emblematic protists.

Protists ○ are unicellular eukaryotes that live alone or in colonies. They come in a wide variety of shapes and sizes, and with various appendages and organelles. All have a nucleus ● with chromosomes ▮, mitochondria ●, and a cytoskeleton consisting of microtubules ▮ and microfilaments ▮. Plant protists also possess photosynthetic chloroplasts ●.

Some protists, like the dinoflagellate and paramecium (*top left*), move quickly using flagella or cilia. Others, like the acantharian (*right*), possess long extensions to capture prey. These animal protists often live in symbiosis with plant protists.

A wide variety of ciliated or flagellated protists proliferate in plankton. They live from prey they capture, from photosynthesis, or from both.

Diatoms, the most abundant protists in phytoplankton, live alone or in colonies of identical cells. Diatoms surround themselves with protective envelopes made of silica.

Protists have complex life cycles and some, like this volvocale (*right*), form spherical colonies of cells, including larger reproductive cells.

Giant protists—
Stentor and *Physarum*,
the blob

Let's go exploring in the woods and ponds for giant protists, like *Stentor* and *Physarum*, famously known as "the blob." In ponds, with the aid of a magnifying glass, you might catch sight of a *Stentor* gracefully stretching. In the forest, you might encounter blobs—those gelatinous amoebas spreading their yellowish pseudopods over undergrowth and decaying wood.

Inside the blob, millions of nuclei coexist within a single shared cytoplasm. Remarkably, if the blob is cut into small pieces—provided that a fragment contains a nucleus—it will regenerate into a complete blob. Despite lacking a brain or nervous system, the blob exhibits a surprising ability to learn, even determining the shortest path through a maze to reach food. It can transmit this acquired knowledge to other blobs by merging with them.

Stentor also demonstrates exceptional regenerative abilities. Regardless of how it is sliced up, each fragment containing a nucleus and a portion of the ciliated cell periphery—the cortex—will regenerate as a new *Stentor* within a few hours, complete with its characteristic, ciliated oral apparatus (a sort of mouth). This remarkable phenomenon suggests that the blueprint for forming a new *Stentor* resides not only within its nuclear genome, but also within the cortex—as if each fragment of the original structure retains a memory of the whole. The cellular and molecular foundations of this mode of information retention—an aspect of epigenetics—remain enigmatic.

**THE BLOB: THE SOCIAL AMOEBA
*PHYSARUM POLYCEPHALUM***

This myxomycete amoeba is in the form of a plasmodium, extending over about 20 cm. The plasmodium is a single cell, containing thousands of nuclei within a single cytoplasm (*see p. 106*). The amoeba moves a few centimeters per hour and doubles in size every day. Using its extensions, called *pseudopodia*, the blob explores its environment in search of the bacteria and fungi on which it feeds.

© Courtesy Audrey Dussutour, CRCA Laboratory, CNRS, University of Toulouse and CNRS Images, France

Right

STENTOR, A SOPHISTICATED ANIMAL PROTIST

Stentor coeruleus is a giant bugle-shaped cell with a ciliated oral apparatus that enables it to feed on microorganisms. This protist attracts and catches prey by creating whirlpools with its ring of cilia (*in red*), swallowing bacteria, archaea, and microalgae into its "mouth" at the center of this apparatus. Each *Stentor* possesses one large nucleus and a multitude of smaller nuclei (macro- and micronuclei *in blue*). All contain chromosomes carrying the *Stentor*'s genes. The peripheral region (the cortex) is rich in microtubules arranged in parallel beneath the membrane.

PROJECTIONS OF CONFOCAL FLUORESCENCE MICROSCOPY SECTIONS
© Courtesy Igor Siwanowicz, Advanced Imaging Center, Howard Hughes Medical Institute, Janelia Research Campus, Ashburn, Virginia

 View *The Blob: A Cell that Learns* by Audrey Dussoutour, CNRS News

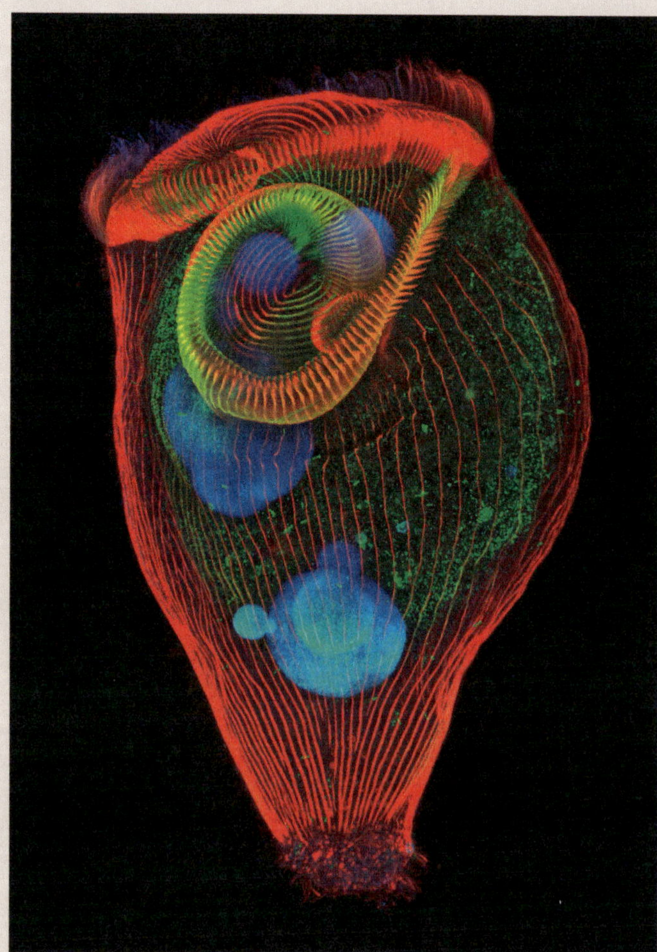

Unicellular eukaryote models—*Saccharomyces* yeast and *Chlamydomonas* microalgae

At first glance, yeasts—most of which are unicellular—might seem to belong to the protists, a category that encompasses all eukaryotic organisms excluding animals, plants, algae, and fungi. In fact, yeasts are closely related to fungi, a diverse kingdom comprising over 100,000 species, both unicellular and multicellular *(see pp. 104–5)*. Among the most-studied yeast species are *Saccharomyces cerevisiae* (brewer's yeast) and *Schizosaccharomyces pombe*, which have long served as model organisms in laboratories due to their ease of cultivation. Since Pasteur's pioneering work on fermentation, these yeasts have contributed to major scientific breakthroughs. In 1996, *Saccharomyces cerevisiae* made history as the first eukaryotic genome to be fully sequenced, comprising around 6,000 genes spread across 16 chromosomes. Remarkably, approximately 1,000 yeast genes have human equivalents (orthologs), allowing researchers to investigate gene functions relevant to various diseases. As a result, yeast models have become indispensable in advancing our understanding of metabolism, gene expression, cell division, and sexual reproduction. Today, tens of thousands of researchers worldwide rely on these organisms for both scientific research and industrial applications.

The green alga *Chlamydomonas reinhardtii*, a plant protist affectionately called "Chlamy" by the large community of researchers who have adopted it as a research model *(see p. 103)*, is one of 500 species of *Chlamydomonas* that live in wet or aquatic environments, including ice and snow.

Chlamy, whose genome (15,000 genes, 17 chromosomes) was sequenced in 2007, is sometimes nicknamed the "green yeast" for its comparably simple structure and ease of genetic manipulation. Research into this species first revealed that chloroplasts have their own genes, and has illuminated many mechanisms of photosynthesis, sexuality, and the growth of flagellae. Our ability to modulate the metabolic capacities of *Chlamydomonas* makes it an ideal organism with which to study the production of biofuels and hydrogen gas.

These model unicellular species have also contributed to our understanding of the transition from single-celled to multicellular life *(see p. 100)*.

SACCHAROMYCES CEREVISIAE

Saccharomyces yeasts are oblong eukaryotic cells, measuring around 10 µm. They reproduce by budding daughter cells, like the one at the top of the picture. The green areas on the mother cells represent scars from repeated budding. These yeasts have been used since antiquity to make bread rise and ferment wine and beer. Ferments have been produced industrially since the 17th century. *Saccharomyces* yeasts lend themselves easily to genetic manipulation, and are mass-cultured for the industrial production of chemical and pharmaceutical substances.

COLORIZED SCANNING ELECTRON MICROSCOPY

© Courtesy Nicole Ottawa and Oliver Meckes, *eye of science*, Germany

Parasitic protists, the cause of malaria and toxoplasmosis

Infectious diseases are transmitted by bacteria and viruses, but also by protists (see p. 190). Parasites of the genus *Plasmodium*, responsible for the transmission and symptoms of malaria, belong to a large family of parasitic protists—the apicomplexans or sporozoans. These parasitic protists are also the infectious agents of other diseases, including toxoplasmosis—transmitted by another kind of sporozoan, *Toxoplasma*. Although less frightening than malaria, toxoplasmosis represents a health hazard, especially for immunocompromised people and pregnant women. It is estimated that the *Toxoplasma* parasite is present in 30 percent of the world's population. *Toxoplasma* is known to be transmitted by cats, via their feces. Curiously, this parasitic protist influences the metabolism and neurotransmitters of infected rodents and humans, triggering changes in their behavior.

Plasmodium and *Toxoplasma* protists share a sophisticated cellular machinery for invading the cells of animals, humans, insects, and even earthworms. In particular, the anterior part of these parasitic cells contains a set of structures and organelles—vesicles, granules, and cytoskeletal elements—which enable them to penetrate the cells of their animal hosts. Then, the protists settle inside the host cells they've infected, stealing some of their membranes to build protective envelopes around themselves. As a consequence, *Plasmodium* and *Toxoplasma* are well protected from phagocytes, the destructive cells of the host's immune system.

The life cycles of these parasitic protists within the various cell types they infect are particularly complex. While enslaving the infected cells and feeding on their proteins, the protists change morphology and behavior. They reproduce inside their hosts' cells, eventually causing them to explode. Once released into the host's bloodstream, the protist parasites spread and infect other cells.

THE MALARIA PLASMODIUM

This so-called *ookinete* form of the malaria-transmitting *Plasmodium berghei* protist measures between 20 and 50 μm long. The parasite possesses an elaborate cytoskeleton consisting of a network of microtubules (*magenta*), located beneath the cell surface. Its cortical structure, together with its ring-shaped organelle—the conoid, visible on one end—enables the protist to penetrate the cells of its host.

CONFOCAL FLUORESCENCE MICROSCOPY
© Courtesy of Eloïse Bertiaux, laboratory of Virginie Hamel and Paul Guichard, University of Geneva, Switzerland

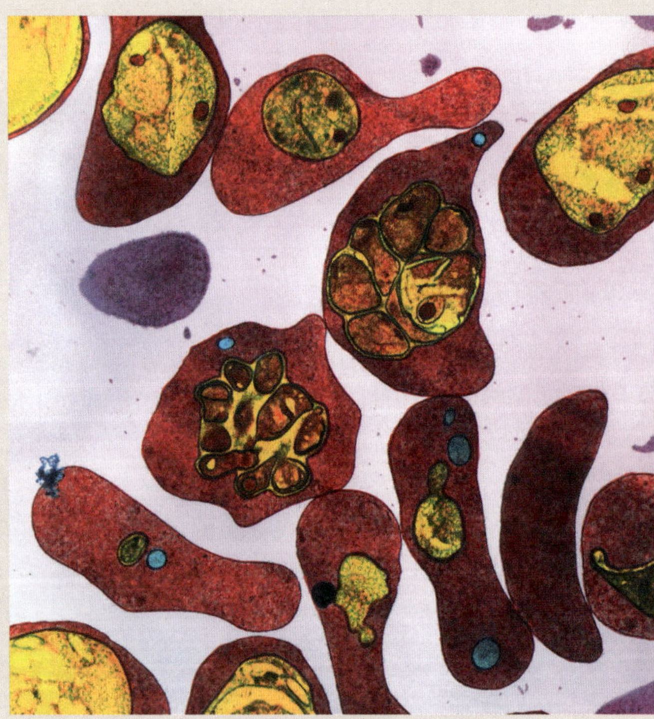

CELL INFECTED WITH PARASITIC PROTISTS THAT CAUSE MALARIA

Several *Plasmodium cathemerium* cells (*brown*) multiplying inside a vesicle (*yellow*) in an infected blood cell (*red*).

COLORIZED THIN SECTION, ELECTRON MICROSCOPY

Solitary and colonial planktonic protists

Planktonic protists drift with the currents in the oceans. Some, like diatoms, are plants, while others, like radiolaria, have an animal lifestyle, and sometimes live in symbiosis with plant protists.

Diatoms, abundant in all aquatic environments, are key photosynthetic organisms, responsible for producing approximately 25 percent of the planet's oxygen. They typically float near the surface, maximizing light capture to optimize photosynthesis. They also sense depth, and seem to communicate using natural fluorescence.

These protists exist either as solitary cells or in colonies—associations that can be permanent or temporary, depending on their reproductive needs and environmental conditions. The tens of thousands of identified diatom species vary in size from 2 μm across to several hundred, each diatom encased in a beautifully ornamented silica shell.

Many species form colonies, arranging themselves into chains, fans, spirals, or other configurations characteristic of their kind. By secreting adhesive molecules, diatom cells bond their shells together, enhancing their collective buoyancy and size, which helps deter predators and optimize light exposure.

Other protists, such as radiolaria and acantharians, form their colonies within a shared gelatinous envelope. This structure enables these heterotrophic cells to float near the surface, where they capture bacteria and microalgae for sustenance.

Notably, many radiolarian species engage in photosymbiosis by hosting microalgal symbionts, such as dinoflagellates, within their cells. In this symbiotic relationship, the microalgal partners perform photosynthesis, providing nutrients to their hosts, particularly in nutrient-poor oceanic regions such as the South Pacific.

Often, the microalgal partner is profoundly transformed by the host, and in some cases this symbiosis resembles slavery more than cooperation (see p. 215).

View *Protists 2, Plankton Chronicles*
by Christian Sardet, Sharif Mirshak,
and Noé Sardet

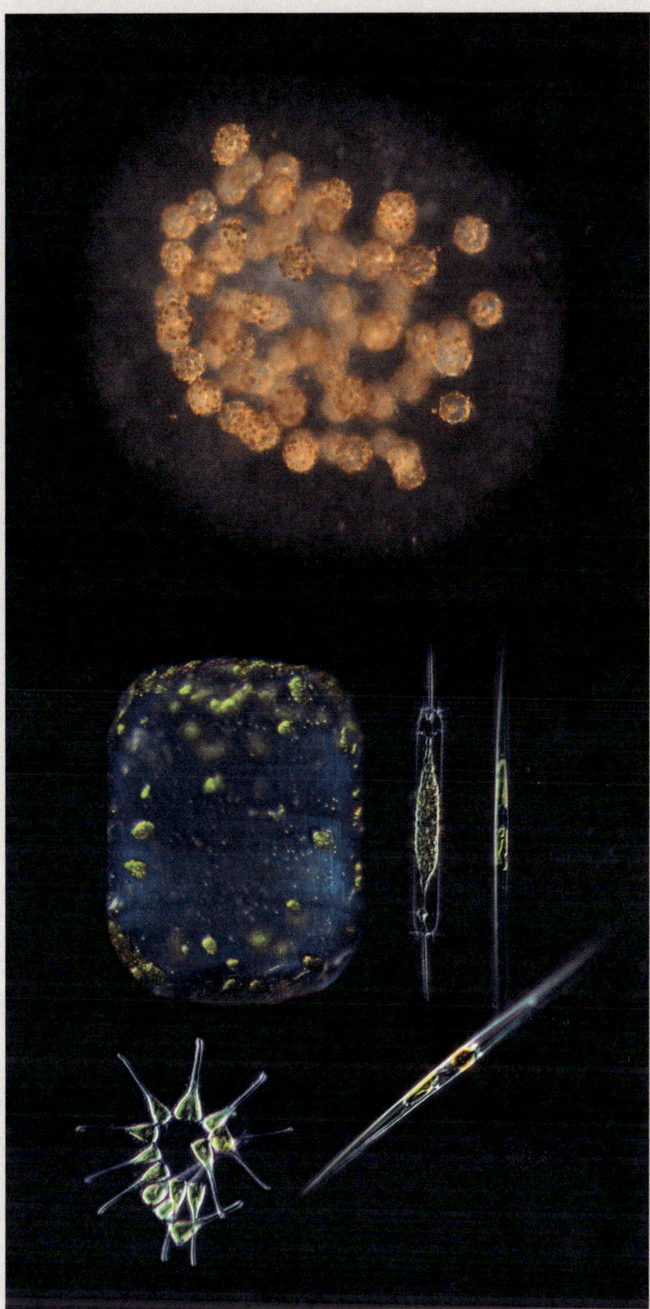

RADIOLARIA AND DIATOMS

Top: A small colony of radiolaria (*Collozum* species), collected on the Mediterranean coast. This colony, measuring 0.5 mm, is made up of around thirty spherical cells that produce and share the same gelatinous envelope. The small, ocher-colored dots inside and outside the cells are symbiotic microalgae.

Bottom: Just a few of the tens of thousands of known diatom species. Some are macroscopic and solitary, while others are colonial, such as the *Asterionellopsis glacialis* shown at bottom left. Only the largest diatoms, such as the one in the center (a *Lauderia annulata* measuring 0.2 mm), are visible to the naked eye. The chloroplasts that carry out photosynthesis in these phytoplanktonic protists are visible as yellow or green granules.

MACRO & DARKFIELD MICROSCOPY

© C. Sardet, excerpted from *Plankton: Wonders of the Drifting World*, University of Chicago Press, 2015

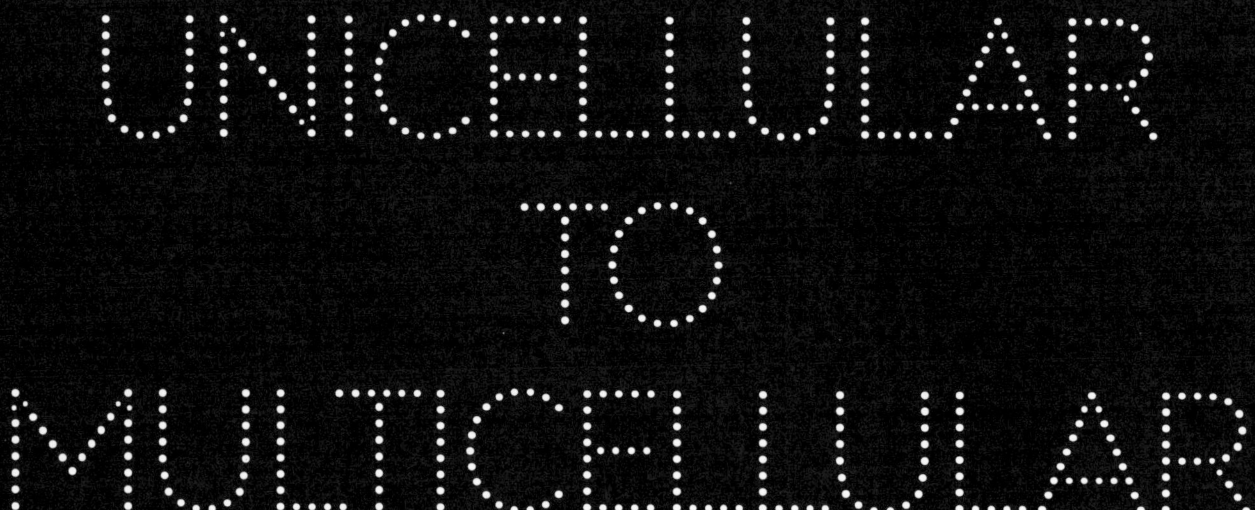

UNICELLULAR TO MULTICELLULAR

From protists to animals, fungi, plants & algae

FROM SINGLE CELLS TO COMPLEX LIFE— THE EUKARYOTES

Most protists are unicellular and live independently as single cells. Some, however—choanoflagellates, for example—are multicellular: They assemble into colonies made of one or two cell types. Ancestral, multicellular protists of this kind evolved into animals, fungi, plants, and algae, made of several types of cells (called *differentiated* cells). This is why we refer to animals, fungi, plants, and algae as *pluricellular* organisms.

The transition from the unicellular state typical of many protists to multicellularity and then pluricellularity likely took place over hundreds of millions of years of evolution. The earliest protists appeared at least a billion years before the first animals, plants, algae, and fungi. The first pluricellular organisms were likely ancestral forms of algae or fungi, with fossil evidence dating back more than 700 million years. Genetic analyses suggest that pluricellularity—which is inherited—evolved more than a dozen times in fungi, as well as in brown and red algae, but apparently only once in animals and once in plants. The trend toward complexity and gigantism among pluricellular organisms may have offered advantages, including enhanced protection, predation, movement, and reproduction—all functions linked to the evolution of numerous specialized cell types. But this shift likely introduced conflict, advantage-taking, and even cheating between different cell types, as well as more instances of deviation from normal cell division and behavior—cancer cells being one example of this.

There are several possible pathways by which a single-celled protist can make the transition to multicellularity. Individual cells might adhere to one another, forming a colony. Or cells might fail to fully separate during division, remaining bound together. Both processes could even occur simultaneously. In either case, the outcome is a cluster of cells resembling an embryo, or a sheet of cells reminiscent of an epithelium. Such colonial formations are observed in various species of choanoflagellates and volvocales—two types of protist thought to be the precursors of animals and plants, respectively *(see p. 103)*. However, these hypotheses remain speculative, as our understanding of these protists' distant ancestors is still limited. Innovative research on multicellularity is currently underway, using budding yeasts—fast-growing members of the fungal kingdom—as model, unicellular eukaryotes *(see p. 104)*.

Remarkably, protists can evolve from unicellular to multicellular in just a few months. In the laboratory, researchers accelerate this transformation by cultivating microalgae in the presence of predator protists, such as paramecia. The union of algal cells into a larger, multicellular organism becomes a powerful strategy for resilience in the face of adversity.

Choanoflagellates are protists ◯ characterized by a corolla, made of microvilli containing contractile microfilaments ▮, and a flagellum, driven by the motion of bundles of microtubules ▮. These structures enable choanoflagellates to move around, so as to capture and feed on bacteria.

When activated by stimulating molecules released by bacteria (the small flagellated cells in the background), choanoflagellate cells divide while remaining attached, forming colonies. Adjacent cells in the colony communicate with each other via cytoplasmic bridges (↔) resulting from incomplete divisions.

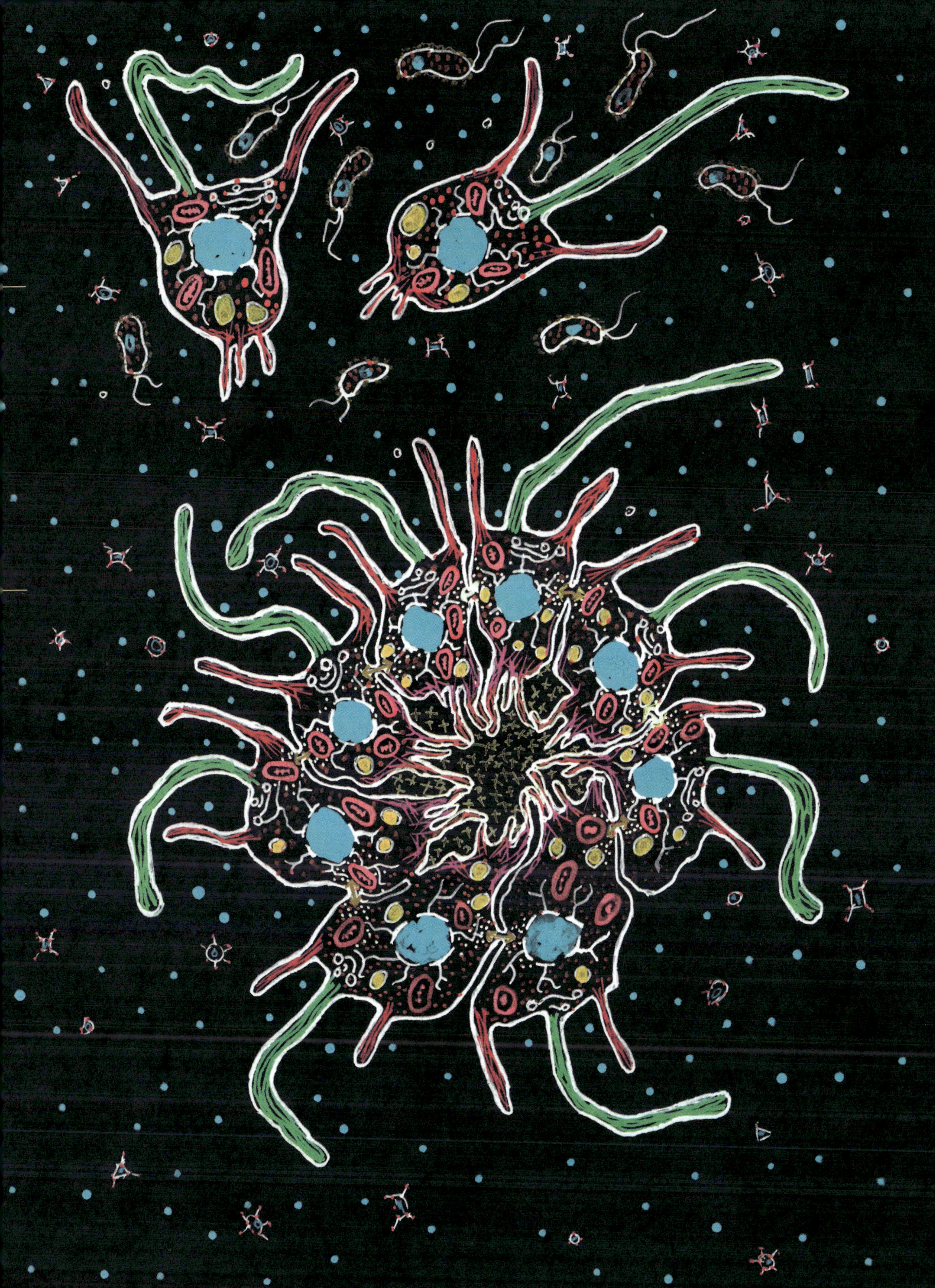

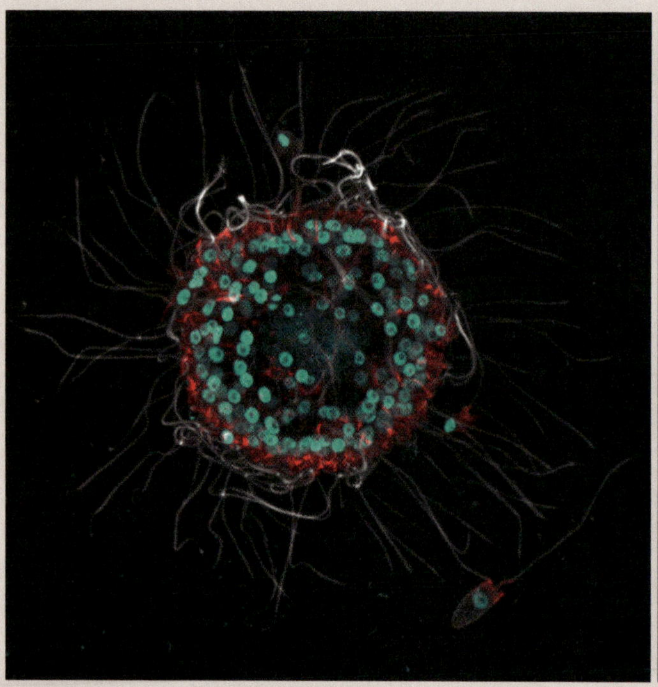

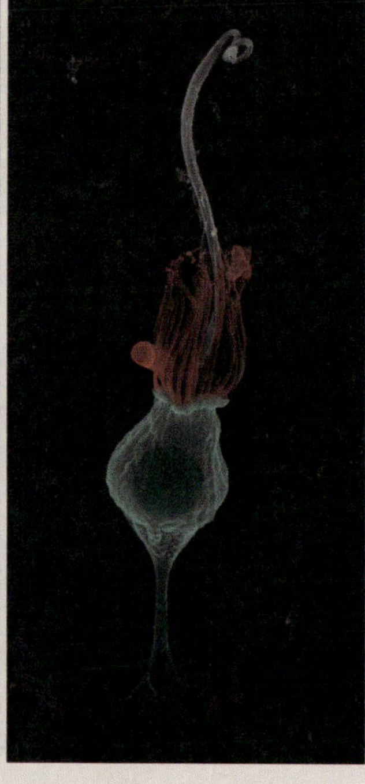

On right: The cell body of this protist is characterized by a corolla of microvilli (*in red*) encircling a long flagellum (*in white*). This cell of the *Salpingoeca rosetta* species is anchored by its stalk.

COLORIZED SCANNING ELECTRON MICROSCOPY
© Courtesy Mark Dayel, www.dayel.com

On left: A large colony of choanoflagellates forming a spherical ball, plus an individual whose nucleus (*in blue*) can be seen. The cells' flagella (*in white*) are filled with microtubules, and their corolla (*in red*) filled with contractile microfilaments.

CONFOCAL FLUORESCENCE MICROSCOPY
© Courtesy Kayley Hake, Nicole King, HHMI, University of California, Berkeley

Choanoflagellates— protists at the origin of animals?

It is highly likely that animals resembling jellyfish, ctenophores, or sponges were already inhabiting the oceans over 600 million years ago *(see pp. 23–25)*. Among these early pluricellular animals, sponges stand out as the oldest and simplest, appearing almost deceptively primitive. Anchored to a substrate and devoid of a nervous system, they feed passively by drawing in the surrounding water and capturing the microorganisms suspended within. Their suction is driven by the rhythmically waving flagella of specialized cells known as *choanocytes*. More than a century ago, biologists observed the striking resemblance between sponges' choanocytes and the single-celled flagellated organisms called *choanoflagellates*. These protists propel their oval bodies with a long flagellum encircled by a funnel-shaped collar, which gives rise to their name (*choano-* derives from the Greek for "funnel"). The movement of the flagellum generates currents that channel food particles, such as bacteria, toward the collar where they are engulfed.

Certain choanoflagellate species are remarkably versatile: They may swim freely with their flagella, anchor themselves to surfaces with a stalk, or assemble into spherical clusters with their flagella protruding outside.

Even more intriguing is the influence of specific bacterial signals in their environment, which can prompt these flagellated cells to either attach by their stalks, or divide. Repeated cell divisions without separation result in a cohesive ball of cells, an elementary multicellular structure reminiscent of an embryo. Rather than forming spheres, some species of choanoflagellate congregate to create cell mats, which exhibit contractile properties thanks to the microfilaments of their cytoskeletons— features remarkably similar to those found in animal cells.

From a genetic perspective, choanoflagellates are the protists closest to animals, sharing a common ancestor some 600 million years ago. In addition, they possess a repertoire of genes and proteins that confer communication, motility, and cell adhesion capabilities comparable to those of animal cells. For these reasons, certain choanoflagellates, like *Salpingoeca rosetta*, have emerged as model research organisms, shedding light on how ancestral single-celled protists might have coalesced to form multicellular organisms—possible precursors of animals as we know them.

Chlamydomonas and Volvox—protists at the origin of algae and plants?

Chlamydomonas is a genus of unicellular green algae in the phylum Chlorophyceae. These agile protists navigate toward light using a pair of flagella, a remarkable adaptation for a photosynthetic organism. In ponds and lakes, they often coexist with volvocales—a family of microalgae which form spherical colonies and can reproduce either asexually or sexually. Among volvocales, the most thoroughly studied species is *Volvox carteri*, which takes the form of a spherical ball. One sphere consists of dozens to several thousand nonreproductive somatic cells (resembling individual *Chlamydomonas* unicellular algae), all waving their flagella at the sphere's periphery. Nestled within this flagellated sphere are 12 to 16 larger reproductive cells. During sexual reproduction, the female *Volvox* produces around forty oocytes, while the male *Volvox* generates sperm packets, each containing approximately a hundred spermatozoa—an evident demonstration of sexuality in protists.

Clamydomonas and *Volvox* diverged several hundred million years ago, leaving researchers to ponder what conditions prompted solitary *Chlamydomonas* microalgae to coalesce into *Volvox* colonies. One plausible hypothesis is that ancient, predatory protists exerted selective pressure, compelling single-celled *Chlamydomonas* to form colonies as a survival strategy. The cells in these groupings—colonies now too large for predators to consume—would have been naturally favored. Researchers have tested this hypothesis by cultivating *Chlamydomonas* in the presence of predatory *Paramecium tetraurelia* paramecia. Astonishingly, within just one year, they observed the emergence of stable aggregates consisting of 4 to 64 *Chlamydomonas* cells, formed through successive divisions within a common outer envelope. This experiment suggests that predation pressure could indeed drive unicellular *Chlamydomonas* to evolve into multicellular forms—perhaps revealing one of the mechanisms behind the emergence of the first multicellular algae and plants.

VOLVOX

Volvox are protist cells with flagella that resemble *Chlamydomonas*. The flagellated cells adhere to each other, forming a spherical colony surrounded by a gelatinous shell. Inside are larger cells that produce egg cells and sperm cells for sexual reproduction.

CONFOCAL FLUORESCENCE MICROSCOPY
© Courtesy Tagide deCarvalho, Keith R. Porter Imaging Facility,
University of Maryland, Baltimore County

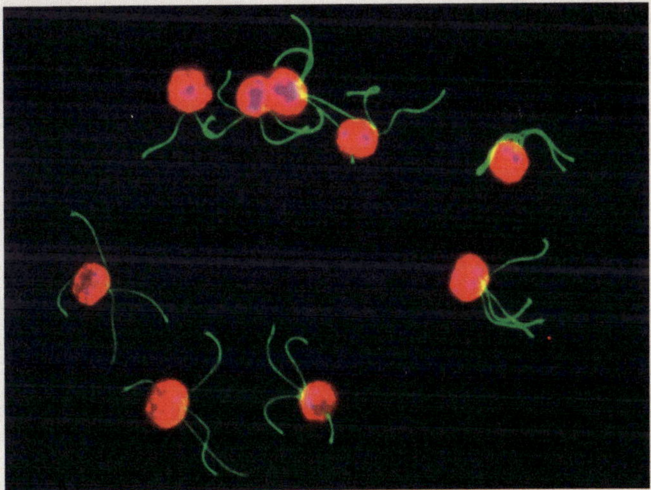

CHLAMYDOMONAS

This phytoplankton protist, a microalgae, contains a voluminous chloroplast that fluoresces red. *Chlamydomonas* moves with two flagella (*in green*). Like *Volvox*, *Chlamydomonas* can reproduce both asexually and sexually. The cells photographed here are mostly zygotes with four flagella, which result from the fusion of two gametes.

FLUORESCENCE MICROSCOPY
© Courtesy George Witman, UMass Chan Medical School,
and Karl Lechtreck, University of Georgia

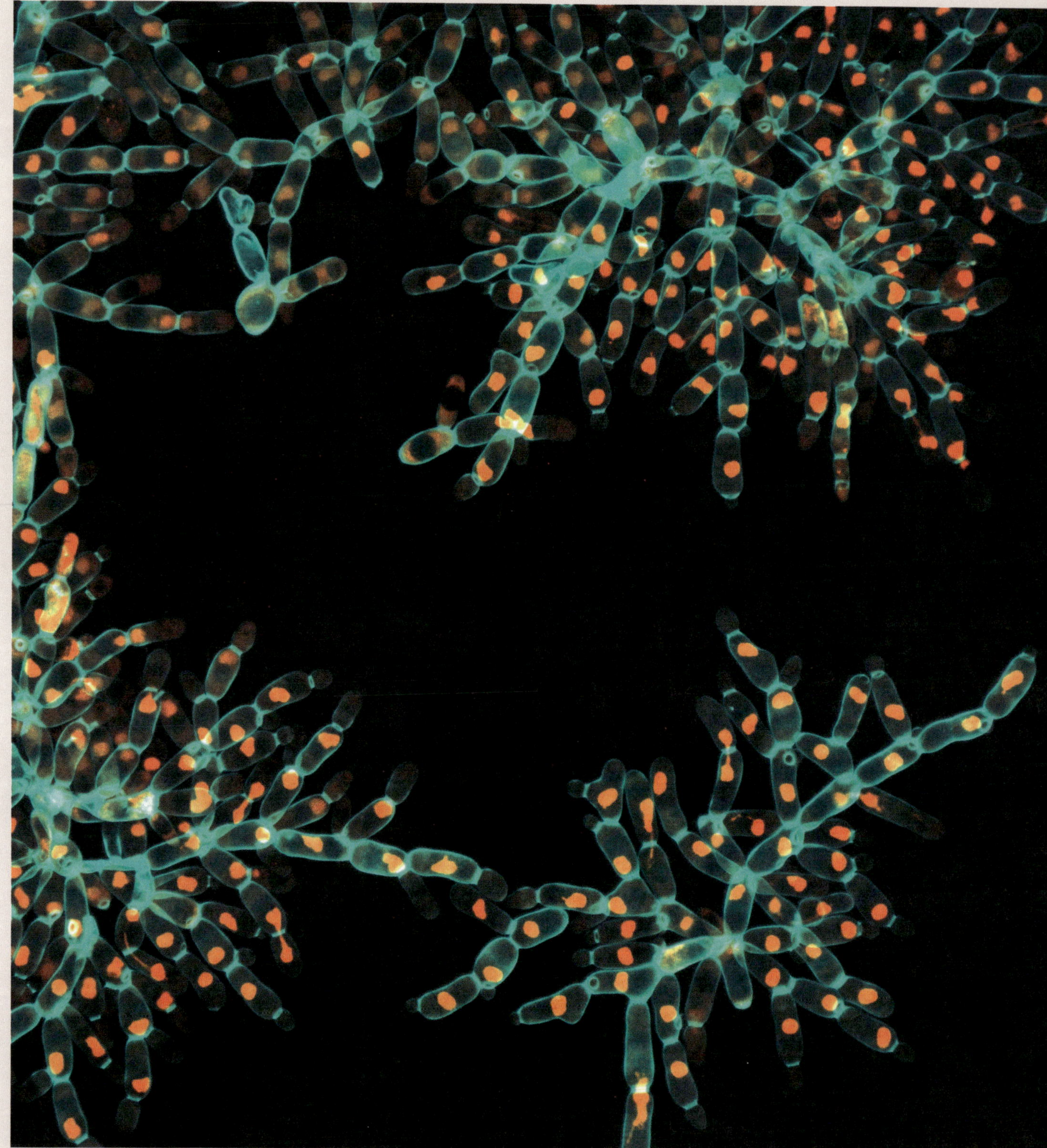

BUDDING YEAST IN MULTICELLULAR ASSOCIATION

When *Saccharomyces cerevisiae* (brewer's yeast) divides in the laboratory, some cells form snowflake-like chains. Some of the chained cells die by apoptosis, causing the flake to fragment. The chains of associated cells in the smaller flakes can then resume their growth. Yeast-cell nuclei fluoresce yellow. On average, yeast cells are 10 μm long.

CONFOCAL FLUORESCENCE MICROSCOPY

© Courtesy Will Ratcliff, Anthony Burnetti, and Ozan Bozdag, Georgia Institute of Technology

Yeast—from solitary cell to colonies and the origin of fungi

Fungi have a complex evolutionary history. They were grouped with plants until the 1960s, when they were recognized as one of the five major kingdoms of life *(see p. 30)*. Molecular analysis has since revealed that fungi are more closely related to animals than to plants, their divergence dating back over a billion years. Unlike animals, which ingest food, fungi take a unique approach to nutrition: They secrete enzymes into their surroundings to break down complex substances and absorb the resulting organic molecules. This mode of feeding is facilitated by their hyphae, microscopic filaments that intertwine to form the mycelium. This extensive network often appears aboveground (as the mushrooms we commonly recognize) and also spreads beneath the surface. Many mycorrhizal fungi engage in symbiotic relationships with plant roots, exchanging essential minerals for sugars, and thereby supporting plant health and soil ecosystems.

Among the vast diversity of fungi, with more than 100,000 identified species, unicellular yeasts exist as a subset. Some, like *Candida albicans*, can be pathogenic, causing infections in humans. Others, notably *Saccharomyces cerevisiae*, play an essential role in fermentation—leavening bread and producing alcoholic beverages—practices dating back to antiquity that have been industrialized over the past two centuries. Beyond its culinary applications, *Saccharomyces cerevisiae* has served as an essential model organism in scientific research since the time of Pasteur *(see p. 95)*.

Under optimal conditions, *Saccharomyces cerevisiae* reproduce by budding: A new daughter cell forms at one end of the mother cell, leading to daily population doubling. Laboratories have harnessed this rapid growth rate to explore evolutionary processes. Recent experiments have subjected yeast cultures to selection pressure by collecting the densest cells daily (which settle at the bottom of test tubes) and recultivating them. Over hundreds of iterations, this method favored the yeast cells that remained attached post-division, in ever-larger multicellular clusters. Successive generations of progeny cells that divide without fully separating create a chain-and-flake formation, making each cluster reminiscent of a snowflake. This type of multicellular state is termed *clonal multicellularity*. Apparently, the mutation of a single gene is sufficient to prevent full separation of mother and daughter cells, creating these microscopic "snowflakes."

Interestingly, when these multicellular clusters grow excessively large, some cells undergo programmed cell death, or apoptosis. This self-sacrificial process fragments the cluster into smaller units, enhancing the remaining cells' access to nutrients and oxygen, and thereby promoting further growth. While this laboratory-induced multicellularity is relatively simple and lacks the specialized cell types found in more complex organisms, it offers valuable insight into the initial steps of cellular differentiation. The speed with which yeast can evolve multicellular traits in experimental settings likely reflects a flexibility inherent to fungi, suggesting that the transition from unicellular to multicellular can occur over a relatively short evolutionary timescale. This may also be the reason why multicellularity seems to have evolved many times over in fungi.

View snowflake yeast, by Will Ratcliff

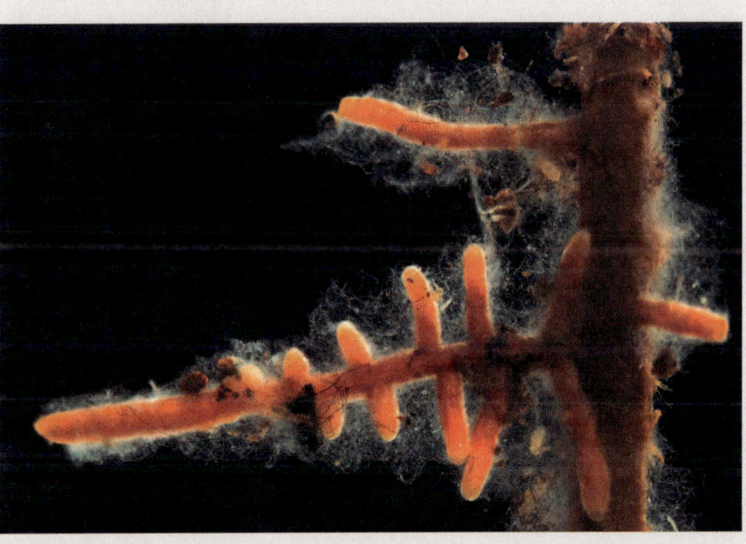

ECTOMYCORRHIZA

Soil-dwelling fungi develop a host of filaments that can attach to the roots of plants.

© Courtesy Nicole Ottawa and Oliver Meckes, *eye of science*, Germany

THE DIVERSITY OF MULTICELLULAR AND PLURICELLULAR ORGANISMS

Protists exhibit a remarkable array of strategies to transition from unicellular to multicellular forms. Many, such as choanoflagellates and *Volvox* species, form colonies where individual cells remain attached and function in unison. In contrast, organisms like the amoeba *Physarum* (commonly known as "the blob"), form a syncytium, a common cytoplasmic space shared by multiple nuclei, without membrane separations. This kind of syncytial organization is also observed in certain fungi and specific animal cells, including muscle, immune, placental, and reproductive cells, such as growing oocytes.

While protist cells may exist independently or in colonies, the cells of pluricellular organisms—such as animals and plants—undergo differentiation during embryonic development and operate as a "team." This process leads to larger organisms comprising thousands to billions of cells. The millions of fungi, plant, and algae species are made up of dozens of different sorts of cells, and animals can possess several hundred types.

A common property of pluricelluar organisms is that the vast majority of cells in tissues and organs adhere to one another, whether they are permanently linked by openings or pores resulting from incomplete splits between dividing cells, or more transiently bonded by adhesive molecules the cells secrete. In plants, algae, and fungi, cells are often connected at the division site by structures known as *plasmodesmata* or *septa*. These channels allow molecules—including metabolites, proteins, RNA, nanomachines, organelles, and sometimes viruses—to move between cells, promoting intercellular communication and coordination.

By contrast, animal cells typically exhibit narrower, direct cytoplasmic connections, as well as connecting structures called *nanotubes* that permit the exchange of larger molecules and organelles between cells (*see p. 205*). There are notable exceptions to this general rule: Oocytes, spermatozoa, and some stem cells use cytoplasmic bridges—broader, lasting openings that enable them to exchange macromolecules and even organelles. These bridges often persist during the early, embryonic stages but usually diminish as cells specialize into the myriad types found in animal tissues. The differentiated animal cells adhere and communicate through various junctions they build and molecules they secrete. They also exchange membrane vesicles containing RNA and protein messages, and secrete molecular signals, such as hormones, to coordinate functions. The diverse mechanisms by which different organisms' cells connect and interact suggest that multi- and pluricellularity evolved independently numerous times, leading to a rich variety of life-forms.

The simplest form of multicellularity is that of organisms (amoebas, certain fungi) whose cells are not physically separated by membranes ▮ but share a single cytoplasm, a "syncytium" containing multiple nuclei ●, mitochondria ●, and other organelles and components. The "blob" (an amoeba) falls into this category (*see p. 94*).

Plant cells (here a planar layer of cells in the epidermis of a leaf) are attached by their membranes and communicate via openings or pores called *plasmodesmata* ↔ that pass through the cellulose walls ▮. The hyphal cells of some fungi also communicate via large openings called *septa* ↔.

On right: Differentiated cells of animal epithelia are delimited by membranes, and passages between cells are regulated by junctions ↔ that open and close.

On left: During growth, oocytes of many animals (here an insect) depend on feeder or nurse cells, with which they form a kind of syncytium, connected by large openings called *cytoplasmic bridges* →.

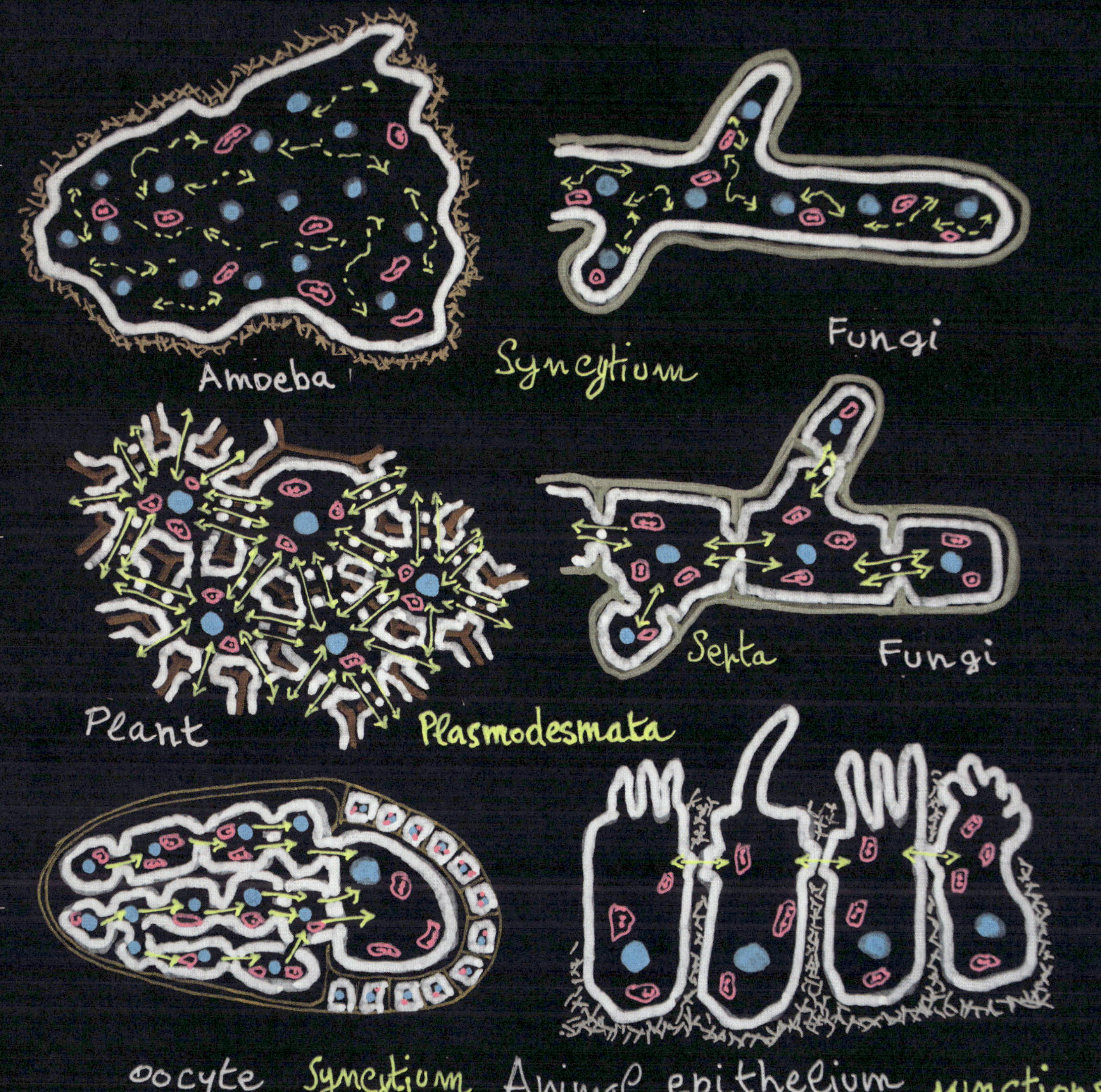

Amoeba
Syncytium
Fungi
Plant
Plasmodesmata
Septa
Fungi
oocyte
Syncytium
Animal epithelium
junctions

DIFFERENCES BETWEEN PLANT AND ANIMAL CELLS

Plants and animals are composed of myriad cells, thousands to billions of them, all intricately specialized to perform distinct functions. In plants, tissues such as roots, stems, leaves, and flowers are constructed from differentiated cell types. These cells are typically stationary, in fixed positions relative to one another, and encased by rigid cellulose walls that provide structure. This rigidity is further sustained by large vacuoles—membrane-bound organelles—that regulate turgor pressure. Animal tissues and organs comprise a greater diversity of cell types, many of which exhibit remarkable dynamism, capable of movement and morphological transformation. This contrast between plants and animals is evident at the microscopic cellular level.

The major distinction between animals and plants is metabolic, as plants convert sunlight to chemical energy via photosynthesis, yielding sugars and oxygen. This is the work of specialized organelles called *chloroplasts*. Although plants and animals share a set of ubiquitous organelles (discussed below), animal cells lack chloroplasts, while plant cells generally lack the centrosomes and lysosomes typical of animal cells.

Another difference is that plant cells are much more interconnected than animal cells. In plants, the plasmodesmata—large channels through the cell walls—let ions, molecules, nanomachines, and even organelles and viruses move between adjacent cells. Animal cells are often more individuated, and tightly regulate intercellular communication through specialized junctions (see p. 204). They typically lack cell walls, and instead are surrounded by a flexible plasma membrane coated with a sugar-rich outer layer.

The similarities between plant and animal cells trace back to their common eukaryotic origins. Typically, all eukaryotic cells have a nucleus containing a set of linear chromosomes, which carry the species-specific genome. Other ubiquitous organelles are the mitochondria, the endoplasmic reticulum membrane network, Golgi apparatus, and peroxisomes, as well as biomolecular condensates such as nucleoli. Both cell types possess a cytoskeleton principally composed of microfilaments and microtubules. However, a notable feature unique to animal cells are the centrosomes—star-shaped structures—which organize the microtubules and thus play a crucial role in cell division and motility (see p. 149). Also common in animal cells, but generally absent in plant cells, are structures associated with cell movement, such as microvilli, cilia, and flagella. In sum, the different life strategies of plants and animals are supported by distinct cellular adaptations.

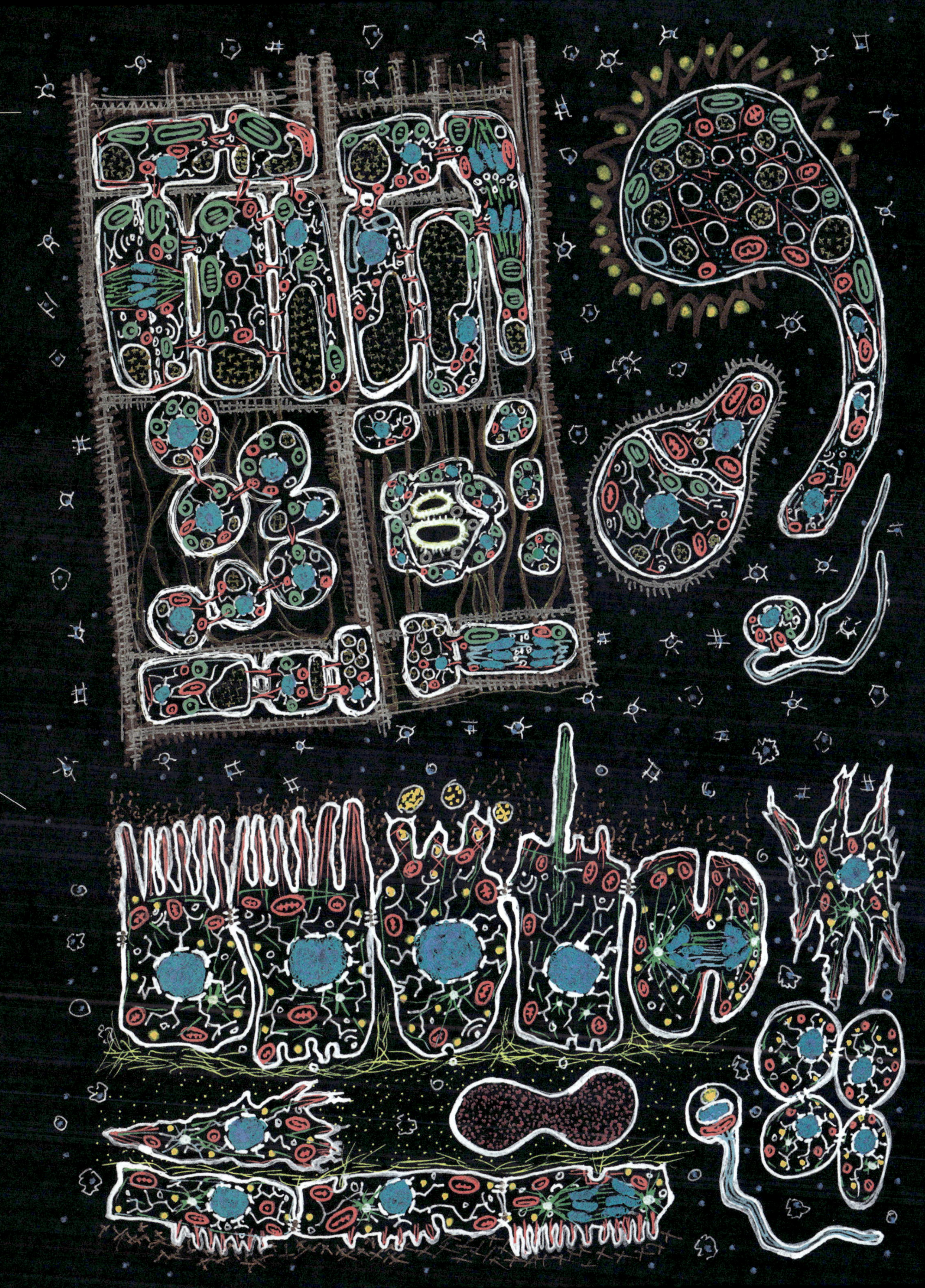

Tissue development in animals and plants

A living tissue is a group of cells collaborating to accomplish specialized functions. The tissues of pluricellular organisms emerge during development, as their cells undergo many successive divisions and differentiate into distinct lineages. In animals, this differentiation primarily takes place during embryogenesis, although additional changes occur during metamorphosis or budding in some species. Cells from different lineages communicate by exchanging signals. They often migrate, positioning themselves according to the messages they receive and decode. As they continue to divide, they develop into the specialized cells of distinct tissues and organs, such as the muscles, epidermis, intestines, and nervous system.

By contrast, basic tissue organization in plants begins during embryogenesis, but the majority of tissue and organ development occurs continuously throughout a plant's life. Roots, stems, leaves, and flowers, along with their respective tissues, emerge from specific regions known as *meristems*. The primary meristem regions (apical meristems) are located at the tips of stems and roots, although lateral meristems, such as the vascular cambium, also contribute to tissue growth in mature plants.

Unlike animal cells, which migrate and form tissues that can fold and reshape, plant cells are bound within rigid walls composed of cellulose fibers and pectins. Although limited in movement, plant cells can change shape by expanding and elongating, driven by turgor pressure. Consequently, the growth and structure of a plant and its organs are controlled by the speed and direction of cell division, as well as variations in cell size.

To study cell lineages, two model organisms are favored: the mouse (*Mus musculus*) for animals and thale cress (*Arabidopsis thaliana*) for plants. By using genetic manipulation to make certain proteins fluoresce, researchers can observe cell division, motility, and differentiation under the microscope. Molecules labeled with fluorescent dyes illuminate cell relationships and behaviors, allowing biologists to analyze the cells' differentiation within tissues and organs with remarkable precision.

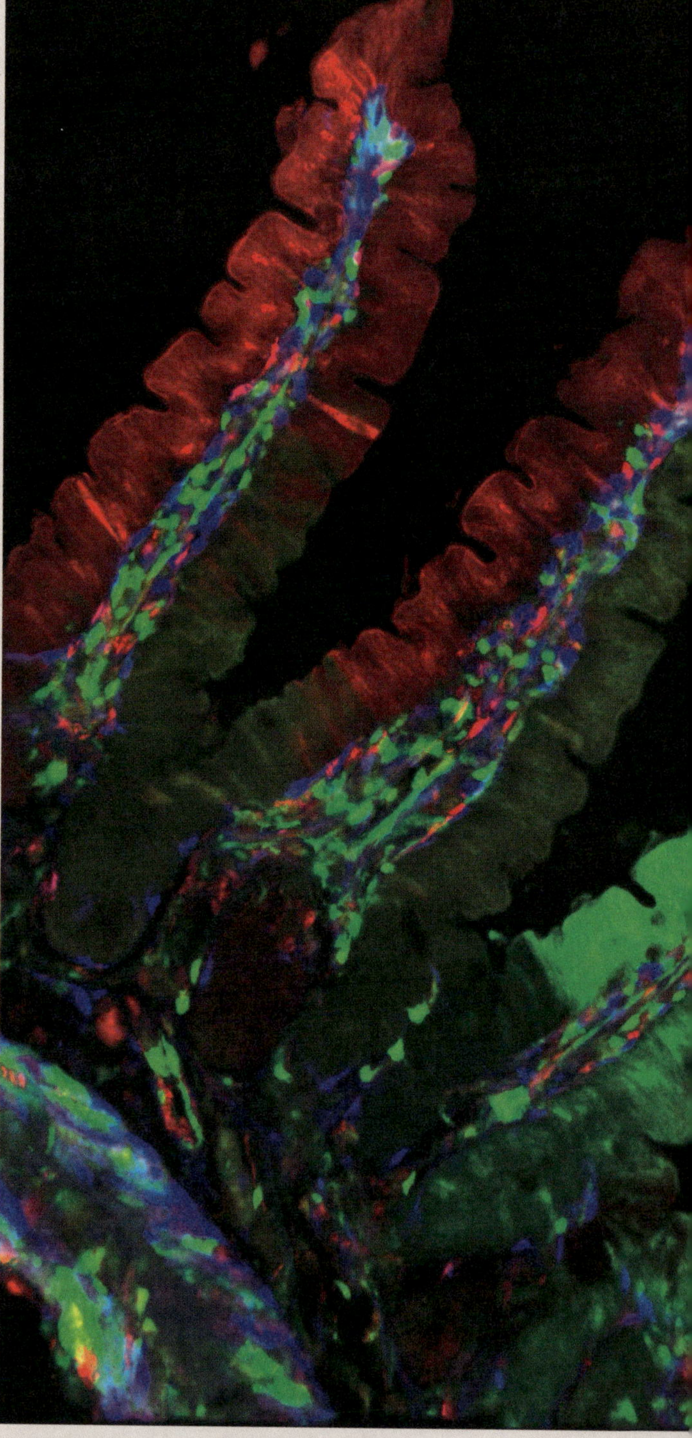

INTESTINAL TISSUE OF A MODEL ANIMAL— THE MOUSE, *MUS MUSCULUS*

The folds of a mouse's intestinal villi, where food absorption takes place. Villi cells can be genetically manipulated and labeled with fluorescent molecules to track their progeny and specialization during the rapid divisions that form different regions of the intestinal lining. This image shows families of cells that have divided, which fluoresce either red, green, or blue.

CONFOCAL FLUORESCENCE MICROSCOPY
© Courtesy Jonathan Nowak, Clément Ghigo, Marc Bajénoff, Marseille-Luminy Immunology Center & CNRS Images, France

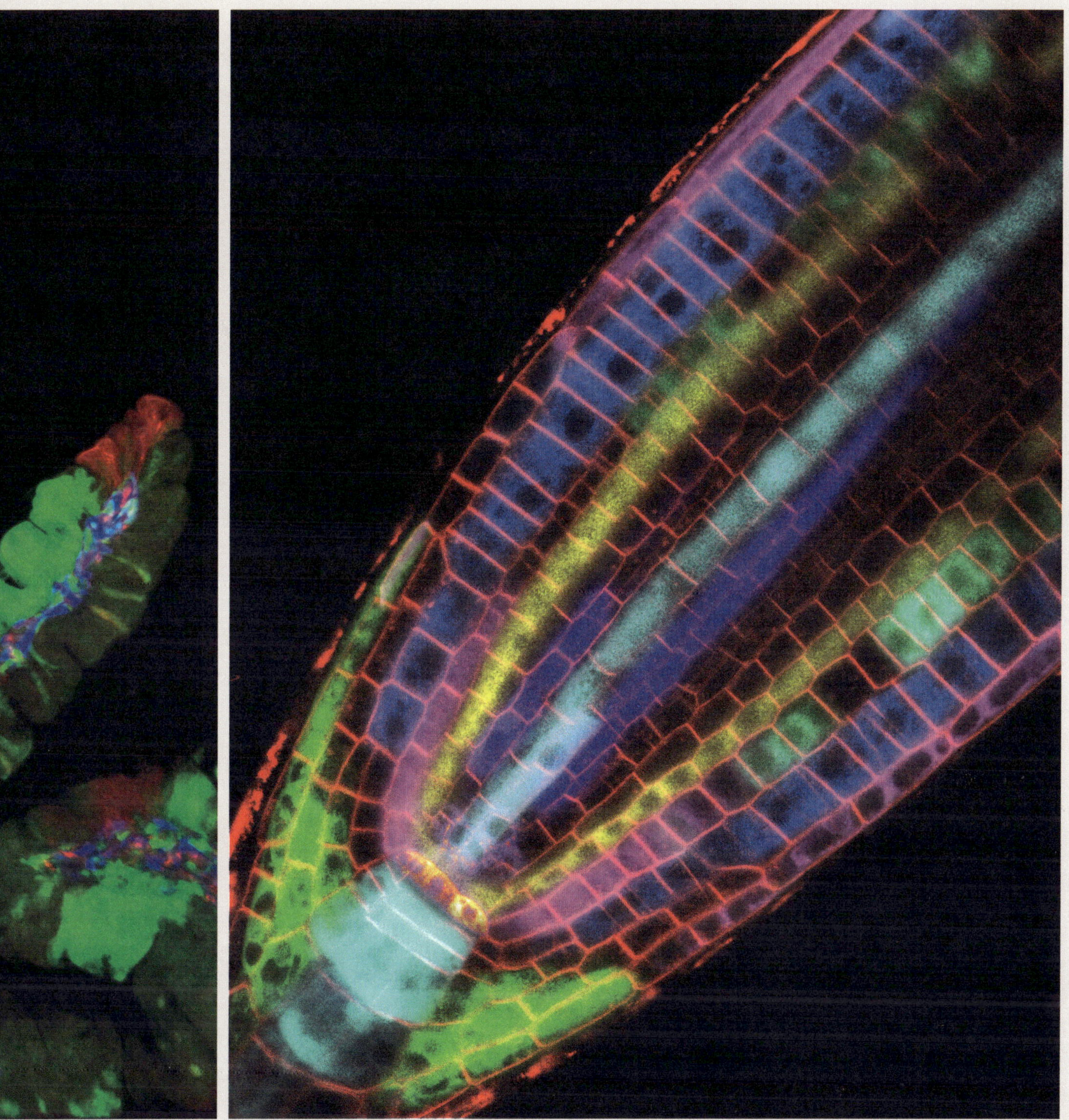

ROOT TISSUE OF A MODEL PLANT—
THALE CRESS, *ARABIDOPSIS THALIANA*

Root growth is influenced by a plant hormone (auxin) which acts as an organizer, determining the position and speed of cellular growth. The cell walls and membranes of this root tip of *Arabidopsis* fluoresce red. The cell lineages are revealed by labeling proteins that flouresce in different colors in each of the root tissues: At the tip, the protophloem in the meristem zone is light blue, the epidermis dark blue, the endodermis yellow, and the cells of the lateral cap and cortex are magenta and green.

CONFOCAL FLUORESCENCE MICROSCOPY

© Courtesy Yvon Jaillais, Plant Reproduction Laboratory, ENS Lyon & CNRS Images, France

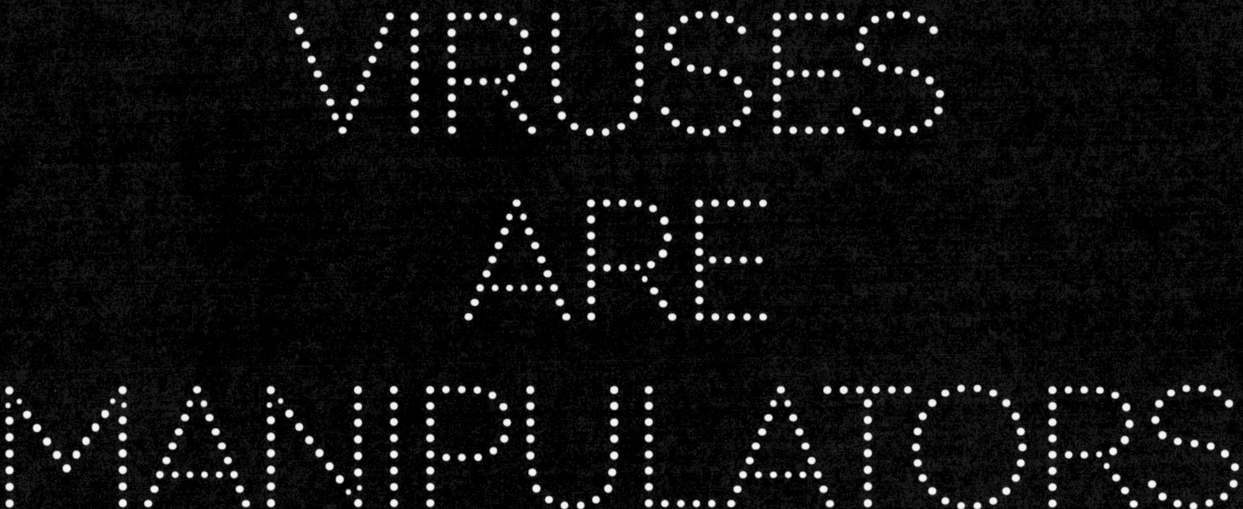

CHAPTER VI

Some are beneficial, some are pathogenic

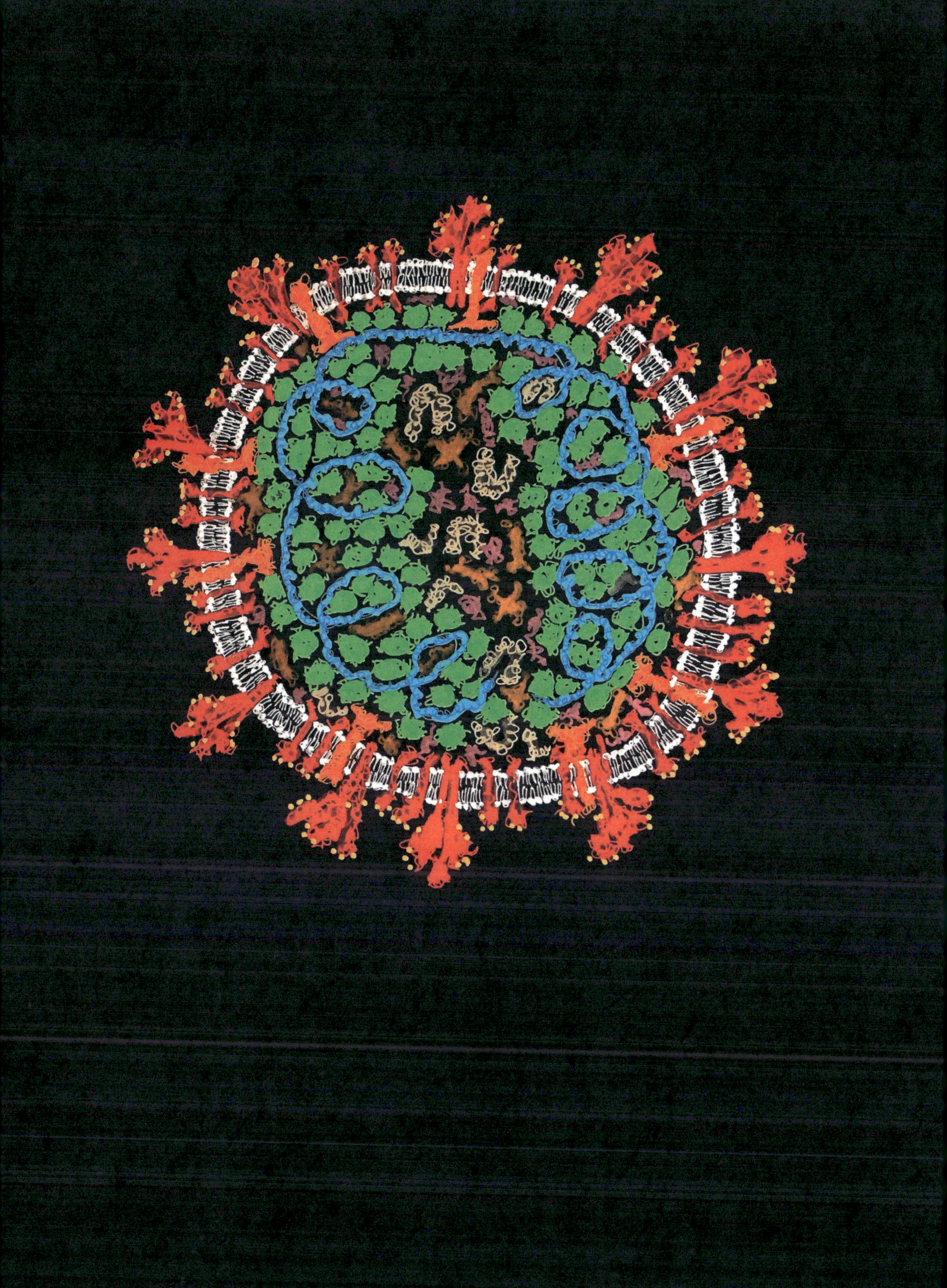

WHAT ARE VIRUSES? ARE THEY ALIVE?

Viruses are omnipresent: found in soil, liquids, and even the upper atmosphere, carried there by dust and water droplets. They rain down on the entire planet, immersing all organisms—from bacteria to humans—in a continuous viral bath. Despite their undeniable abundance and their fundamental role in the origins and evolution of life, viruses remain conspicuously absent from the tree of life *(see pp. 30–31)*. We still struggle to determine their place, to estimate their true number, and even to decide whether they are genuinely alive. Their very existence challenges our understanding of life itself.

Unquestionably, viruses are the most abundant biological entities in the world. They outnumber the cells immersed in Earth's viral soup by a factor of ten. Viruses' estimated quantity surpasses that of the stars in the universe! While the precise origin of viruses remains uncertain, some researchers hypothesize that they existed even before the emergence of primordial protocells.

Defining what a virus is—and whether it is "alive"—is no simple task, since viruses take on many forms. Outside of cells, they exist as virions—inanimate particles composed of DNA or RNA encased within a protein shell known as a *capsid*. Certain virions, such as those of HIV or COVID-19, surround themselves with an added lipid membrane, acquired from a host cell as they exit. Clearly, virions are not alive in the traditional sense: They lack an energy source, have no metabolism, and cannot reproduce on their own. Virions become living viruses only when they infect, penetrate, and parasitize cells. Once inside, they hijack the host's cellular machinery, forcing it to produce countless copies of the original virion. In this sense, virions serve as vehicles for viral reproduction and dispersal, much as seeds, spores, and gametes do for plants, algae, fungi, and animals.

First detected around 150 years ago as a mysterious "poison" or "toxin," viruses were initially defined as entities smaller than bacteria. This perception changed some twenty years ago with the discovery of giant viruses—called *mimiviruses*, for <u>mi</u>micking <u>mi</u>crobe <u>virus</u>. These colossal viruses can be larger than some small bacteria or archaea, and can even fall victim to their own parasitic viruses, known as *virophages*. With their imposing size and highly changeable genomes, giant viruses represent a vast reservoir of genetic diversity and offer immense potential for manipulating eukaryotic organisms. Alongside bacterial and archaeal viruses, viruses targeting eukaryotes—including giant viruses—must be regarded as influential regulators and manipulators of living systems. Their effect on evolution and ecosystem dynamics is far greater than once thought, positioning them as central players in the complex web of life.

Viruses and virions are made up of DNA ▌ or RNA ▌ surrounded by a "capsid" made of proteins (other colors). They are generally classified into seven major families, based on the nature and structure of their genetic material (RNA ▌ or DNA ▌).

At top right is a partial view of the first giant virus discovered—a mimivirus 400 to 600 nm in size, itself infected by a "virus of a virus" called Sputnik (70 nm).

Viruses vary in size and shape. Some are tiny spheres (circovirus, 17 nm; hepatitis D virus, 30 nm) or threadlike rods (tobacco mosaic virus, 300 × 15 nm; Ebola virus, 900 × 80 nm), while others are bullet shaped (rabies virus, 140 × 70 nm), ellipsoid, or icosahedral (smallpox virus, 300 × 200 nm; various adenoviruses).

Some viruses, such as those causing influenza, herpes, COVID, AIDS, Zika, and other diseases, are enveloped by lipid membranes ▌. The membrane envelope is taken from an infected cell when a virion's capsid (enclosing its genetic material) buds out of the cell.

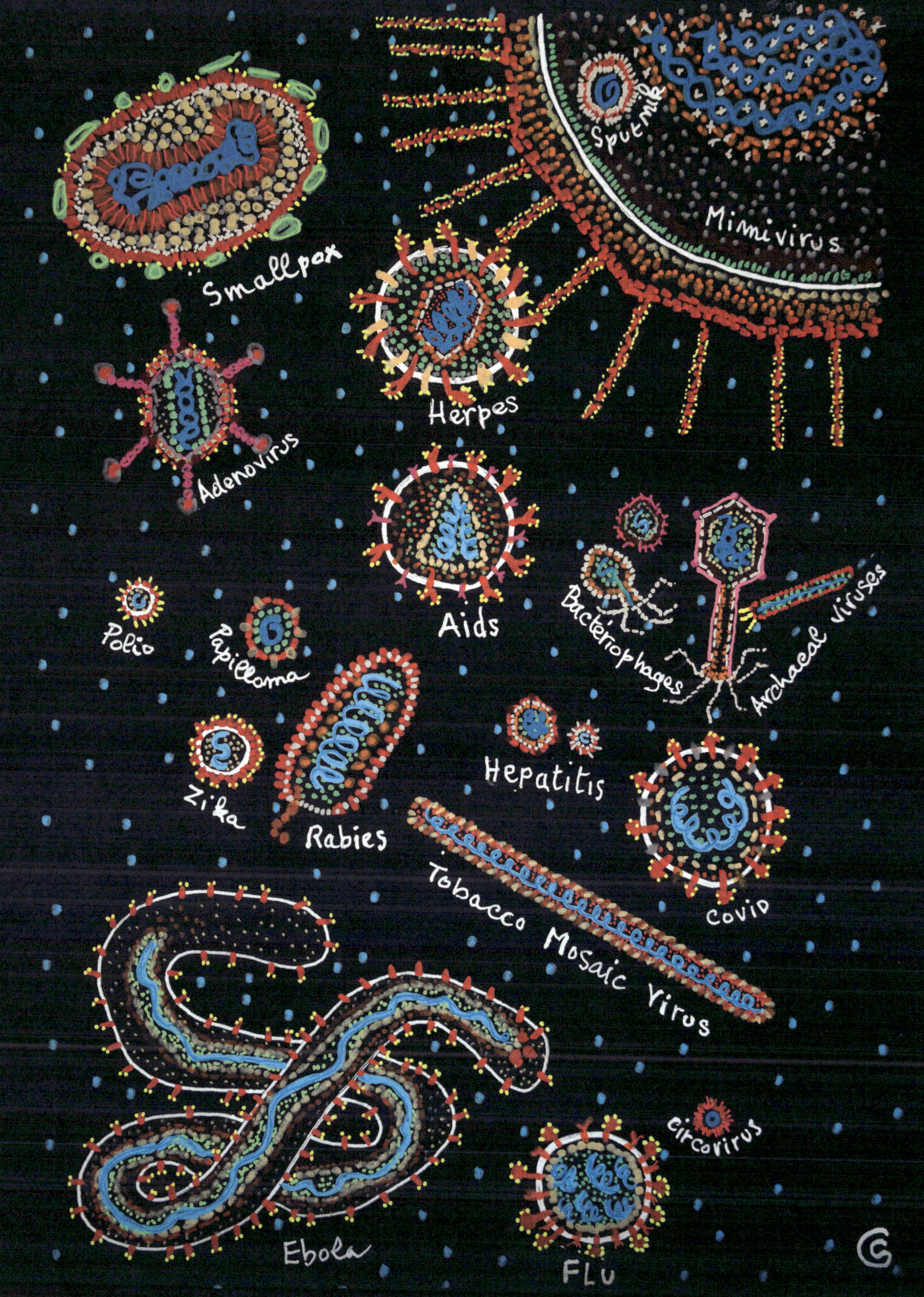

Smallpox
Spuermik
Mimivirus
Herpes
Adenovirus
Aids
Bacteriophages
Archaeal viruses
Polio
Papilloma
Zika
Rabies
Hepatitis
Covid
Tobacco Mosaic Virus
Circovirus
Ebola
FLU

The first virus was discovered in tobacco plants

The first virus ever identified infected plants, causing tobacco mosaic disease. This disease, which presents as discolored and dying leaves, was wreaking havoc on Europe's rapidly expanding tobacco fields as early as the 1850s. At the time, germ theory, advanced by Robert Koch and Louis Pasteur, had established that certain diseases—such as tuberculosis, cholera, and the plague—were caused by infectious bacteria. But Pasteur had also observed that rabies must be caused by an infectious agent much smaller than any known bacterium.

In 1884, Pasteur's collaborator Charles Chamberland developed a technique to isolate bacteria for identification and cultivation, using a simple porcelain filter with pores fine enough to retain microbes. Using such a Chamberland filter, Russian biologist Dmitri Ivanovski (working in Crimea) discovered that the infectious agent responsible for tobacco mosaic disease was present in the plant's sap. Remarkably, the agent passed through a Chamberland filter, leading Ivanovski to conclude that infection must be caused by some kind of toxin or poison. Seeking a solution to stop the spread of tobacco mosaic disease in Holland, microbiologist Martinus Beijerinck recognized that this infectious agent, a "liquid poison," required living tobacco leaves to reproduce. He called it a *virus*—the Latin word for "poison"—coining a term that would forever change our understanding of pathogens.

It would take nearly forty more years of research before Wendell Stanley succeeded in isolating (or "crystallizing") the infectious agent—TMV (<u>t</u>obacco <u>m</u>osaic <u>v</u>irus) in 1935. Using the first electron microscopes and X-ray diffraction, researchers discovered that TMV virions were in fact tiny, spiral-shaped rods. A decade later, they succeeded in reconstituting the virion in the laboratory from purified viral RNA and proteins, marking a milestone for the nascent field of molecular and structural biology. Since then, advanced techniques have revealed the three-dimensional shapes of thousands of viruses, and their complex molecular structures, mapping the relative positions of every atom with unprecedented precision.

The discovery of bacterial viruses known as *bacteriophages* (literally "bacteria eaters") followed during World War I. In Great Britain, Frederick Twort observed that a substance capable of passing through porcelain filters could destroy staphylococci bacteria. Shortly afterward, Félix d'Hérelle demonstrated that different types of phages specifically infected and burst distinct bacterial species, destroying them. He advocated the use of these viruses to treat bacterial infections in people—a practice known as *phage therapy*. Today, as increasing numbers of bacteria develop resistance to multiple antibiotics, phage therapy is regaining attention as a promising approach to combating severe infections caused by antibiotic-resistant bacteria.

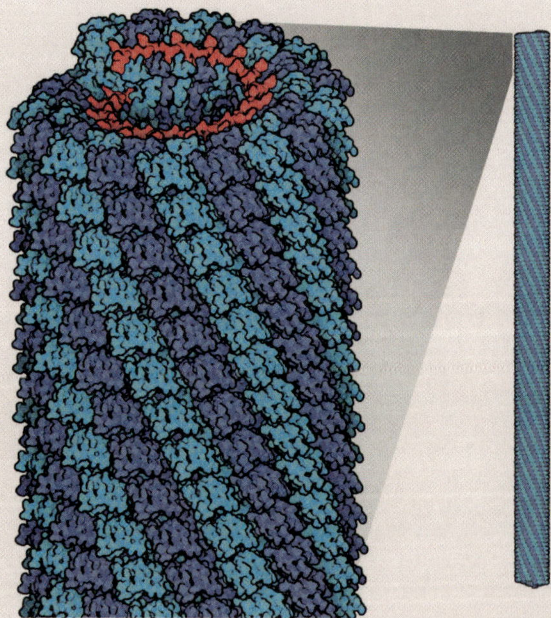

TOBACCO MOSAIC VIRUS—THE FIRST VIRUS DISCOVERED

The RNA of tobacco mosaic virus consists of 6,400 nucleotides in a spiral (*in red*), surrounded by a capsid of 2,130 proteins arranged in a helix (the two forms of the protein are shown in blue and purple). The virus's RNA encodes only four proteins, all it needs to complete its life cycle.

© Artwork by David Goodsell for the RCSB PDB "Molecule of the Month"

TOBACCO MOSAIC—A VIRAL DISEASE

Tobacco mosaic virus is a disease of tobacco plants that spreads in patches, causing necrosis of the leaf. It is transmitted by the TMV virus, which infiltrates the sap and passes from plant to plant in the soil.

© Courtesy Karen-Beth G. Scholthof, Texas A&M University

The saga of giant viruses

In the early 2000s, large viruses such as the smallpox virus were known to exist, but they remained within the established norm: They passed through the 200-nanometer pores of Chamberland porcelain filters designed to retain bacteria. This paradigm was challenged by the investigation into an outbreak of pneumonia at an English hospital. Researchers initially suspected the culprit was a bacterium residing within amoebae that had proliferated in the hospital's cooling system. However, researchers in Marseille revisited the samples in 2003 and made a groundbreaking discovery: The infectious agent within the amoebae was not a microbe but a giant virus, which they named *mimivirus* (*mi*micking *mi*crobe *virus*). The virion, shaped like an icosahedron, contained over 1,000 genes—100 times more than the genome of HIV. Measuring around 500 nm, mimivirus exceeded the size of the smallest bacteria. It was a revelation!

The laboratory discovery of mimiviruses sparked a successful search to find more giant viruses. Amazingly, they were discovered all over the world—along the coasts of Chile, in Austrian sewage-treatment plants, at the bottom of the oceans, and even in Siberian permafrost! One of the most astonishing discoveries was made recently at the bottom of hot springs in Japan, where researchers identified the medusavirus. Aptly named after the Medusa of Greek mythology, this virus transforms an infected amoeba into a hardened, mineralized cyst— turning it to "stone." New giant viruses continue to be discovered almost daily, boasting increasingly impressive dimensions and genomes. The current record for gigantism is held by amphora-shaped viruses known as *pithoviruses*, which can reach 2 μm.

All giant viruses identified so far carry their genes on double-stranded DNA. They belong to the category of NCLDVs (<u>n</u>ucleo<u>c</u>ytoplasmic <u>l</u>arge <u>D</u>NA <u>v</u>iruses) and constitute the phylum *Nucleocytoviricota*. These viruses infect both unicellular and multicellular eukaryotes, replicating within specialized virus factories inside the parasitized cells—transforming them into so-called *virocells*. Even more surprisingly, some NCLDVs, such as mamaviruses and mimiviruses, possess their own viruses called *virophages*. The particular virophages named Sputnik and Zamilon carry DNA genomes capable of

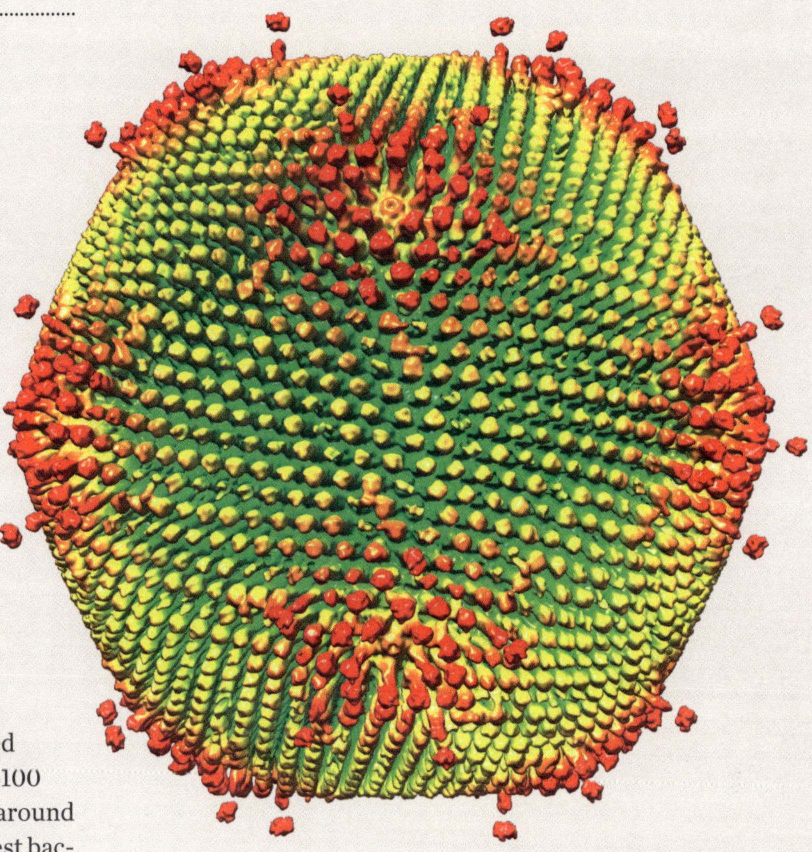

MEDUSAVIRUS TURNS INFECTED AMOEBA INTO STONE

The medusavirus is a giant DNA virus, measuring 260 nm, which preys on the amoeba *Acanthamoeba castellanii* in hot thermal springs. This digital reconstruction reveals the icosahedral structure of its outer protein shell, made up of thousands of protrusions.

© Courtesy Hiroyuki Ogata, Kyoto University; Kazuyoshi Murata, National Institute for Physiological Sciences, Okazaki; and Masaharu Takemura, Tokyo University of Science

inactivating or even killing the giant viruses they infect. This intricate parasitic relationship highlights the complex interplay between viruses and their hosts.

Due to frequent DNA exchanges between giant viruses and the cells they infect, these viruses have acquired a vast array of genes, including genes involved in metabolism, protein synthesis, cytoskeletal formation, and chromosome compaction (by which linear DNA is densely folded up into its chromosome superstructure). These genes endow giant viruses with remarkable potential to reprogram the cells they infect, profoundly influencing cell populations' viability and adaptability, and manipulating them to the virus's advantage. Despite these astonishing discoveries, we are only at the dawn of the giant-virus saga. To date, only a small number of giant viruses in existence have been identified, and it is suspected that their diversity could well exceed that of bacteria.

ORIGINS AND LIFE CYCLES OF VIRUSES

Viruses have been an integral part of life since its very beginnings, and likely predated the emergence of the ancestral cell LUCA (last universal common ancestor). The earliest viruses may date back to the RNA–peptide world, when protocells first appeared *(see pp. 60-64)*. One bold theory posits that viruses are relics of protocells that gradually lost their reproductive autonomy, evolving into parasites of bacteria and archaea, and later of eukaryotes. In this view, viruses would represent a fourth branch of the tree of life. Other biologists, however, consider viruses more likely to be fragments of DNA or RNA which escaped from cells, or the result of cellular reductions that gave rise to strictly parasitic entities.

These contrasting perspectives reflect our evolving understanding of viruses and their origins. Initially defined at the end of the 19th century as entities smaller than bacteria, viruses are now recognized as a vast and diverse category of macromolecular parasites. They range from simple plasmids—small fragments of DNA or RNA carrying only a few genes—to giant viruses with more than 2,000 genes (one tenth the size of the human genome)! Some of these giant viruses even exceed small bacteria or microalgae in size.

Unlike most living organisms, viruses follow a unique life cycle characterized by intermittent parasitism. Without a host, a virus is an inanimate particle made of DNA or RNA encapsulated by proteins, and sometimes lipids. This dormant form is called a *virion*. A virion only becomes a living virus when it penetrates and takes control of a host cell, transforming it into a virocell—a kind of "zombie cell" reprogrammed as a virus factory. The massive production of virions within the infected cell often leads to its destruction, giving rise to infections and diseases such as influenza, hepatitis, polio, COVID-19, AIDS, Ebola, and others *(see p. 190)*.

But this harmful aspect of viruses is not a universal rule. Some viruses remain dormant, while others enter and exit cells without causing any pathology. There are even some viruses with beneficial effects! Most viruses act as agents of gene exchange and transfer, and over the course of evolution, some of these exchanges have sparked beneficial adaptations. Far from being mere harbingers of disease, viruses are major players in evolution—rewriting genomes, driving diversity, and even conferring advantages to the organisms they infect.

1 – Bacteria ○ are infected and destroyed by viruses (called *phages*) ▮. Like tiny syringes, the phages inject their DNA ▮ inside the bacteria, instructing them to make copies of the phages. Then the phages lyse (rupture) the infected bacteria's membrane, releasing countless copied phages into the external environment.

2 – SARS-CoV-2 virions of COVID-19, with their spike proteins ↑ and their RNA ▮ inside.

3 – *On left:* The COVID-19 virus enters and replicates in an animal cell by introducing its RNA ▮ to the cell, whose own DNA ▮ is also represented.

4 – *On right:* New viruses are produced and bud out of the infected cell, enveloping themselves with a membrane ▮ taken from the cell.

MANIPULATIVE AND BENEFICIAL VIRUSES

There are good reasons to believe that ancestral viruses supported the emergence of larger, more complex cells—eukaryotes—characterized by chromosomes (enclosed in a nucleus) and a diversity of organelles. For one thing, the viral factories that form within infected cells closely resemble nuclei, and may have been their precursor. Also, viruses have introduced genes to cells that have shaped cells' structures and functions, including their ability to fuse together. Although the precise mechanisms remain obscure, it appears that around 1.5 to 2 billion years ago, eukaryotic cells arose from complex interactions between archaea, bacteria, and their viral companions, evolving over hundreds of millions of years *(see chap. IV)*.

Our perception of viruses is often distorted by the infectious diseases they cause. But contrary to popular belief, most viruses are not malevolent! In fact, many have beneficial effects. They regulate microbe populations—destroying organisms that proliferate to the point of exhausting resources and threatening other life-forms—and recycle their organic matter. From the very beginning, viruses have acted as major vectors of gene transfer between cells, fostering necessary innovations for adaptation and evolution. These exchanges are particularly intense among bacteria and protists, where viruses facilitate the transfer of genes that confer resistance to antibiotics and parasites.

For humans, as for all sexually reproducing organisms, transferred viral genes can integrate within the DNA of gametes (sex cells). Although this is rare, integrated viral genes have accumulated in our DNA over millions of years, and today they account for around 8 percent of our genome. These genes bear witness to the various viruses that infected our animal ancestors (mammals and primates), driving significant evolutionary changes—including essential genes linked to cognition and reproduction.

Viruses' ability to manipulate the genes of infected cells carries profound ecological implications. One striking example is found in insect populations whose behaviors are altered following viral infection, with cascading effects on ecosystems. On a global scale, viruses significantly influence biodiversity in the oceans. The *Tara Oceans* expedition revealed that the diversity of viruses is highest at the Arctic and Antarctic poles. Inversely, the biodiversity of marine organisms infected by viruses—from fish and jellyfish to krill, protists, and bacteria—peaks in equatorial and temperate regions. Could viruses be responsible for this striking difference? It remains a mystery for now, but when it comes to viruses, we are accustomed to surprises!

RNA
DNA

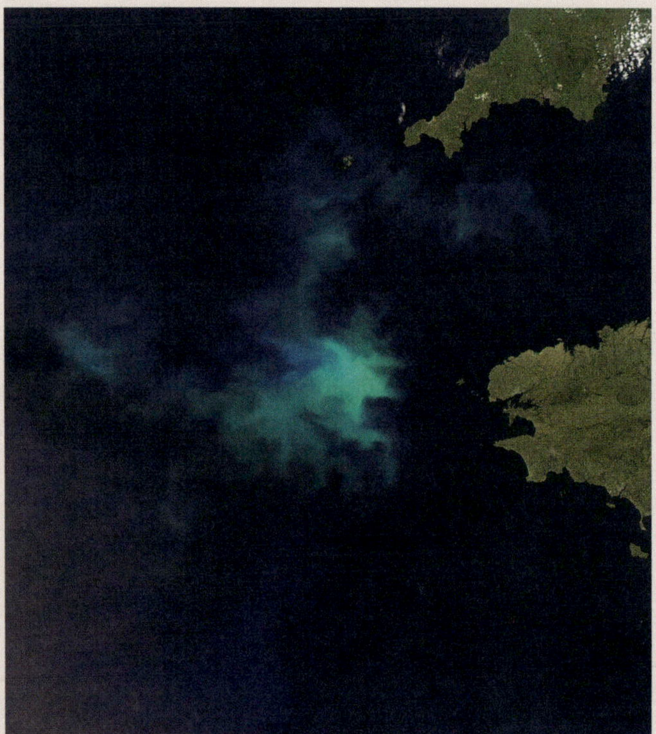

COCCOLITHOPHORE BLOOMS SEEN FROM SPACE
Coccolithophores are tiny, phytoplankton cells covered with calcium carbonate plates that reflect light. When their density reaches millions of organisms per liter—in proliferations called *blooms*—they reflect enough light to be seen from space, as in this satellite image of an ocean region near the English and French coasts.

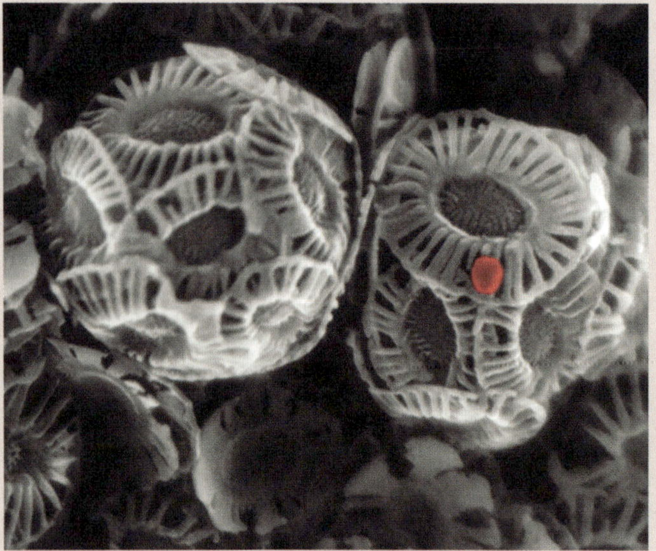

COCCOLITHOPHORE VIRUSES
Two coccolithophores of the species *Emiliana huxleyi*, about 5 µm in diameter, are covered with calcium carbonate plates called *coccoliths*. The cell on the right features a giant DNA virus (EhV, *Emiliana huxleyi virus, colored pink*), measuring 180 nm in diameter.

SCANNING ELECTRON MICROSCOPY
© Courtesy Willie Wilson and Keith Ryan, Marine Biological Association, Plymouth, Great Britain

Master manipulators— from insects to plankton

Baculoviruses are master manipulators. These viruses can infect over 800 species of invertebrates and are sometimes harnessed as biological insecticides for their ability to decimate pest populations. This kind of virus can turn caterpillars into zombie-like creatures. After infecting the caterpillar's gut, the virus integrates its DNA into the host cells' genome. By doing so, it seizes control of the cell machinery, directing the infected caterpillar to produce countless copies of the baculovirus. This viral takeover triggers a dramatic change in the insect's behavior: Instead of descending to bury itself and pupate as usual, the infected caterpillar, irresistibly drawn to sunlight, climbs to the top of the plant. The baculovirus drives this peculiar ascent by manipulating the amino acid sequence of two light-sensitive proteins in the caterpillar's eyes. Once at the top, the virus completes its deadly mission: The caterpillar's body liquefies, releasing a torrent of virions that flood the entire plant, contaminating other caterpillars below.

Single-celled organisms are not spared from viral manipulation, either. In the oceans, coccolithophores—a type of phytoplankton protist covered with calcium carbonate scales—are crucial players in the biological process that sequesters carbon on the ocean floor. Coccolithophores can proliferate massively under optimal conditions of light, temperature, nutrients, and minerals. They multiply into immense microalgal blooms, stretching across thousands of square kilometers, and often visible from space. However, these blooms are vulnerable to viruses—particularly giant DNA viruses which encode hundreds of proteins. When a virus invades a coccolithophore population, it triggers intensive virion production that ultimately delivers the coup de grâce to the bloom. Such an immense population collapse profoundly impacts geochemical and biological balances, as coccolithophores also emit volatile sulfur compounds that influence cloud formation, such as dimethyl sulfide. In general, marine viruses are remarkably abundant and carry a wealth of metabolic genes stolen from the bacteria and protists they infect. They ingeniously repurpose these borrowed genes to manipulate the physiology of their hosts, enhancing photosynthesis and the ability to "fix" captured carbon or nitrogen into new, organic molecules. In this way, viruses play an essential yet overlooked role in enhancing planktonic life in the oceans, subtly shaping marine ecosystems and global biogeochemical cycles.

Memory and gestation— gifts from beneficial viruses

For hundreds of millions of years, DNA and RNA viruses have been infecting animals, leaving indelible marks on their genomes. Among these, RNA viruses—particularly retroviruses—are uniquely able to insert copies of their genes into the DNA of parasitized cells. Remarkably, around 8 percent of the human genome consists of these acquired viral sequences. When retroviral genes integrate into sperm or oocytes, they become part of our germline and are passed on from generation to generation—inherited relics of ancient infections.

Over the course of evolution, many of these viral genes have been recycled and repurposed, giving rise to new functions, some with reproductive and cognitive benefits.

One of the oldest and most striking examples of genetic recycling is that of syncytins—retroviral genes captured and reused by various mammals over more than 80 million years. Syncytin genes are essential to forming the placental barrier, which regulates exchanges between a fetus and its mother. This barrier, known as the *syncytiotrophoblast*, consists of giant cells with multiple nuclei, formed from fused-together embryonic cells. The cells fuse thanks to syncytin proteins, which induce membrane fusion—the same type of proteins that enable viruses to enter host cells (like the spike proteins of the COVID-19 virus). Without these ancient retroviral genes, mammals would lack a placenta and would probably still be laying eggs! Syncytin proteins also play a key role in fusing muscle precursor cells, to form the multi-nucleated cells of muscle fiber.

Another extraordinary example of our viral legacy is the Arc gene, expressed in the brain. This gene plays a major role in learning and memory. These cognitive abilities depend on nerve cells (neurons) and their communication via contact sites called *synapses*, where signals are transmitted and processed. These synapses constantly adapt and remodel themselves to encode and store information, creating a molecular memory of our experiences. The Arc gene encodes both messenger RNA and the Arc protein, which, surprisingly, assemble into a capsid structure reminiscent of a virus. This Arc capsid surrounds the messenger RNA and buds off from the neuron, taking a fragment of cell membrane with it. Once released, the capsid can fuse with the membrane of a neighboring neuron or muscle cell, delivering its

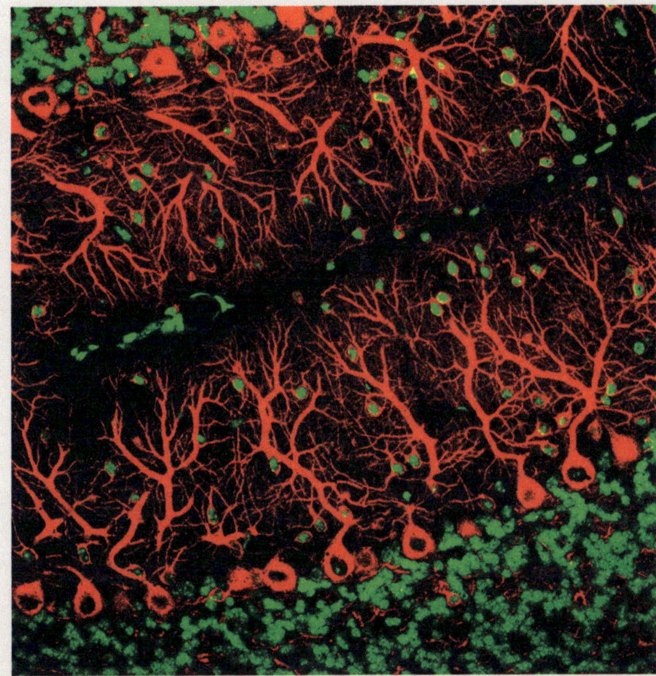

NEURAL NETWORKS

Neurons are nerve cells that store information and carry signals in the brain and throughout the body. Made up of a cell body, plus threadlike extensions called *axons* and *dendrites*, a neuron has abundant contact sites, called *synapses*, with other cells. Neurons form networks connected by these synapses, and also communicate via extracellular vesicles and particles, such as the Arc capsid, inherited from viruses. Pictured here are Purkinje cells, large neurons named for their discoverer.

CONFOCAL FLUORESCENCE MICROSCOPY
© Courtesy Timothy Vartanian and Yinghua Ma, Cornell University

ARC CAPSID STRUCTURE

The icosahedral Arc capsid consists of 240 copies of the Arc protein arranged in hexamers and pentamers (not unlike the sides of a soccer ball), and bears remarkable similarity to retroviral capsids, such as that of the AIDS virus. Like a retroviral capsid, the Arc capsid is designed to protect and deliver messenger RNA. The Arc proteins allow the RNA to enter the cell via endocytosis.

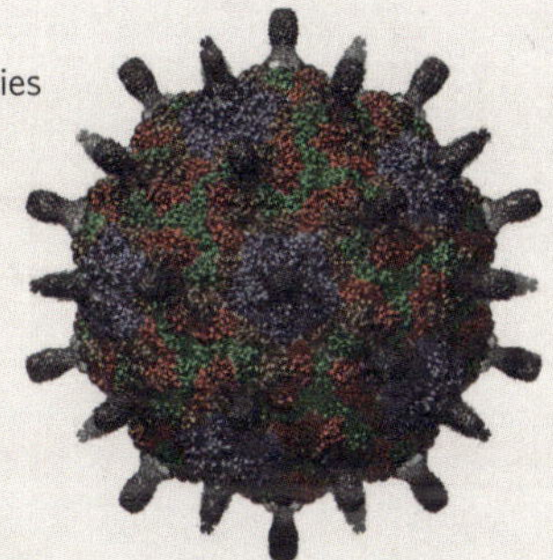

© Courtesy Simon Erlendsson, University of Copenhagen, with permission of *National Geographic*

genetic message. This process is strikingly similar to viral mechanisms, revealing how neuron cells have repurposed viral strategies to communicate with each other. This form of communication, born of a recycled viral gene, parallels the exchange of extracellular vesicles—membrane-bound bubbles that ferry molecular messages between cells *(see p. 200)*. These vital processes highlight the profound influence of viruses on the evolution of complex life.

The malevolents— contrasting portraits of COVID-19 and AIDS

Humanity has faced two major pandemics in recent decades: AIDS (acquired immune deficiency syndrome), since the 1980s, and COVID-19, which took the world by surprise at the end of 2019. Both diseases are zoonoses, meaning they originated via transmission from other animals to humans. AIDS was transmitted by monkeys in Africa, while COVID-19 likely emerged from bats in Asia, possibly via other small mammals. Over the past forty years, the AIDS virus has claimed between 0.5 and 1 million lives each year, while, in its first three years alone, COVID-19 caused around 18 million deaths, based on excess mortality estimates (which assess death rates above the usual average).

The virus responsible for AIDS—HIV (human immunodeficiency virus)—can be transmitted via intimate contact. HIV invades certain body fluids, including blood and semen, and suppresses our immune response by destroying lymphocytes, white blood cells that fight infection.

The virus responsible for COVID-19, SARS-CoV-2 (severe acute respiratory syndrome from corona virus 2), is a highly transmissible respiratory virus that spreads through airborne droplets exhaled by infected individuals. When inhaled, this virus attacks the respiratory mucosa, and may penetrate more deeply to infect other cell types, including neurons and immune cells. This can lead to severe complications, hospitalization, and even death, particularly for the elderly and those weakened by other health conditions.

The AIDS and COVID-19 virions are surprisingly similar in size and structure. Both measure roughly 100 nm across and are surrounded by a lipid membrane acquired from an infected cell. And both are studded with proteins that protrude from their surface, popularized as *spike proteins* in the case of SARS-CoV-2. These proteins enable the virion to attach and cling to a host cell, fuse its membrane with the cell's, and finally penetrate the cell. Both the HIV and COVID-19 virions contain RNA, encased in a protein shell called a *capsid*. This capsid protects the RNA, as well as the enzymes required to hijack the host cell's machinery and replicate the virus.

Despite their structural similarities, these viruses operate by fundamentally different mechanisms. HIV is a retrovirus, meaning it integrates its genome into the infected cell's DNA as the first step of replication. By contrast, the SARS-CoV-2 genome—which is approximately three times longer than that of HIV—is replicated directly within the host cell as RNA, without first integrating into the cell's DNA.

Both HIV and SARS-CoV-2 evolve and mutate rapidly. Especially in immunocompromised individuals, SARS-CoV-2 quickly develops mutations that change the amino acid sequences of its spike proteins. These mutations can enhance the virus's ability to attach to host cells or evade detection by antibodies. During the pandemic, variants of SARS-CoV-2 (such as alpha, delta, and omicron) succeeded one another, spreading across the globe in waves of infection.

Combating pathogenic viruses is a formidable challenge, although vaccination campaigns have eradicated smallpox and nearly contained polio. PrEP (pre-exposure prophylaxis), an ongoing regimen, effectively prevents AIDS, but unfortunately we still lack a true AIDS vaccine, and those living with AIDS must undergo lifelong antiretroviral therapy to suppress their viral load. As for COVID-19, rapidly developed and distributed RNA-based vaccines have helped mitigate the pandemic's impact, preventing countless hospitalizations. However, the threat of new viruses looms perpetually on the horizon. Despite our progress, we still lack comprehensive knowledge of how to preempt or thwart the emergence of new, pathological viruses. The fight against viral diseases remains a relentless pursuit, demanding constant vigilance and innovation.

View the HIV life cycle, by Janet Iwasa

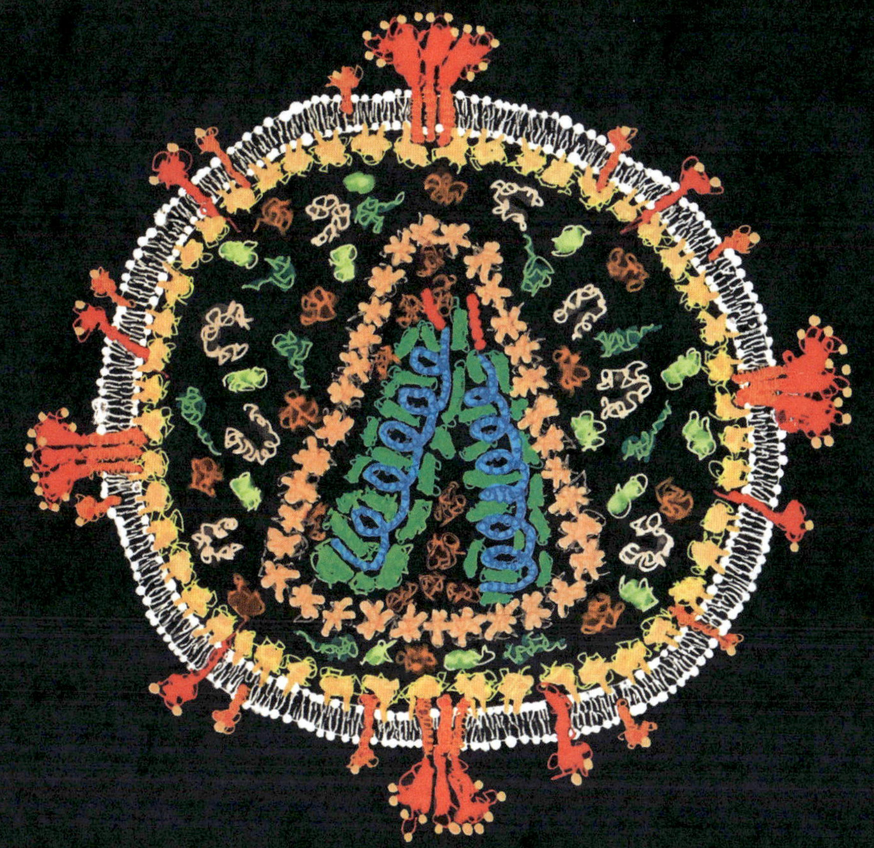

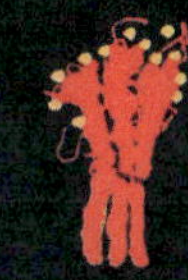

The surface protein: Around 15 copies of this protein allow the virion to attach to and fuse with a cell's membrane, then penetrate inside.

The capsid protein: 1,500 to 2,500 copies surround the two helical strands of viral RNA.

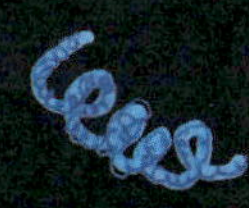

The RNA genome: The two strands contain 15 to 19 genes, which encode 19 proteins to be manufactured inside an infected cell.

The viral membrane : The virion acquires this double layer of lipids when it buds out of an infected cell.

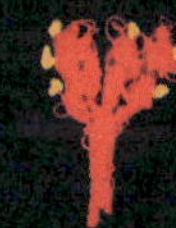

The spike protein: A COVID-19 virion carries around 25 to 100 copies of its spike protein.

The capsid protein: Around 1,000 copies surround the long, helical strand of viral RNA.

The RNA genome: The long strand of RNA contains 13 genes, which encode some 29 proteins to be manufactured inside an infected cell.

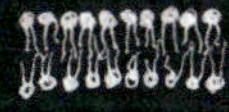

The viral membrane.

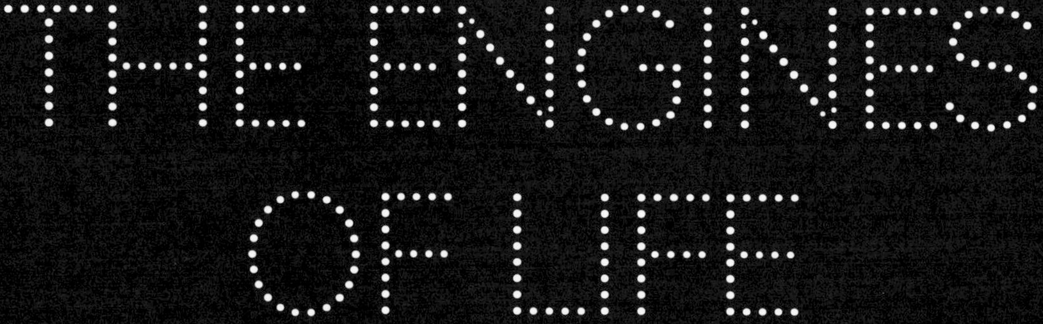

CHAPTER VII

THE ENGINES OF LIFE

Energy & metabolism

ENERGY
AND METABOLISM

From its beginning, life has relied on a continuous flow of energy and matter. All over our planet's surface, plant cells harness light energy from the sun, fueling a vast web of existence. This energy sustains photosynthetic bacteria, plants, algae, and phytoplanktonic protists, allowing them to thrive and create organic compounds—some of the myriad molecules of life. In turn, these molecules nourish a rich diversity of terrestrial and aquatic life. In the light-deprived depths of the oceans and subterranean realms, life follows a different path. In the abyss, bacteria and archaea extract energy from metals and gases, sustaining intricate ecosystems of remarkable creatures—first discovered in the 1960s at the bottom of the ocean *(see pp. 68–69)*.

Organic matter is transformed and recycled by communities of organisms —from bacteria to animals—with complementary metabolic abilities. We animals survive by eating, drinking, and breathing. Our food supplies proteins, sugars, fats, vitamins, and minerals, which our cells transform into energy and essential organic compounds through respiration and fermentation. Fungi, nature's decomposers, absorb nutrients from their surroundings, and in intricate exchanges with plant roots. Oxygen, the vital element of respiration, fills our atmosphere thanks to photosynthesis—produced half by phytoplankton and algae, and half by trees and land plants. For them, oxygen is a by-product. Photosynthesis is made possible by chlorophyll pigments, found in the membranes of cyanobacteria and chloroplasts of plants. As these organisms transmute sunlight into chemical energy by photosynthesis, they capture atmospheric CO_2; synthesize sugars, fats, and macromolecules (like starch and cellulose); and release O_2. Ultimately, plant and animal residues, broken down by microbes, provide fuel and nutrients for a vast array of protists and fungi. All these different organisms have complementary metabolisms, the chemical processes of the body that transform organic matter.

At the heart of metabolism—across the whole tree of life—lies a universal force, the proton-motive force. This force, both electrical and chemical, arises from the flow of protons—positive hydrogen ions (H^+)—across cell membranes. This fundamental energy mechanism, embedded in membranes, is as ancient as life itself. First established in bacteria and archaea, it later evolved within chloroplasts and mitochondria—eukaryotic cells' powerhouses—which have cyanobacterial and bacterial ancestry, as detailed in chapter IV. Thus, the universe's primordial and most abundant element—hydrogen—emerges as the ultimate driver of life, the silent force behind the ceaseless transformations that sustain the living world.

On the surface of the planet, energy comes from the sun. Photons of light (denoted hv) are captured by cyanobacteria with photosynthetic membranes ◯, as well as by eukaryotes—phytoplankton protists, plants, and algae—equipped with chloroplasts ●, the membrane-bound organelles that capture CO_2 and carry out photosynthesis to make organic compounds.

Photosynthetic organisms, from cyanobacteria ◯ and protists ◯ to plants, are eaten or absorbed by uni-, multi- and pluricellular organisms—animals and fungi—which produce energy from the organic matter they digest. Protists and pluricellular animals can do this thanks to the mitochondria ●, energy-producing organelles which oxidize food molecules, using O_2 to break the hydrogen bonds (and produce H_2O).

In the darkness of the abyss and the depths of the Earth, bacteria ◯ and archaea ◯ extract chemical energy from metals and gases such as hydrogen sulfide (H_2S), and use CO_2 and other carbon-rich molecules to produce organic matter.

N2
photon hv
O2
CO2
CO2
H2O
O2
H+
OH-
H2O
CH4
CO2
H2S
H2
NH3
NH4

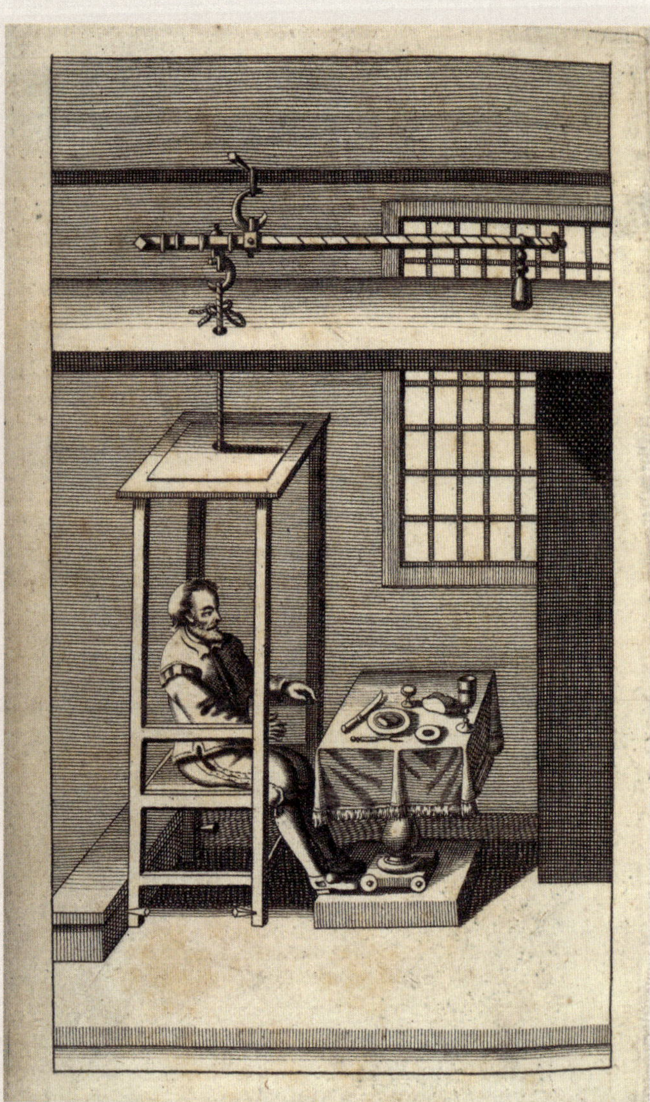

Right
A MAP OF METABOLISM
The chemical reactions represented on this map constitute the main metabolic pathways and cycles of life. It resembles a subway map, with stations (the molecules being metabolized) and lines from one station to the next (the reactions, catalyzed by a host of different enzyme proteins). At the center, in yellow, is the Krebs cycle, which metabolizes sugars, fats, and proteins to release energy and create simpler molecules. These molecules (metabolites) are used in other metabolic pathways—chemical chain reactions that create more complex molecules the body needs. In black, starting from top left, we see the pathway of glycolysis (the breakdown of sugars), which provides a citrate molecule—an essential metabolite—to the Krebs cycle.

© Courtesy the author, Richard Wheeler, Cambridge, Great Britain

The discovery of metabolism and organelles

The concept of animal metabolism—the transformation of food into vital energy—dates back to antiquity. But this idea was not tested experimentally until the 17th century. Santorio Santorio, a pioneer of physiological study, meticulously weighed himself before and after each meal, drink, and activity, seeking to quantify the body's hidden transformations. Soon after, the study of respiration in animals, and fermentation in yeast, began to unravel the mysteries of metabolism. However, it was only in the 19th century, with the advent of modern chemistry, that scientists identified the crucial element in the air we breathe: oxygen (O_2). Oxygen is the essential element that allows our mitochondria to oxidize food, producing water (H_2O), carbon dioxide (CO_2), and heat (*see p. 40*).

At the dawn of the 20th century, European researchers turned their attention to fermentation—the process by which cells convert sugars. By then, fermentation's role as catalyst of *other* important reactions was well established, thanks to the foundational work of Anselme Payen, Jean-François Persoz, and Louis Pasteur. Researchers had also recognized that certain diseases, such as scurvy, stemmed from dietary deficiencies. Gradually, they isolated essential nutrients—vitamins A, C, and D—as well as enzymes with remarkable biochemical properties, including urease, catalase, and protease. With the understanding that these enzymes, in concert with vitamins, facilitate the breakdown of sugars, researchers embarked on mapping the intricate

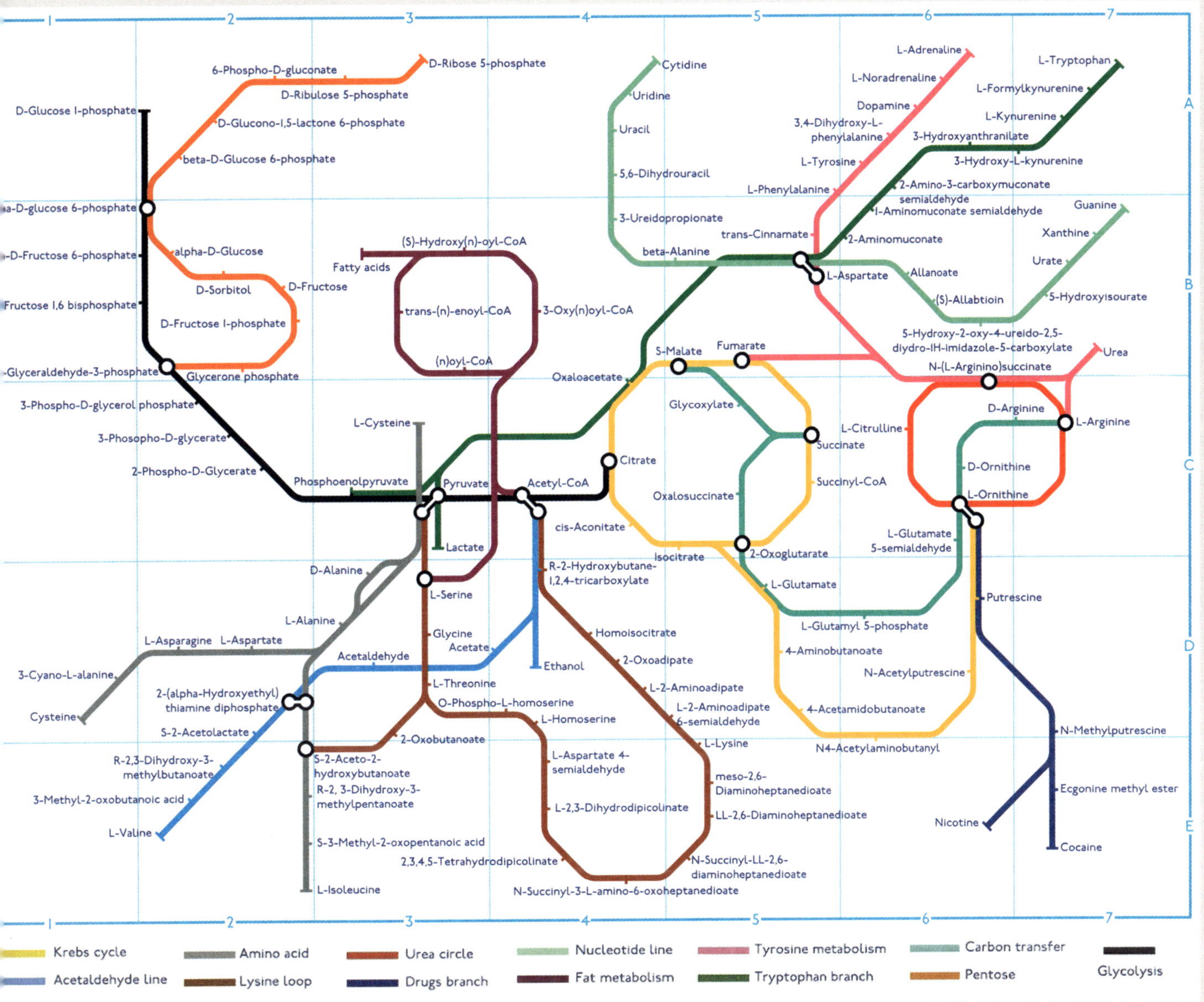

network of metabolic reactions.

In the 1920s, Gustav Embden, Otto Meyerhof, and Jakub Karol Parnas delineated the first metabolic pathway: glycolysis—a sequence of enzymatic reactions that converts glucose into pyruvate, releasing energy in the process. Soon after, Karl Lohmann unveiled the existence of ATP (adenosine triphosphate), a universal energy currency molecule. In the following decade, Albert Szent-Györgyi and Hans Krebs uncovered the citric acid cycle, which plays a pivotal role in governing the metabolism of sugars, fats, and proteins. These groundbreaking discoveries, made through studies of both muscle tissue and yeast cultures, revealed metabolic pathways that underpin the function of all living cells, highlighting their universal nature.

This growing understanding naturally led to the question: Where within the cell do these metabolic reactions occur? By the mid-19th century, the microscope had already revealed the existence of mitochondria and chloroplasts, although their functions remained enigmatic. In the 1940s, Albert Claude, George Palade, and their colleagues devised a revolutionary technique—cell fractionation. By first crushing cells open, they were able to isolate their larger components using a centrifuge. Aided by the newly invented electron microscope, they sorted and identified different organelles, unlocking the secrets of their roles in metabolism. Mitochondria and chloroplasts, once mere structures under a microscope, were now understood as bioenergetic powerhouses, even capable of functioning outside the cell. Today, spectroscopic techniques allow scientists to study metabolic reactions running *in vivo*, bringing ever-greater clarity to the intricate transformations that drive life within cells.

FROM THE PROTON-MOTIVE FORCE TO ATP

The term *metabolism* originates from the Greek word meaning "transformation." It encompasses the vast network of chemical reactions that sustain life in cells and organisms. While plants, animals, and fungi rely on a limited set of metabolic processes—photosynthesis, respiration, and fermentation—unicellular organisms possess an astonishingly diverse metabolic repertoire. Microorganisms, including bacteria, archaea, and protists, can harness energy from an array of molecules in their environment. Some even thrive in extreme, lightless conditions, deriving energy from minerals or gases.

Despite their metabolic diversity, all cells share the extraordinary ability to convert light or chemical energy into the universal energy currency molecule—ATP (adenosine triphosphate). This molecule, brimming with energetic phosphate bonds *(see p. 47)* is synthesized thanks to a fundamental electrochemical mechanism known as the *proton-motive force.* Proposed by Peter Mitchell in 1961, this concept, though initially met with skepticism, ultimately revolutionized our understanding of bioenergetics. It became accepted as the universal principle of energy production after twenty years of controversy.

A proton is the charged form of hydrogen (H⁺). The proton-motive force is generated by a gradient—meaning there are more protons on one side of a membrane than the other. This gradient is created by nanomachines that actively pump protons in one direction across the membrane. Because the protons are positively charged, they naturally repel each other and seek a balance. As a result, protons flow back across the membrane in the other direction, exiting through another nanomachine (ATP synthase) and powering it just like a turbine so that it can produce ATP. In prokaryotic cells, the proton-motive force takes place across the membrane that surrounds each bacterium or archaea. This activity of the membrane provides all the energy necessary for metabolism. For some bacteria and archaea, the proton-motive force also drives the rotation of flagella that allow movement *(see pp. 76–77).*

Eukaryotic cells have vastly expanded on this energy-generating capacity, after first inheriting it through a remarkable evolutionary event over a billion years ago: the internalization and domestication of bacteria *(see chap. IV).* These ancient bacterial symbionts evolved into mitochondria and chloroplasts—organelles that now serve as cellular powerhouses. In daylight, plants, algae, and plantlike protists harness energy from the sun, using chloroplasts to generate ATP via photosynthesis. And even when night falls, the mitochondria continue their ceaseless work, supplying energy just as they do in animals, fungi, and non-photosynthetic protists. Across the membranes of life, molecular pumps and turbines direct an unending dance of protons, sustaining the perpetual production of ATP that fuels all forms of metabolism, and ultimately life itself.

The mitochondrion's ● outer membrane is permeable to protons (H⁺), while its inner membrane is impermeable. The inner membrane folds to form enclosures called *cristae*, each with a duo of nanomachines on top: ATP synthases, which produce the universal energy currency molecule, ATP. They do so by adding a phosphate group to its precursor molecule, ADP.

The ATP synthase nanomachines work like turbines, powered by protons flowing through them. The protons are pumped inside the crista by other nanomachines called *respirasomes*. Because the positively charged protons seek electrical and chemical equilibrium across the membrane of the crista, they resist accumulating inside the crista, and exit through the ATP synthases, creating the proton-motive force.

Respirasomes are powered, in turn, by the flow of electrons (e-). Special proteins in the respirasome (I, III, IV) are organized into an "electron transport chain." As electrons flow from protein to protein along the chain, and finally reduce O_2 producing H_2O, they power the proteins to pump protons inside the crista.

Respirasomes extract the needed electrons from the molecule NADH, breaking its hydrogen bond to create NAD. NADH, in turn, is supplied by the universal metabolic cycle known as the *Krebs cycle*.

mitochondria

ADP
ATP
ATP SYNTHASE
ADP
ATP
H+
H+
H+
H+
H+
CRISTA
Respirasome
O2
H2O
e-
NADH
NAD
Krebs Cycle
H+
internal mitochondrial membrane
INTER MEMBRANE
external mitochondrial membrane

The proton-motive force —the hydrogen engine of life

Vital energy is generated through oxidative phosphorylation, the chemical reaction that grafts a phosphate molecule onto ADP (_adenosine diphosphate_) to produce ATP (_adenosine triphosphate_).

Oxidative phosphorylation, which generates ATP, takes place in membranes, where the hydrophobic double layer of lipid molecules acts as an insulator _(see p. 48)_. ATP production depends on the unequal electric charge on either side of the membrane, the power of which is comparable to that of a thunderbolt. The cell maintains this difference by creating a gradient of protons—hydrogen ions (H^+), or more rarely sodium ions (Na^+)—across the membrane.

The discovery of this electrochemical force, the proton-motive force, was the visionary achievement of Peter Mitchell, an eccentric English gentleman who pursued his research independently, financing the experiments himself. From the solitude of the Cornish manor house he transformed into a laboratory, Mitchell meticulously studied mitochondria extracted from the liver cells of rats, manipulating their electrical properties to regulate ATP production. In 1961, he introduced his groundbreaking "chemiosmotic" theory, igniting a fierce scientific debate known as the "ox phos war." His theory proposed that the cell concentrates protons on one side of the membrane, and that as the protons return to the other side—seeking equilibrium—their motion drives the synthesis of ATP.

At the same time, another scientific revolution was unfolding in Great Britain, one that would forever transform our understanding of heredity and evolution. Molecular biologists Francis Crick, James Watson, Sydney Brenner, and Rosalind Franklin, along with their colleagues, were deciphering the molecular basis of the genetic code, revealing that the nucleotide sequences of the DNA double helix encode life's blueprint _(see pp. 158–59)_. This discovery rapidly gained worldwide recognition, culminating in a series of Nobel Prizes in 1962.

By contrast, Mitchell's chemiosmotic theory and the concept of the proton-motive force faced a far more arduous path to acceptance. It was not until 1978, 17 years after his proposal, that Mitchell was awarded the Nobel Prize, granting his work long-overdue recognition, though it remained largely uncelebrated. In truth, ATP synthase—the proton-powered nanomachine that

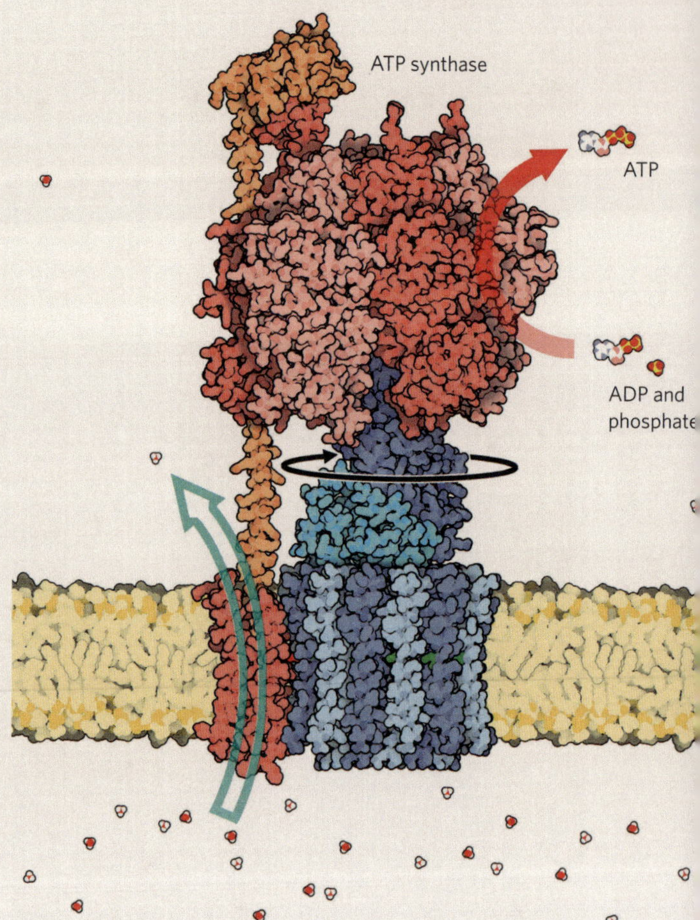

ATP SYNTHASE—A TURBINE THAT GENERATES ENERGY IN THE FORM OF ATP

ATP synthase is an amazing nanomachine that produces ATP. Embedded in the membrane, it is actually a turbine with an axle and rotor, assembled from some twenty proteins (represented here in different colors). A turbine to generate energy makes sense. Like water flowing over the top of a dam, protons that are concentrated on one side of the membrane (here, at the bottom in the intermembrane space) flow through the ATP synthase nano-turbine (_blue arrow_) and turn its rotor. This rotation produces the energy needed to make ATP by grafting an additional phosphate group onto its precursor molecule, ADP. In one second, the flow of protons turns the rotor more than 300 times, generating about a hundred ATP molecules. Our mitochondria produce hundreds of thousands of ATP molecules every minute, which are immediately consumed to maintain metabolism (they revert to ADP, in an ongoing cycle). Each day, an animal produces and consumes its own equivalent weight in ATP.

© Artwork by David Goodsell from _The Machinery of Life_, Springer, 2009

generates ATP—had all the potential to become as iconic as the DNA double helix. Yet its remarkable rotating turbine structure was not unveiled until 1998, a historic delay that likely contributed to the theory's lack of popularization. This gap of nearly four decades between Mitchell's 1961 proposal and the 1998 revelation of ATP synthase may well explain why this fundamental nanomachine of life remains far less renowned than DNA's double helix.

Energy-generating nanomachines

It is likely that the ancestral cell LUCA *(see chap. III)* harnessed energy through a primitive form of the proton-motive force, coupled with an electron current flowing across its membrane. This ancient mechanism was inherited and refined by bacteria and archaea, which further evolved its complexity. Over time, the molecular nanomachines at the heart of this process—ATP synthase and the protein complexes of the electron transport chain—became enriched with additional proteins and cofactors, which fine-tuned their efficiency and expanded their energy production capacities.

A leap in evolution occurred when other unicellular organisms internalized and domesticated bacteria and cyanobacteria, giving rise to eukaryotic cells with mitochondria and chloroplasts. It is believed that the abundant energy supplied by these new organelles fueled the transition to more sophisticated eukaryotic cells, characterized by greater complexity.

At the core of ATP production lies an extraordinary molecular turbine—ATP synthase. However, for this nanomachine to operate, it requires a powerful proton gradient—established by other nanomachines that actively pump H^+ ions across the membrane. This feat demands both specialized proteins called *proton pumps* and a continuous energy supply. Plants, algae, and phytoplankton protists are self-sufficient in this regard, seamlessly coupling their proton pumps with photosynthesis. When bathed in sunlight, they convert solar energy into electron currents, which in turn power the proton pumps. In darkness, their mitochondria take over, running the electron transport chain to generate ATP. By contrast, animals, fungi, and many non-photosynthetic protists are entirely dependent on their mitochondria for ATP production.

In all cells, energy generation requires three interconnected steps: electron current coupling, proton pumping, and ATP synthesis. These processes take place within membranes, orchestrated by a suite of nanomachines known as respiratory complexes I, II, III, and IV, which work in concert with smaller molecular partners such as cytochromes and ubiquinones. Precisely arranged, these protein complexes form the highly efficient electron transport chain: They interlock like the components of an electrical circuit, channeling electrons from one to another as if through a microscopic wire. This universal mechanism allows the cells to generate ion gradients—whether of H^+ or Na^+—to drive

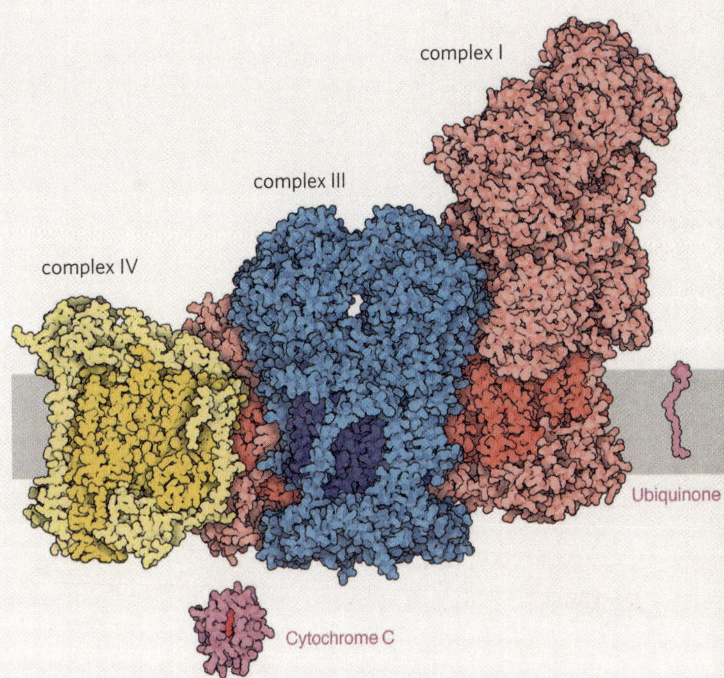

THE RESPIRASOME–THE NANOMACHINE THAT GENERATES THE PROTON GRADIENT

The respirasome is a (relatively) huge nanomachine that spans a mitochondrion's inner membrane (*shaded area*) and pumps protons from its interior into its cristae, dedicated spaces that exist between the mitochondrion's two membranes. The respirasome is made up of three types of protein complex—associations of several dozen enzyme proteins of the electron respiratory chain: one complex I (*pink*), two complexes III (*blue*), and one complex IV (*yellow*). The dark-colored proteins are those encoded by mitochondrial DNA and manufactured inside the mitochondria. The respirasome captures electrons from the NADH molecule produced by the Krebs cycle and uses them to pump H^+ ions across the membrane, creating an electrochemical proton gradient. This gradient drives the motion of the protons that turn ATP synthase's nano-turbine to produce ATP. Also shown are the small protein cytochrome C and, in the membrane, a ubiquinone molecule. The respirasome requires both to be present for successful electron transport.

© Artwork by David Goodsell for the RCSB PDB "Molecule of the Month"

their essential metabolic processes. In mitochondria, this intricate process occurs within the inner membrane *(see p. 133)*. In bacteria and archaea, the electron transport chain and ATP synthase turbine are embedded in the plasma membrane, powering life at its most basic level.

View ATP synthase in action (Harvard Online)

ORGANELLES AND CONDENSATES— CRUCIBLES OF METABOLISM

Eukaryotic cells, along with a few bacteria and archaea species, exhibit a remarkable degree of compartmentalization. The largest compartments, best observed under the microscope, are membrane-bound organelles. These include nuclei, chloroplasts, mitochondria, lysosomes, the endoplasmic reticulum, and many others. Membrane-bound organelles each have their own metabolic functions. Within mitochondria, the Krebs cycle—a cyclical sequence of chemical reactions—metabolizes sugars, fats, and proteins, extracting energy and producing vital metabolites in the process *(see pp. 130-31)*. Meanwhile, inside chloroplasts—the photosynthetic powerhouses of plant cells—enzymes of the Calvin cycle fix CO_2, synthesizing sugars using energy generated by photosynthesis (wherein eight photons of light yield six molecules of ATP). Since the 1970s, we have known that these two essential organelles, complete with their metabolic pathways, are evolutionary inheritances from bacteria and cyanobacteria *(see chap. IV)*.

Beyond membranous organelles, cells also contain a diverse array of biomolecular condensates—granular compartments without membranes that play crucial roles in metabolism and the synthesis of macromolecules. The most striking and largest of these biomolecular condensates is the nucleolus, a prominent granular region within the nucleus where ribosomes and their RNA components are assembled. Biomolecular condensates emerge from the dynamic interplay of proteins, RNA, and other molecular partners. Many of these molecules, including the "intrinsically disordered proteins"— renowned for their structural flexibility—self-assemble into transient, highly dynamic aggregates.

Biomolecular condensates spontaneously assemble in response to cellular conditions, emerging from their surroundings via a sorting process known as LLPS (liquid–liquid phase separation)—just as oil and water, mixed into an emulsion, separate. These dynamic droplets merge, fragment, or vanish depending on environmental cues, orchestrating complex biochemical interactions. For instance, when yeast or human cells experience oxygen deprivation, the proteins and RNA involved in glycolysis (sugar breakdown) organize into specialized condensates known as *G granules*. Similarly, proteins, metabolites, and ATP congregate in condensates to structure the cytoplasm and regulate key metabolic pathways, including glycolysis. These condensates not only form freely within the cytoplasm but also interface with membranes, chromosomes, and organelles, serving as hubs for molecular interactions. In essence, biomolecular condensates function as crucibles, where chemical reactions, information exchanges, and nanomachine interactions unfold. As our understanding of these enigmatic structures deepens, they promise to revolutionize our perception of cellular organization, despite the fact that the principles governing their self-assembly and function remain largely mysterious.

A cyanobacterium whose photosynthetic membranes ▌ capture photons. In the cytoplasm, its single chromosome ▌ is represented, as well as four carboxysomes ◆. These are biomolecular condensates, principally made of Rubisco proteins, which capture CO_2 molecules.

A partial view of a eukaryotic plant cell, featuring membranous organelles: two chloroplasts ●, two mitochondria ●, and the endoplasmic reticulum ▌ surrounding the nucleus. Half a dozen biomolecular condensates in the form of granules are represented in the cytoplasm.

Inside the nucleus, we distinguish several chromosomes ▌ and a nucleolus *(red and ocher granules)*. The nucleolus is a large biomolecular condensate, made of different compartments inside which ribosomal RNA and ribosomes *(see pp. 56–57)* are produced and assembled.

The biomolecular condensate revolution—fluid droplets

The interior of a eukaryotic cell—its cytoplasm—is a liquid environment with the viscosity of honey, teeming with life at the molecular scale. It is densely packed with membrane-bound organelles and other compartments, all immersed in a dynamic soup of billions of nanomachines, molecules, and ions. Among these structures, membrane-bound organelles—such as mitochondria, chloroplasts, and nuclei—represent relatively stable compartments. In contrast, biomolecular condensates are more ephemeral, membrane-free assemblies of macromolecules that convene and disperse in response to cellular conditions. Iconic examples include germ granules in the cytoplasm of embryos, and nucleoli within the nucleus of eukaryotic cells *(see p. 175)*. Unlike membranous organelles, biomolecular condensates are changeable by nature: They expand, shrink, merge, vanish, and reappear depending on the cell's state or phase within the cell cycle.

A breakthrough in the discovery of these condensates came in 2009, when Anthony Hyman and Clifford Brangwynne, working in Woods Hole and at the Max Planck Institute in Dresden, discovered that P granules—germline condensates of proteins and RNA in the embryo of the nematode *Caenorhabditis elegans*—behave like fluid droplets. Initially termed "liquid droplets" by Hyman and Brangwynne, similar structures were soon identified elsewhere, leading researchers to coin the broader term *biomolecular condensates*.

Biomolecular condensates self-organize through the principles of LLPS (liquid-liquid phase separation), a process that enables proteins, RNA, and their molecular partners to concentrate into nanoscale droplets, fostering interactions precisely when and where they are needed. They form in response to various factors, including signals received at the cell surface, the presence of cytoskeletal elements (microfilaments and microtubules)

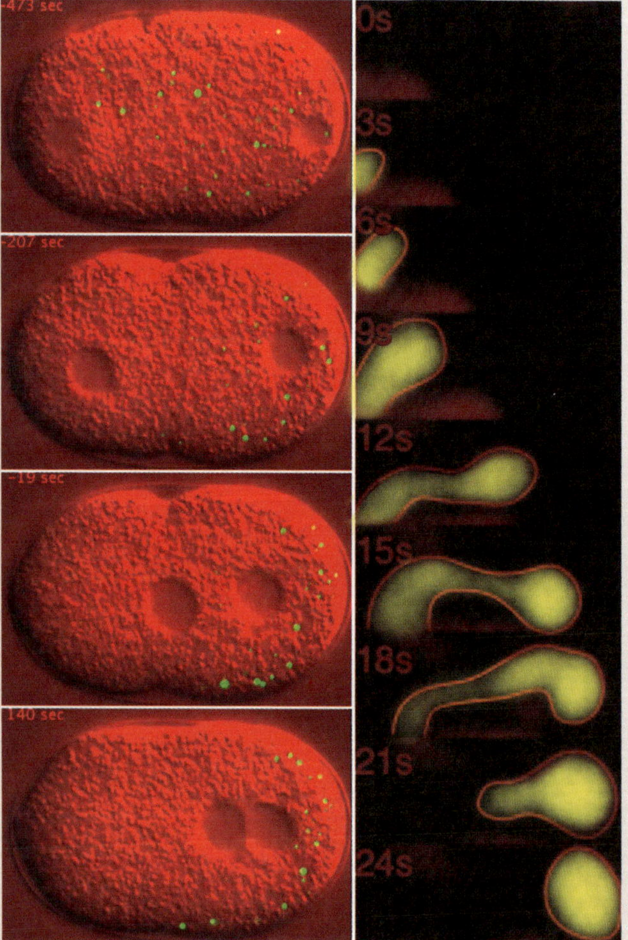

THE DISCOVERY OF BIOMOLECULAR CONDENSATES—P GRANULES AS LIQUID DROPLETS

The discovery in 2008: During the annual physiology course at the Woods Hole Marine Biological Laboratory, research instructors Anthony Hyman and Clifford Brangwynne and two students observed germ granules called *P granules* under the microscope (*yellow*), and filmed them in the minutes following fertilization of a nematode worm's oocyte (*sequence on left*). They noted that the P granules behave like liquid droplets that stretch, coalesce, split, and dissolve as they migrate toward the posterior pole of the fertilized oocyte (*close-up sequence on right*). Back in their laboratory at the Max Planck Institute in Dresden, the researchers analyzed the phenomenon in detail, and in 2009 wrote a landmark article in *Science*: "Germline P granules are liquid droplets that localize by controlled dissolution/condensation." This research first revealed the importance of liquid droplets, aka biomolecular condensates, and phase-change phenomena.

Some details: The P granules present in the oocytes and embryos of the nematode worm *Caenorhabditis elegans* are aggregates of proteins and RNA molecules. They are part of a family of granules called *germ granules*, which direct the development of germ cells (reproductive cells) in the developing embryo. P granules, revealed here by a yellow-fluorescing molecule, are initially dispersed in a newly fertilized oocyte (*top left*). About ten minutes after fertilization, as the two nuclei (the sperm's and the egg's) migrate toward the center of the cell and merge, the P granules concentrate at the embryo's posterior pole (*bottom left*). During the embryo's first cell division, since the germ granules are all located together, they are all contained in the posterior cell, which gives rise to the germline of the new nematode worm *(see p. 175)*.

SELECTED IMAGES (FROM VIDEO FOOTAGE)
© Courtesy Anthony Hyman and Clifford Brangwynne, Max Planck Institute, Dresden, Germany; published in *Science*, vol. 324, pp. 1729-1732 (2009)

and membranes, or the cyclic upheavals of cell division. Not unlike a flash mob, macromolecules rapidly gather in response to a signal, assemble to perform a task, and then swiftly disperse once their work is done. The droplets' properties are finely tuned to the cell's ionic balance, metabolic state, and molecular composition. They range from highly fluid to gel-like or solid, and their size varies dramatically, from the imposing nucleoli to minuscule "nuclear speckles" *(see p. 165)*. They adapt their shape to their surroundings, often appearing alongside one another, nested within one another, or positioned at interfaces with membranes, microfilaments, or microtubules. It is now clear that biomolecular condensates serve as molecular crucibles, essential for metabolism, protein assembly, and the operation of nanomachines. Their essential role in cell signaling, chromosome organization, and gene expression is increasingly recognized, revealing yet another layer of complexity in the symphony of cellular life.

The discovery of phase transitions within cells and their compartments has sparked an experimental gold rush. Scientists are racing to reconstitute condensates *in vitro* from various more- or less-structured proteins, RNA, and their partners. Physicists long familiar with the principles of deformation in soft matter, capillary action, and surface tension are now finding their expertise relevant to the inner workings of cells. Despite these advances, the precise molecular codes governing condensate dynamics—their assembly and dissolution—remain largely mysterious. The implications, however, are profound. Understanding these processes could offer critical insight into age-related diseases such as Alzheimer's and cancer, in which aberrant phase transitions may play a fundamental role. A revolution is underway! It promises to reshape our understanding of cellular organization and disease.

View P granules' localization in an embryo (Hyman Lab)

A BIOMOLECULAR CONDENSATE IN THE BACTERIUM *CAULOBACTER*

Caulobacter crescentus is a bacterium that lives in two forms: a mobile form that swims with a flagellum at the posterior pole (*green, at bottom*), and a differentiated, immobile form that anchors itself by a stalk. To swim, the bacterium grows a flagellum, assembling a biomolecular condensate at its pole around an aggregate of "intrinsically disordered," flexible proteins (*yellow*). These proteins recruit other partner proteins (*orange and magenta*) to effect the change from the bacterium's immobile form, to mobile. In the cytoplasm, the bacterium's chromosome (*violet*) and ribosomes (*blue*) are also represented.

WATERCOLOR/DIGITAL ILLUSTRATION
© Artwork by David S. Goodsell and Keren Lasker, RCSB Protein Data Bank and Scripps Research (doi: 10.2210/rcsb_pdb/goodsell-gallery-046)

Learn more about the creation of this digital watercolor from David S. Goodsell's series "Molecular Landscapes"

*Activation, signaling
& division*

ACTIVATION AND SIGNALING— AWAKENING THE CELL

In constant dialogue with their environment, cells receive, decode, transmit, and respond to countless signals. Some signals are physical, such as a change in temperature or pressure. Others are molecular, such as hormones produced and dispatched from distant cells. Signals can also arise from direct contact with proteins on the surface of an interloping virus or a neighboring cell.

Some signaling molecules are too large or too hydrophilic to cross the target cell's membrane unaided. These, the cell detects and receives using specialized receptor proteins embedded in the membrane and protruding outside *(see p. 48)*. Upon contact with a signaling molecule, the receptor protein transforms—generally changing shape—and associates with other proteins in both the membrane and the cytoplasm. Together, they form a complex called a *signalosome*. Once received and decoded at the cell cortex (the peripheral zone around the cell membrane), the signal is relayed to deeper within the cell. This marks the initial stage of the intricate process known as *cell signaling*.

The immediate consequence of this signal relay is "cell activation"—a dynamic cellular awakening characterized by a surge of calcium ions and/or the production of messenger molecules. Frequently, when a cell encounters a neighbor or receives a hormonal cue, its internal calcium concentration skyrockets. A striking example occurs during fertilization: Upon contact with a spermatozoon, an oocyte is activated by a calcium wave that sweeps across it like a tsunami *(see pp. 42 and 144)*.

Specialized membrane lipids act as molecular messengers, coordinating with a vast network of enzyme proteins to transmit these signals throughout the cell. Among them, kinase and phosphatase proteins play a crucial role, activating or inhibiting target proteins and messenger molecules by either adding or removing phosphate groups. Other proteins create messenger molecules from ATP, its close relative GTP (guanosine triphosphate), or other key metabolites. These molecular signals travel from the cell's cortex to its cytoskeletal structures and organelles, including the mitochondria and nucleus. Within the nucleus, the messenger molecules recognize and bind to activator or repressor proteins (within dynamic biomolecular condensates that also include RNA). These proteins, in turn, bind to specific target genes, either to repress the gene or "express" it by transcribing it into messenger RNA. This mRNA, together with other RNA, directs the synthesis and distribution of proteins necessary for the cell's response to the original signal.

Signal molecules are received from outside the cell by receptor proteins spanning the cell membrane. These protein receptors associate with enzyme proteins (as biomolecular condensates) to add or remove phosphates $\bigcirc$ to macromolecules, so as to control enzymatic activities and produce messenger molecules.

The cell is activated by a massive increase of calcium ions (Ca^{2+}) in the cytoplasm, released by channel proteins spanning the membrane of the endoplasmic reticulum, a primary reservoir of calcium within the cell.

Messenger molecules deliver the signal from the cell cortex to the biomolecular condensates that envelop and regulate the chromosome ▌. These condensates—made of repressor and activator proteins, as well as other proteins and RNA—turn specific target genes "on" or "off."

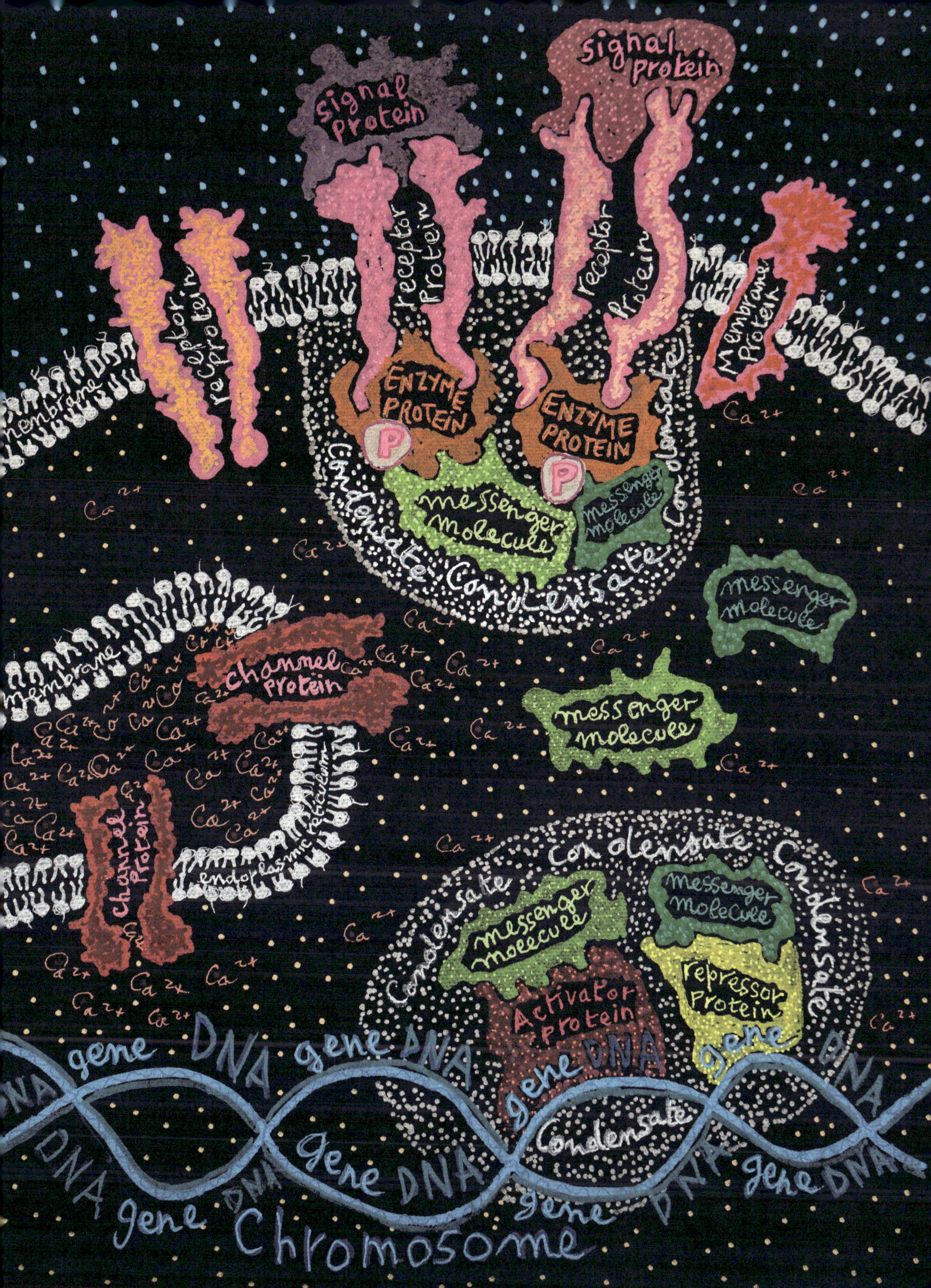

signal Protein
signal Protein
Receptor Protein
Receptor Protein
Membrane Protein
receptor protein
receptor protein
membrane
ENZYME PROTEIN
ENZYME PROTEIN
P
P
Condensate
messenger molecule
messenger molecule
Condensate
messenger molecule
Ca 2+
membrane
Channel Protein
Ca 2+
messenger molecule
messenger molecule
endoplasmic reticulum
Channel Protein
Condensate
condensate
Condensate
messenger molecule
messenger molecule
Activator Protein
repressor Protein
gene
Condensate
DNA gene DNA gene DNA
DNA
gene DNA
gene DNA
gene
gene DNA
gene DNA
DNA
DNA gene
Chromosome

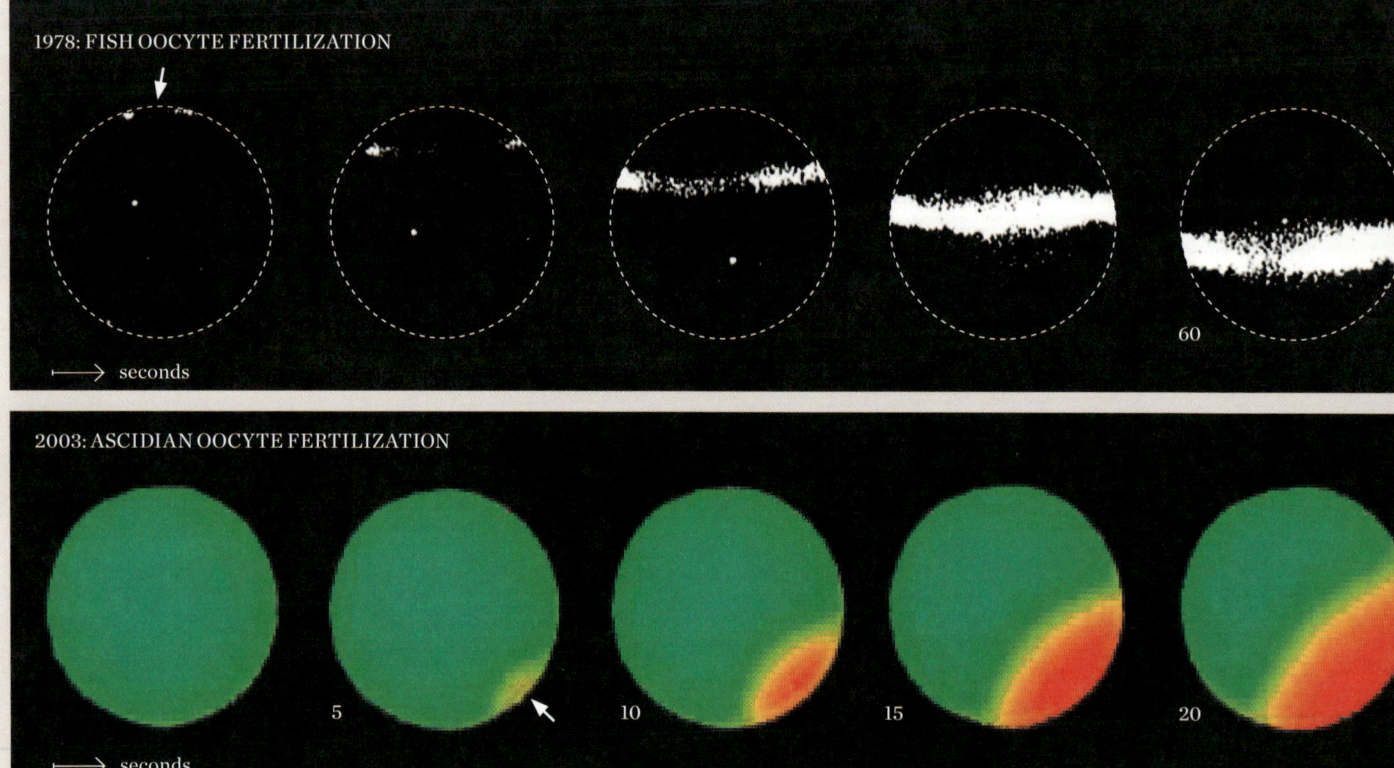

Cell activation—life begins with a calcium tsunami

The genesis of a new individual—whether human or animal—begins with a striking event: A sudden surge of calcium ions sweeps through the oocyte, starting from the site of fertilization. This sudden influx, often described as a calcium tsunami, is a universal response to various stimuli that cause all sorts of dormant cells to awaken, whether triggered by hormones or by contact with another cell (the spermatozoon, in the case of the oocyte). The calcium surge sets off a powerful cascade of intracellular events with metabolic and genetic consequences.

This remarkable phenomenon was discovered in the 1970s by Lionel Jaffe, John Gilkey, Richard Steinhardt, David Epel, and their American colleagues, who captured the earliest images of calcium explosions and waves in fertilized sea urchin and fish oocytes. A wave of calcium travels across an oocyte within mere seconds. In species such as fish, sea urchins, and amphibians, a single calcium wave that starts from the fertilization site is sufficient to awaken the oocyte. By contrast, in ascidians (sea squirts), mollusks, and mammals, fertilization sets off a series of rhythmic calcium oscillations that can last for minutes—or in mammals, even several hours (see p. 42).

Most cells, including primary or "immature" oocytes, are diploid (2n), containing two sets of chromosomes. Once it is triggered by the fertilizing sperm—and the ensuing calcium tsunami—the oocyte finally concludes meiosis to reach the haploid (1n) state, with only one set of chromosomes. (The extra set is jettisoned in a "polar body.")

The sperm delivers its own haploid set of chromosomes to the oocyte, along with a centriole located at the base of its flagellum. This paternal centriole recruits proteins from the oocyte to form a centrosome—also known as an MTOC (microtubule organizing center). Microtubules radiate from this center to create a star-shaped structure called the *sperm aster*. In animals and certain plants, these microtubules function as tracks, guiding the paternal nucleus toward the maternal nucleus. The nuclei converge, merging the two haploid sets of chromosomes (maternal and paternal) to restore the diploid state (1n + 1n = 2n), giving the fertilized oocyte its own complete and unique genetic identity as a new individual (see p. 177).

Once in possession of both paternal and maternal chromosomes, the fertilized oocyte—now a *diploid zygote* cell in biologist's jargon—embarks on a series of mitotic divisions, becoming an embryo whose cells progressively specialize into tissues and organs. The destiny of each embryonic cell hinges primarily on its location within the embryo, relative to its anterior–posterior axis (head to tail), dorsal–ventral axis (back to front), and later, left–right axis. These axes are established either before, during, or after fertilization, depending on the species. We will explore the specific pathways of embryonic development across the animal kingdom in a later chapter *(see p. 180).*

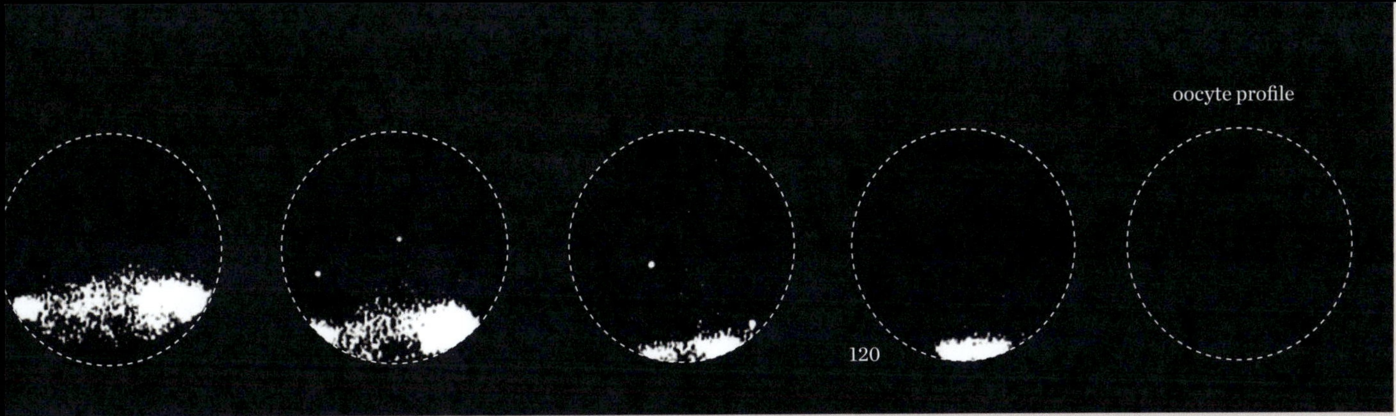

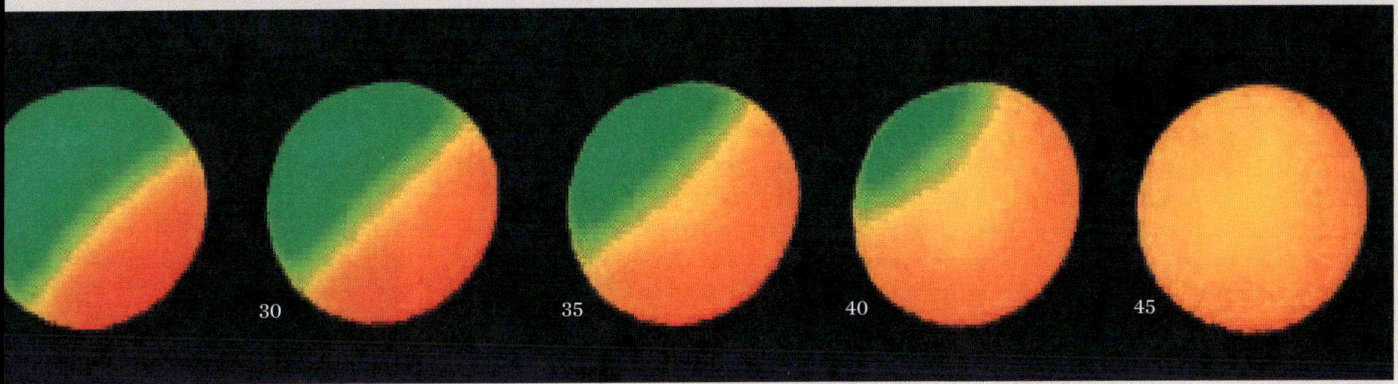

View calcium waves traversing medaka fish eggs (IWF Göttingen)

Top

THE DISCOVERY OF CALCIUM WAVES IN FISH EGG FERTILIZATION

In 1977, Lionel Jaffe and his student John Gilkey flew to Princeton to join Ellis Ridgway and physicist George Reynolds. They brought small Japanese *medaka* fish, which laid large oocytes that were easily fertilized and observed under a microscope. Gilkey injected the oocytes with aequorin, a jellyfish protein that emits light when the concentration of calcium ions rises around it. At the time, Reynolds was one of the few researchers with an ultrasensitive military camera that could detect and film the emission of light photons (luminescence), which revealed the oocyte's rise in calcium under the microscope. Still images from the recorded sequence are shown here. The calcium wave, visualized as a white band, propagates down the fertilized egg's cortex from the sperm's entry point (indicated by the arrow on top). The wave crosses the 1 mm egg in two minutes, at a speed of approximately 10 μm per second. The video soundtrack is full of exclamations that convey the researchers' joy at discovering the first calcium wave ever observed.

STILL IMAGES FROM VIDEO RECORDING
© Courtesy Lionel Jaffe, coauthor of the original article "A free calcium wave traverses the activating egg of the medaka, *Oryzias latipes*," J. C. Gilkey, L. F. Jaffe, E. B. Ridgway, G. Reynolds, *J. Cell Biol.* 1978; vol. 76(2), pp. 448–466

Bottom

MULTIPLE CALCIUM WAVES TRIGGERED BY FERTILIZATION IN AN ASCIDIAN

New techniques using fluorescent molecules have replaced luminescence techniques. In this sequence, an ascidian (sea squirt) oocyte has been injected with a molecule that fluoresces in response to increased calcium concentration (*green* indicates the lowest concentration, and *red* indicates highest). A calcium wave triggered by the fertilizing sperm (indicated by the arrow at five o'clock) passes through the oocyte in 45 seconds.

CONFOCAL FLUORESCENCE MICROSCOPY
© Courtesy A. McDougall, CNRS & Sorbonne University, IMEV, Villefranche-sur-Mer, France

THE CELL CYCLE— GROWTH, MITOSIS, AND DIVISION

A cell's response to an external signal is dictated by the nature of the signal and the cell's inherent abilities. The activated cell may respond by changing its shape or motility, by specializing or multiplying. When stimulated to divide, a cell embarks on its transformation in a sequence known as the *cell cycle*: a preparatory phase of activation and growth (interphase), the separation of chromosomes and centrosomes (mitosis), and finally division into two distinct daughter cells (cytokinesis).

The timing of this intricate dance is partly directed by activation or suppression of specific cell-cycle genes, first discovered in yeast during the 1980s and 1990s. Signals received by the genome are deciphered and translated into instructions—transmitted by various proteins and RNA—that direct the synthesis and localization of any proteins and nanomachines necessary for the task at hand *(see pp. 164–66)*. Even more immediate is the response of enzyme proteins, which are swiftly activated or inactivated through transient chemical modifications or targeted degradation events. Over the past four decades, scientists have revealed the central role of proteins bearing evocative names such as *cyclins*, which regulate these cascades of enzymatic activity. They govern not only the orderly progression of the cell cycle, but also the decision to exit the cycle and embark on differentiation into a specific cell type: nerve, epithelial, blood, muscle cells, or beyond.

Before division, during its growth phase, the cell duplicates its chromosomes and membrane-bound organelles, while reorganizing its biomolecular condensates such as nucleoli and centrosomes. At the same time, the cell deploys its intricate cytoskeleton—a dynamic scaffold of microtubules and microfilaments—to oversee a dramatic internal reconfiguration known as *mitosis*, parceling the chromosomes and organelles into two separate groups, one for each daughter cell. Cytokinesis, the final phase, unfolds as a spectacular constriction of the cell cortex, cleaving the mother cell into two daughters—each inheriting similar, or sometimes distinct, contents.

All these cell-cycle events depend on the astonishing self-organization and coordination of countless proteins, nanomachines, organelles, and dynamic molecular structures, all in perpetual motion and finely tuned by ions and metabolites. Today, we know many of the key players involved in cellular activation, signaling, and division, but the deeper mechanisms that coordinate their interactions remain largely mysterious. Still, emerging tools in modeling and artificial intelligence are illuminating the intricate laws that govern the astonishing complexity of life.

Once activated to divide, a cell goes through three major successive phases: **Interphase**, **Mitosis**, and **Cytokinesis**.

Interphase includes two stages of cell growth (G1 and G2), punctuated by synthesis (S), when the chromosomes I and centrosomes ● are duplicated, and the nucleoli ● disperse.

In **Mitosis** (M phase), the nuclear membrane I fragments, and the chromosomes I separate into two batches concentrated at opposite poles—moved into position by the microtubules I of the cytoskeleton.

Cytokinesis, which divides the cell in two, is also run by cytoskeletal structures. A spindle of microtubules I between the two poles distributes the cell's organelles and components into two groups. And a network of microfilaments I that form an equatorial ring at the cellular cortex contracts to "pinch" the mother cell into two daughter cells.

Differentiation into specific cell types (neuron, fibroblast, lymphocyte, muscle, etc.) occurs when cells exit the cycle at G0 phase.

MITOSIS
CYTOKINESIS
INTERPHASE
G1
G2
S
Go
DIFFERENTIATION

From mitosis to the cell cycle—a century of discoveries

In 1882, Walther Flemming immortalized the process of mitosis in a drawing of a dividing cell with moving chromosomes—a name inspired by the Greek words *chroma* (color) and *soma* (body). During mitosis, these chromosome filaments migrate toward two opposing poles of a spindle-shaped structure. In 1890, German biologist Theodor Boveri observed that each pole appeared as a spherical mass with dense granules at its center, which he named a *centrosome* and *centrioles*, respectively. Boveri demonstrated that what begins as one centrosome duplicates itself, separates, and journeys to opposite poles of the cell, separating the chromosomes into two sets and positioning them at opposite poles.

A century later, the study of mitosis was revitalized. From the 1960s onward, advances in light and electron microscopy, boosted by image-processing techniques, revealed that the spindle is composed of bundles of highly dynamic microtubules. Biochemical studies soon revealed that these are built from proteins called *tubulins*, which assemble and disperse to lengthen or shorten the microtubules, allowing them to move and reposition nanomachines, organelles, or chromosomes (see p. 56).

Starting in the 1960s, Yoshio Masui, a Japanese researcher working first at Yale and then at the University of Toronto, sought to understand the mechanisms behind mitosis. By transferring cytoplasm between large *Xenopus* (toad) oocytes at various stages of the cell cycle, he discovered a mysterious agent capable of triggering cell division. He named it MPF (maturation promoting factor) since it caused the injected oocytes to divide, undergoing meiosis (also known as *maturation*). Meanwhile, in Seattle, Leland Hartwell identified mutant *Saccharomyces cerevisiae* yeasts that failed to divide properly, leading to the discovery of specific genes that control the cell cycle. He named them CDC (cell division cycle) genes. By the late 1980s, biochemists had successfully isolated MPF (now renamed mitosis promoting factor) and determined it to be a kinase protein—an enzyme that activates other proteins by attaching phosphate groups. Around the same time, in Edinburgh, Paul Nurse identified the CDC genes first described by Hartwell, revealing that they, too, coded for kinase proteins closely related to MPF. This breakthrough established a fundamental principle: that the cell cycle is governed by a delicate balance of phosphorylation and dephosphorylation. Completing this transformative picture, Tim Hunt—through elegant studies of sea urchin embryos, remarkable for their synchronized divisions—discovered proteins he termed *cyclins*. While teaching at the Marine Biological Laboratory in Woods Hole, Massachusetts, and later in his Cambridge laboratory, Hunt showed that cyclin proteins are synthesized just before mitosis and rapidly degraded afterward. Their rhythmic rise and fall set the timing of cell division. Ultimately, the mechanism was laid bare: An enzymatic kinase protein, active only when bound to a cyclin, "kicks off" the cell division process. For their groundbreaking contributions, Hartwell, Nurse, and Hunt were awarded the Nobel Prize in Physiology or Medicine in 2001. Since then, the landscape has grown more complex, with the identification of numerous cyclin and kinase protein variants, and an expanding cast of regulatory molecules. Yet, the essential principle endures: The cell cycle is driven by the rhythmic synthesis, modification, and degradation of interactive molecules—choreography as elegant as it is vital.

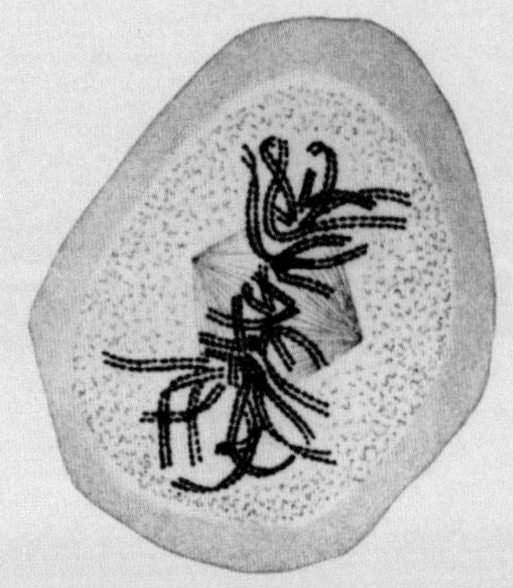

HISTORICAL DRAWING OF MITOSIS

In 1882, observing large salamander skin cells, Walther Flemming drew filaments separating longitudinally into two batches. These filaments were named *chromosomes* a few years later. This stage of mitosis, when the chromosomes separate, is called *metaphase*.

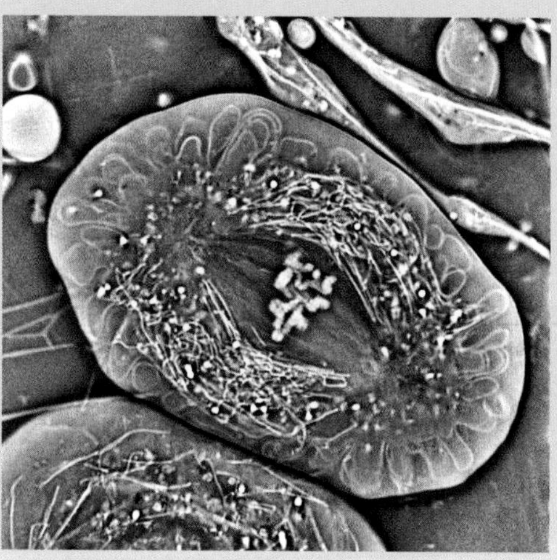

MITOSIS OBSERVED WITH POLARIZED LIGHT

A dipteran (fly) cell undergoing meiosis, in metaphase, is viewed under a polarized light microscope. The chromosomes are at the center of the microtubule spindle, surrounded by threadlike mitochondria. Bubble-like structures (probably biomolecular condensates) surround the centrosomes at both poles of the spindle.

Centrosomes and centrioles

During cell division, separation of the chromosomes and cellular components involves thousands to millions of dynamic microtubules made of proteins called *tubulins (see p. 56)*. Tubulin proteins add on to one end of the microtubule (called the "+ end"), lengthening it, and depart from the other end (the "– end"), shortening it. In animal cells, microtubules radiate from organizing centers (centrosomes), forming asters, which change in size over the course of the cell cycle. Dynamic microtubules grasp the chromosomes in their central location, then slide along one other to separate the chromosomes into two sets, concentrating one set near each pole. Then, the microtubules stimulate an equatorial region of the cortex, arranging a ring of microfilaments around the cell's periphery. Finally, this ring contracts, cleaving the cell in two. The microfilaments, made primarily of actin proteins, are maneuvered by motor proteins called *myosins*. This same actin–myosin protein pair, energized by ATP, enables the cells of our muscle fibers to contract, powering our movements.

While the core choreography of mitosis is universal, variations abound across the kingdoms of life—from protists to animals, plants, and fungi. Yet all forms of mitosis rely on an MTOC (microtubule organizing center): This is the centrosome in animals, or its structural equivalent in other organisms. Both centrosomes and kinetochores—the structures that anchor microtubules to chromosomes—are in fact specialized biomolecular condensates: dynamic aggregates of dozens of distinct proteins, nanomachines, and RNA. These condensates—involving tubulin proteins and enzymes that govern microtubule growth, stability, motion, and interactions—vary in size and consistency throughout the cell cycle.

Among the most captivating structures in animal mitosis are the centrioles, tiny cylindrical bodies that resemble miniature gyroscopes. Found in pairs at the heart of each centrosome, centrioles duplicate and part ways during the synthesis (S) phase, when DNA replication and chromosome duplication occur. In the mitosis (M) phase, the mother-daughter pair of centrioles nucleates ("seeds" the growth of) a fully functional centrosome. Endowed with the capacity to organize and animate microtubules, the centrosome acts as the conductor of the mitotic symphony.

Outside of mitosis, during interphase, the centrioles migrate toward the cell periphery, embedding themselves in the cortex and becoming so-called *basal bodies*. Sitting just beneath the plasma membrane, basal bodies give rise to cilia or flagella. In some cells, entire families of cilia emerge where the centriole implants. Even more mysteriously, centrioles can arise spontaneously, assembling as if by magic.

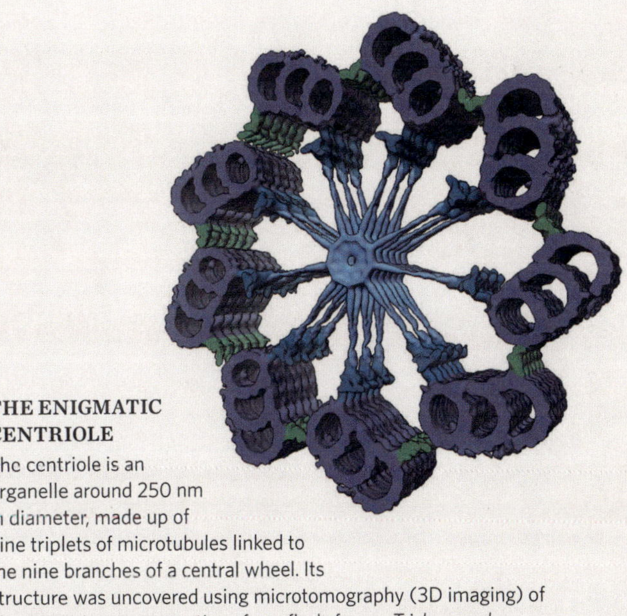

THE ENIGMATIC CENTRIOLE

The centriole is an organelle around 250 nm in diameter, made up of nine triplets of microtubules linked to the nine branches of a central wheel. Its structure was uncovered using microtomography (3D imaging) of electron microscope sections from flash-frozen *Trichonympha*, flagellated protists which live symbiotically in the gut of termites.

3D RECONSTRUCTION FROM ELECTRON MICROSCOPE SECTIONS
© Courtesy Pierre Gönczy EPFL/ISREC, Lausanne, and Paul Guichard, Department of Cell Biology, University of Geneva, Switzerland

MICROTUBULES AND CENTROSOMES DURING MITOSIS
A fertilized oocyte of the nematode worm *Caenorhabditis elegans* during chromosome separation. The microtubules (*green*) of the mitotic spindle emanate from the centrosomes (*yellow*), forming large asters at both poles. The microtubules move and separate the chromosomes (*blue*) while sliding along the cell cortex.

CONFOCAL FLUORESCENCE MICROSCOPY
© Courtesy Pierre Gönczy, Swiss Federal Institute of Technology in Lausanne (EPFL), Swiss Institute for Experimental Cancer Research (ISREC)

EVERY CELL IS BORN FROM ANOTHER

Since every cell arises from a preexisting one, life perpetuates life. In prokaryotic organisms—archaea and bacteria—cells reproduce primarily through binary fission, one mother cell yielding two daughters. In their natural environment, *Escherichia coli*, the bacteria inhabiting our intestines, divide approximately every two hours. Yet in laboratory conditions, where nutrients abound, an *E. coli* population can double every twenty minutes, producing millions of descendants in a single day. To divide, the bacterium replicates its single, circular chromosome, which is anchored to the membrane. It then elongates its cell wall, doubles its membrane, and finally splits into two daughter cells. These daughters typically inherit the same internal components as their mother, though exceptions exist. Not all bacteria and archaea divide through fission—some reproduce through budding, or differentiate into distinct cell types (*see pp. 75 and 87*).

By contrast, all eukaryotic cells—from unicellular protists to the cells of plants, algae, fungi, and animals—undergo the successive stages of the cell cycle: growth, mitosis, and cytokinesis. The complexity and duration of these phases vary widely. Eukaryotic division is slower and more intricate than bacterial fission. Before dividing, a eukaryotic mother cell must replicate several bulky linear chromosomes and duplicate its many organelles and condensates. Nonetheless, some unicellular eukaryotes—particularly certain yeast species—are able to complete a full division cycle in under an hour. The early embryonic cells of sea urchins, mussels, and amphibians, which bypass the growth phase entirely, also exhibit rapid division. Among embryos, the record holders are appendicularians (a swimming type of zooplankton), whose cells divide every few minutes. In stark contrast, differentiated cells within animal tissues and fluids require much more time to divide. Some, like neurons and muscle cells, even cease dividing altogether, and others, such as red blood cells, never divide. Cancer cells, however, defy these norms. Unshackled from the regulatory constraints that govern their tissue of origin, they divide relentlessly and uncontrollably.

In cell division, the positioning of the mitotic spindle and the orientation of the cleavage plane crucially influence the destiny of the daughter cells. Most divisions are symmetrical: With the spindle centered, the mother cell cleaves evenly down the middle, usually producing two identical daughters. Yet when the mitotic spindle is eccentrically placed—with one centrosome nearer to the cell's periphery than the other—the mother cleaves into asymmetrical daughter cells, differing in size, content, and fate. Asymmetric division is employed as a developmental strategy by embryos and stem cells alike. A mother stem cell, for instance, divides into one differentiated daughter and one self-reproducing daughter—still a stem cell—ensuring the lineage continues.

Division creates two daughter cells from one mother (here, an animal cell), dividing the cell membrane ❙ and its contents: the chromosomes ❙ (housed in the nucleus ●), organelles (including mitochondria ●), and cytoskeleton (made up of microfilaments ❙ and microtubules ❙, radiating from centrosomes ●). These cytoskeletal elements move and act as an anchor for the organelles, chromosomes, and cortex.

During mitosis, the nucleus ● temporarily disappears. Two identical sets of chromosomes ❙ are moved toward the centrosomes ● by bundles of microtubules ❙. Microfilaments ❙ contract at the cortex to split the mother cell into two daughter cells, each of which re-forms a nucleus ● around the chromosomes ❙ and nucleoli.

It's different in bacteria (*below right*): The single circular chromosome ❙ duplicates while anchored to the membrane ❙. The cell grows, duplicates its other contents, and then divides into two daughter cells, each one inheriting a chromosome ❙ and a collection of cellular components.

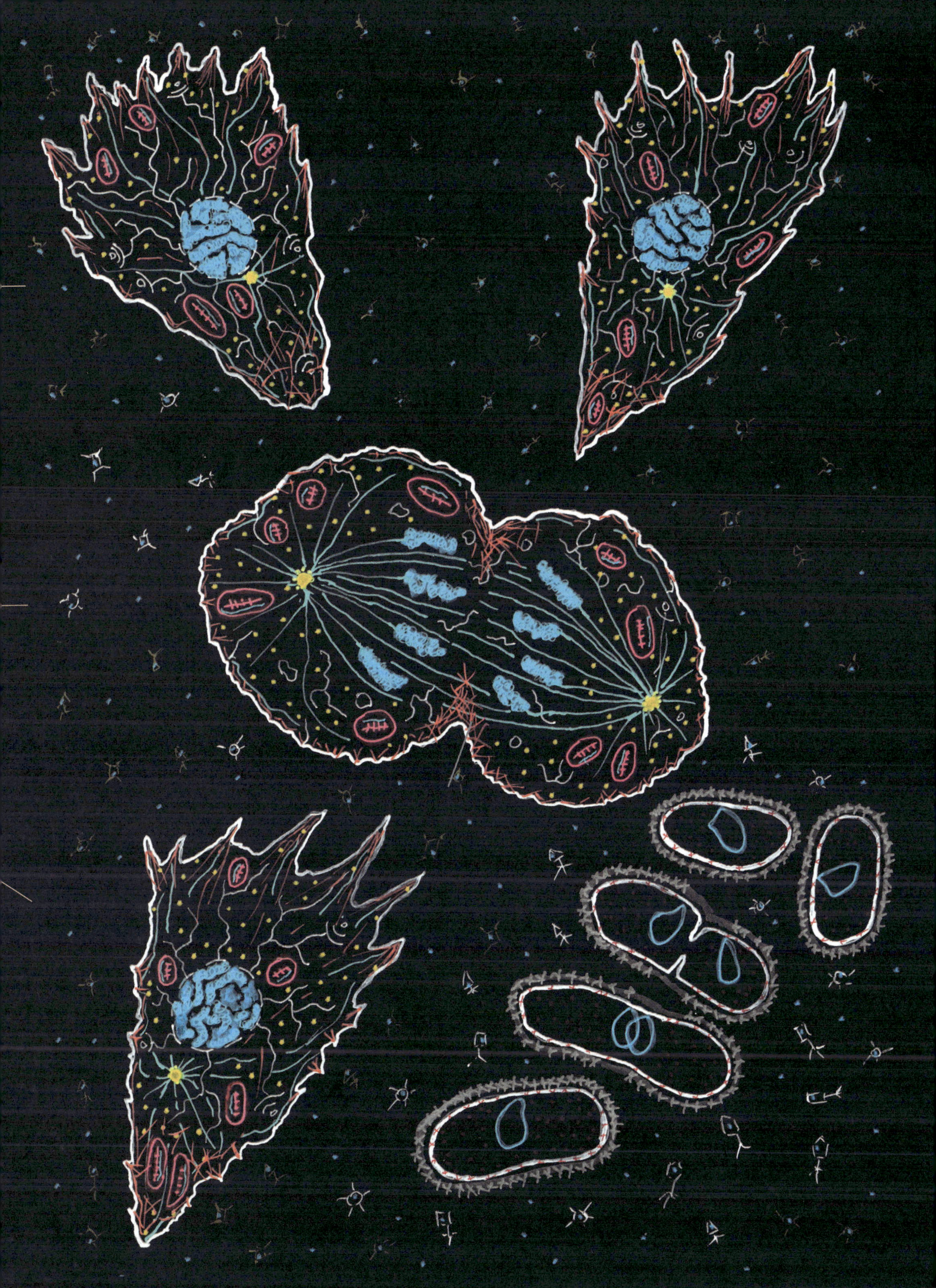

Phases of the cell cycle and mitosis in animals

The phases of the cell cycle track the life and reproduction of a cell *(see p. 147)*. Cells spend most of their time in interphase, a period of quiescence, growth, and finally duplication of the cell components in preparation for mitosis, the phase of separation. These components are distributed between two nascent daughter cells, which split apart in the final phase, cytokinesis.

All the architectural changes in the cell that enable the mechanics of division are brought about by modifications to hundreds of proteins, primarily via phosphorylation. And this is accomplished by MPF, whose discovery we have already reported *(see p. 148)*.

INTERPHASE

Interphase is the longest phase of the cell cycle. During interphase, the chromosomes (*in blue*) reside in the nucleus in their elongated (decondensed) form. When the cell is ready to divide, it enters a growth phase called G1 (for gap/interval 1). Once the cell reaches a critical size, synthesis (S phase) begins, when the chromosomal DNA is duplicated. Once complete, each chromosome is made up of two identical DNA strands. Next comes G2 phase (for gap/interval 2), during which the cell replicates its membranous organelles and condensates, such as centrosomes and nucleoli. Then, microtubules (*in green*) begin to radiate around two centrosomes that act as organizing centers.

MITOSIS

Compared to interphase, mitosis (M phase) is relatively short. During mitosis, the chromosomes and other cell components separate and distribute themselves between the two centrosomal "poles."

Mitosis itself is a succession of four subphases: prophase, metaphase, anaphase, and telophase. In animal cells, they are as follows:

1. Prophase: Prophase immediately follows the conclusion of G2. It begins as each duplicated chromosome is condensed into two sister chromatids, compact structures that remain joined at the middle, so that each pair is shaped like an X. The nuclear membrane disperses, freeing the chromosomes. Meanwhile, the two organizing centrosomes move apart, becoming two poles from which bundles of microtubules radiate, forming a mitotic spindle surrounding the chromosomes.

2. Metaphase: The dynamic microtubules of the mitotic spindle attach to the chromosome pairs and position them on an equatorial plane, midway between the two centrosomes at the poles.

3. Anaphase: During this rapid phase, the microtubules maneuver the chromosomes, separating the chromatids to distribute one copy toward each pole. Then, as the cell elongates, the microtubules stimulate the cell cortex around the perimeter of the equatorial plane.

4. Telophase: A nuclear membrane forms around each chromatid set near the poles. The chromosomes decondense within the nuclei, while bundles of actin microfilaments form a ring at the equatorial region of the cortex.

CYTOKINESIS

Once mitosis has concluded with telophase, cytokinesis begins, triggering cell division. At its onset, the ring of microfilament bundles at the cellular cortex contracts. This constricts the plasma membrane, ultimately splitting the cell into two daughter cells, each containing an identical set of chromosomes in its own nucleus.

View "Inside the Cell" (Digital Studio SA, Paris)

PHASES OF ANIMAL CELL DIVISION
Epithelial cells from a rat kangaroo (from the common experimental cell line PtK2) divide within two to three hours. Chromosomes fluoresce blue, mitochondria red, and microtubules green.

CONFOCAL FLUORESCENCE MICROSCOPY
© Courtesy Eric Clark, John Griffin, Nathan Claxton, Michael Davidson, Molecular Expressions, National High Magnetic Field Laboratory, Florida State University, Tallahassee

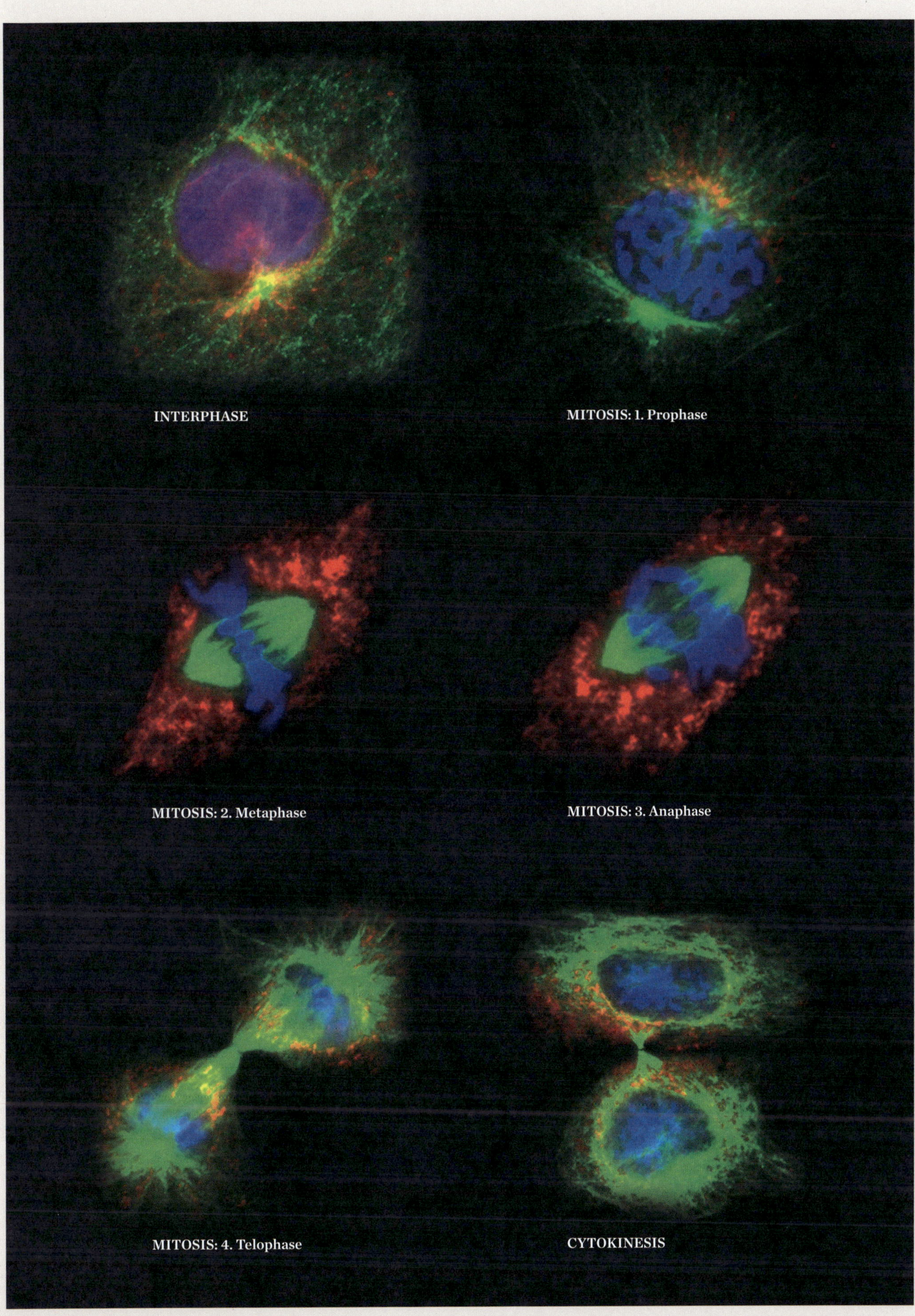

INTERPHASE
MITOSIS: 1. Prophase
MITOSIS: 2. Metaphase
MITOSIS: 3. Anaphase
MITOSIS: 4. Telophase
CYTOKINESIS

CHAPTER IX

INFORMATION & HEREDITY

From genes to proteins

DNA
RNA
Protein
TRANSCRIPTION
TRANSLATION

DNA, GENES, AND CHROMOSOMES

DNA has held iconic status ever since the 1950s and 1960s, when Rosalind Franklin's famous "photo 51" allowed Crick and Watson to unveil its elegant double-helix structure. DNA (deoxyribonucleic acid) was revealed as the key to the genetic function of all living things. This helical marvel consists of two intertwined strands, composed of four types of nucleotide—adenine (A), thymine (T), guanine (G), and cytosine (C)—whose specific sequences define the genes *(see p. 52)*. In eukaryotic cells, the DNA double helix wraps around protein cores known as *histones*, forming structures called *nucleosomes*. These nucleosomes—which also include RNA molecules and accessory proteins—are further coiled and folded into chromatin filaments. These, in turn, are compacted into distinct chromosomes. In human cells, this extraordinary feat of architecture squeezes roughly 2 meters of linear DNA into a nucleus a hundred times thinner than a human hair. These 2 meters of DNA are faithfully duplicated and distributed between daughter cells during mitosis.

The transmission of hereditary information from one generation to the next is orchestrated by thousands of genes that encode proteins and functional RNA molecules. While an *E. coli* bacterium carries approximately 4,200 genes on a single circular chromosome, a *Homo sapiens* cell contains around 21,000 genes spread across 23 pairs of chromosomes. That humans possess only five times more genes than *E. coli* is astonishing—especially when one considers that a mere 2 percent of our genome consists of protein-coding sequences. These coding regions are interspersed with vast stretches of "non-coding" DNA, which account for the other 98 percent of the human genome. This is in stark contrast to bacterial genomes, where non-coding DNA represents a modest 10 percent. Bacteria supplement their primary circular chromosome with small, autonomous DNA fragments known as *plasmids*. These plasmids are readily exchanged among bacteria, archaea, and even viruses, facilitating horizontal gene transfer—including the spread of antibiotic resistance. Such exchanges transcend species boundaries and can even extend to certain eukaryotic cells.

Far from being "junk," most of the non-coding regions in our genome are transcribed into a rich array of RNA molecules that orchestrate and fine-tune cellular processes. Our DNA also harbors mysterious repetitive sequences and mobile genetic elements called *transposons*—segments that can leap from one chromosome region to another, contributing to genomic diversity. Remarkably, about 8 percent of our DNA originates from RNA viruses that infected our human and animal ancestors. This viral legacy forms a genetic reservoir—an archive of foreign sequences that, over evolutionary time, has been repurposed to serve our own biology. Indeed, certain viral genes have been co-opted to produce innovative proteins, including those crucial for placenta formation in mammals and enhancing the memory function of neurons *(see p. 123)*.

  Information & heredity

The DNA double helix **I** is made up of two strands of nucleotides, occurring in four types: A, T, G, C. A gene is defined by its specific sequence of nucleotides, which is transcribed into an RNA message that is finally translated into a protein.

The DNA double helix **I** is wound into coils (nucleosomes), principally made of histone proteins (H).

The coils fold and twist, forming chromatin filaments that wind and compact into chromosomes **III**. Histone proteins and DNA **I** are subject to chemical modification, including the addition of methyl groups ●, which modulate gene expression.

Within the nucleus ● of the cell, bounded by the nuclear envelope **I**, different chromosomes **III** occupy distinct regions.

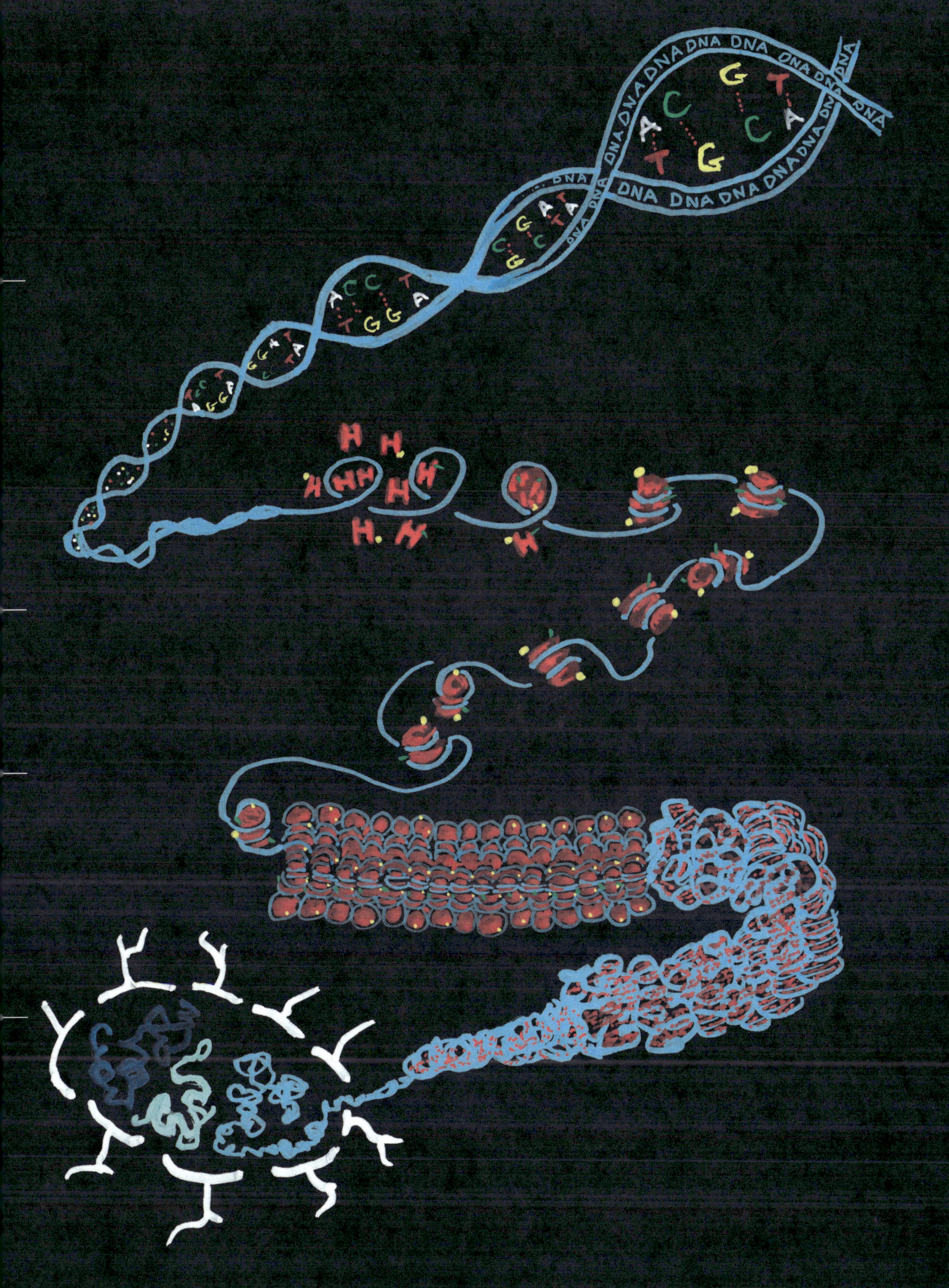

DNA
A C G T
T G C A
C G A T
G C T A
A C C T
T G G A
H H H
H H H H H
H H H

From heredity to DNA— 200 years of discoveries

Two hundred years ago, the term *heredity* referred solely to the transmission of property, titles, or official roles. The concept of biological heredity only emerged in the mid-19th century, born at the confluence of three discoveries: Darwin's theory of evolution and his vision of the tree of life connecting all organisms; Mendel's formulation of hereditary laws, derived from his meticulous crossbreeding of pea plants in the quiet of his monastery garden in Brno; and the cellular theory pioneered by Schwann and Schleiden, which asserted that all plants and animals are composed of similar cell types, each bearing a nucleus that houses chromosomes *(see chap. I)*. It was also during this period that scientists began isolating the first proteins and mysterious, phosphate-rich fibers—dubbed "nucleins"—purified by the Swiss chemist Friedrich Miescher. These would later be recognized as DNA, but, at the time, the molecular underpinnings of heredity remained completely unknown.

Curiosity about the cell nucleus and the cryptic nature of chromosomes became the key that gradually revealed the secrets of hereditary transmission. Between 1860 and 1900, European biologists peered through their rudimentary microscopes and tested countless chemical dyes to scrutinize the cell's interior, observing the transformations of nuclei in developing tissues and embryos. They noted that chromosomes revealed themselves during cell division—as filaments that condensed into rods, split along their length, and were distributed to daughter cells in the process we now call *mitosis*. They also discovered another form of division—meiosis—in which chromosome numbers are halved, giving rise to specialized reproductive cells: sperm and oocytes. These gametes represented the end product of what August Weismann theorized as the germline, a distinct cellular lineage responsible for heredity. Like Weismann, several prominent thinkers of the time—including Ernst Haeckel and Wilhelm Roux in Germany, and the American biologist Edmund B. Wilson—began to suspect that chromosomes carry the very substance of inheritance. Studies of fertilization in sea urchins and starfish by Oscar Hertwig and Hermann Fol lent weight to this hypothesis. Fol, a young Swiss researcher, observed chromosomal anomalies caused by polyspermy—penetration of single oocyte by multiple sperm. Building on these insights, German biologist Theodor Boveri and his wife, American biologist Marcella O'Grady, conducted ingenious experiments on sea urchin embryos subjected to polyspermy. They also investigated how chromosomes are distributed in the cells of the parasitic tapeworm *Ascaris*. Together with American scientist Walter Sutton, the Boveris established the individuality of chromosomes and their fundamental role in the transmission of hereditary traits. At the dawn of the 20th century, the Dutch botanist Hugo de Vries rediscovered and championed Mendel's laws, introducing the concept of mutation as

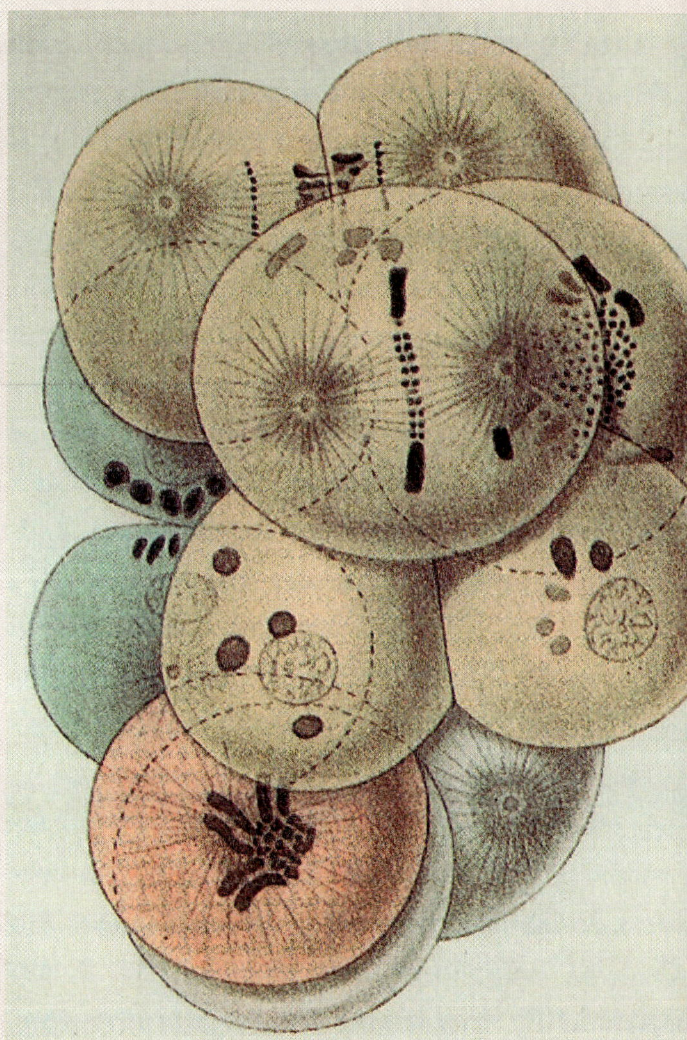

THEODOR BOVERI, THE TAPEWORM, AND THE ROLE OF CHROMOSOMES

At the end of the 19th century, Theodor Boveri determined the roles of chromosomes and centrosomes, mainly using oocytes and embryos of the horse tapeworm *Ascaris megalocephala*. The transparent embryos have only four large chromosomes (two paternal and two maternal) and large centrosomes, which are easy to observe during the first embryonic divisions. This enabled Boveri to characterize the differences between somatic cells (*yellow, green, red*) and germ cells (*grey, in the lower part of the embryo*) and to understand the essential role chromosomes play in hereditary transmission. Theodor Boveri published this lithograph in a 1910 article written in honor of his colleague Richard Hertwig's sixtieth birthday. Boveri carried out most of his experiments with his American-born wife and fellow biologist Marcella O'Grady, although she was never officially associated with his publications. Theodor Boveri died at the age of 53, probably from a chronic infection of the *Ascaris lumbricoides* worm with which he had also experimented.

the source of genetic variation. Around the same time, Danish biologist Wilhelm Johannsen coined the word *gene* to designate the hereditary factors that Mendel had postulated—now understood to reside on chromosomes. The Englishman William Bateson, inspired by the Greek word *ghenetikos* (meaning "origins"), baptized the emerging science with a new name: *genetics*.

America's rise to the forefront of genetic research began with the groundbreaking work of Thomas Hunt Morgan and his collaborators, who elevated *Drosophila melanogaster*, the humble fruit fly, to the status of the first genetic model. In 1911, they succeeded in mapping the fly's genes along its chromosomes, offering a new vision of heredity. Yet it was not until 1944 that Oswald Avery, through pioneering experiments with bacteria and viruses, definitively proved that DNA is the carrier of genetic information. In the 1950s and 1960s, the discovery of DNA's iconic double-helix structure—and ensuing efforts to decipher its genetic code—by Watson, Crick, Sydney Brenner, and their colleagues—ushered in a new era in biology. Since then, the mechanisms of Mendelian inheritance and their molecular foundation in DNA have been firmly established. However, other elements of heredity, beyond the DNA sequence, have emerged, especially the crucial role of RNA molecules. In addition, epigenetics—the study of how environmental factors influence gene expression—has revived the long-debated idea that acquired traits can be inherited.

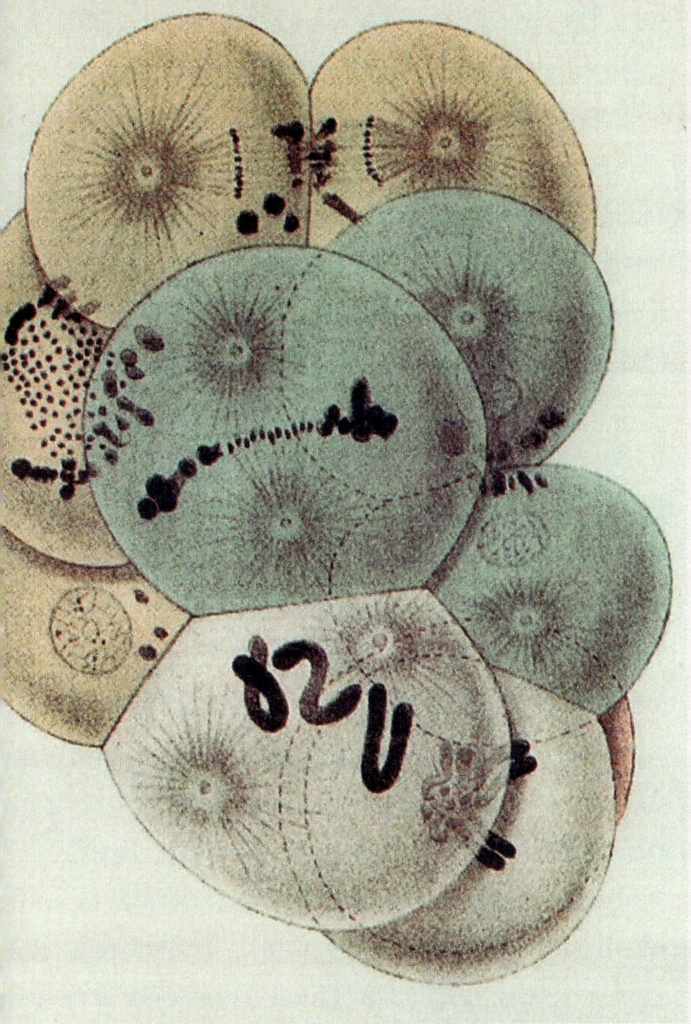

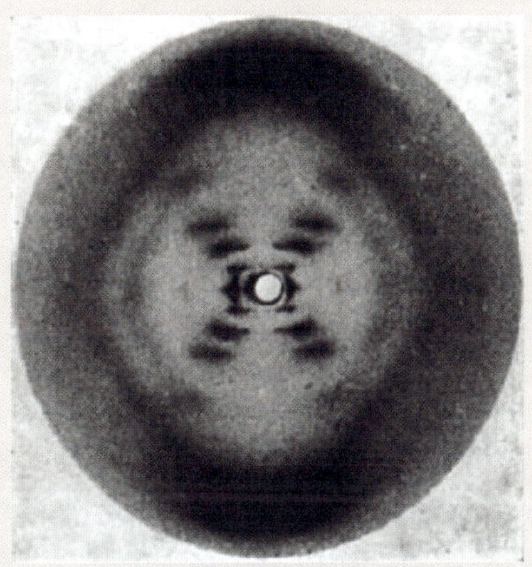

THE DISCOVERY OF THE DNA DOUBLE HELIX

By the early 1950s, DNA was recognized as the macromolecular carrier of heredity. At King's College London, Rosalind Franklin and her student Raymond Gosling obtained the first diffraction images of purified DNA fibers examined by X-ray, and in particular the famous photo 51 (*above*), whose dark spots reveal DNA's double-helix structure. Upon Franklin's departure from King's College, Gosling showed the photo to his new supervisor, Maurice Wilkins. Without Franklin's knowledge, Wilkins shared it, in turn, with James Watson and Francis Crick, who were seeking to elucidate the structure of DNA. Watson and Crick published DNA's famous double-helix structure, of which they had built a metal model (*photo, left*) in *Nature* in 1953. In 1962, Wilkins, Crick, and Watson were awarded the Nobel Prize, without Franklin and Gosling's essential contribution being recognized. This is considered an injustice to Rosalind Franklin, who died of cancer in 1956 at the age of 37.

THE GENETIC CODE —FROM DNA TO RNA TO PROTEIN

The genetic code is the set of rules cells follow to convert the information stored in DNA into the proteins they need. To make the protein encoded by a specific gene, the DNA (deoxyribonucleic acid) sequence is first transcribed into RNA (ribonucleic acid). DNA's double helix is built of two complementary strands of nucleotides—A always pairs with T, and C with G *(see pp. 52–53)*. During transcription, enzyme proteins "unzip" the DNA helix and assemble RNA by taking the complement of each nucleotide on one strand, producing a sequence that matches the other strand (with U standing in for T). The result is messenger RNA (mRNA).

Long ignored by the public, RNA—the essential molecule between DNA and proteins—finally gained notoriety during the COVID-19 pandemic. Messenger RNA is just one member of the family of RNA types that regulate gene expression and protein-making. Ribosomal RNA (rRNA) is an active part of ribosomes, where it works together with transfer RNA (tRNA) to shuttle amino acids. Still other sorts of RNA facilitate chemical reactions, as do enzyme proteins. Innumerable RNA molecules of various shapes and sizes associate with proteins to form biomolecular condensates. These semi-fluid environments at nanoscale are crucibles of metabolic reaction and gene expression, and for the assembly or disassembly of cell structures *(see p. 138)*.

The production of proteins—encoded in DNA and relayed by mRNA—relies on intricate molecular machinery, with ribosomes in a pivotal role. These remarkable nanomachines are each composed of two subunits built from rRNA and an array of associated proteins *(see p. 57)*. Several ribosomes can bind to the same mRNA strand at once, allowing many copies of the same protein to be made simultaneously. Ribosomes operate as translators, decoding the four-letter nucleotide language of mRNA into the twenty-letter language of amino acids that compose proteins. This translation follows a universal genetic code, where each set of three nucleotides—called a *codon*—specifies a particular amino acid. As the amino acid chain emerges, it begins to fold progressively into its functional form—a process driven partly by the chain's own chemical properties, and partly by specialized helper proteins known as *chaperones*. The resulting 3D structure gives the final protein its unique shape and function. Interestingly, certain segments of the amino acid chain remain "intrinsically disordered" (flexible), which enables molecular interactions, especially within biomolecular condensates *(see p. 139)*.

Across all known life-forms, DNA is first transcribed into mRNA, which is then translated into protein. An exception arises in cells infected by RNA viruses such as Ebola or HIV (the AIDS virus). These viruses carry their genetic instructions in the form of RNA, which is reverse-transcribed into DNA by a viral enzyme protein. The newly formed DNA of viral origin is then integrated into the host cell's genome. This process—reverse transcription—may be traceable to the earliest chapters of life's history, as RNA most likely preceded DNA in the primordial world *(see chap. III)*.

TRANSCRIPTION

TRANSLATION

DNA

RNA

Protein

THE ORGANIZATION OF THE NUCLEUS— CHROMOSOMES AND NUCLEOLI

The nucleus is a prominent organelle, enclosed by a double membrane that is continuous with the endoplasmic reticulum—a vast membrane network whose tubular architecture extends throughout the cell, from its periphery (the cortex) to its center *(see pp. 90-91)*. Within the nucleus, the chromosomes— long strands of DNA—reside in well-defined regions known as *chromosomal territories*. Within each territory, the genes engage in an ongoing dialogue with regulatory proteins that act as activators or repressors. These proteins collaborate with RNA to form dynamic biomolecular condensates at the surface of the chromosomes. Basically, these condensates form on cue when mRNA transcription begins, and dissolve as transcription completes. Modulated by other factors, such as local concentrations of ATP, metabolites, or ions, these nuclear condensates serve as biochemical crucibles in which genes are either switched on or silenced, in response to the cell's internal and external conditions.

Protein production depends not only on mRNA, but also on diverse transfer RNA (tRNA), which delivers amino acids to the ribosomes—the nanomachines responsible for protein synthesis *(see p. 57)*—and on ribosomal RNA (rRNA), which forms their structural and functional core. The ribosomes' components, including rRNA, are produced and largely assembled within the nucleoli, multilayered spherical compartments once thought to be inert granules. Thanks to advances over the past decade, we now understand that nucleoli are, in fact, paradigmatic biomolecular condensates. They change their numbers, shape, and malleability in response to stress, and play a key role in regulating the cell cycle, aging, and differentiation.

During interphase—the active growth and preparation phase of the cell cycle—numerous genes are transcribed into mRNA, while the nucleoli swell with accumulating ribosomes and rRNA. When mitosis begins, the chromosomes condense, transcription slows dramatically, and the nucleoli disassemble, seemingly vanishing from view. Yet this disappearance is only temporary. As mitosis concludes, in the nuclei of each daughter cell, the chromosomes decondense into their new territories, and the nucleoli re-form. There, they resume their essential role, producing the ribosomal machinery needed to translate mRNA into proteins—thus setting the stage for a new round of the cell cycle.

In eukaryotic cells, protein production is triggered in the nucleus ● and takes place in the cytoplasm.

The nucleus ● of an animal cell is delimited by a membrane ▌ pierced by nuclear pores ▓. The nuclear membrane is continuous with the membrane network of the endoplasmic reticulum.

The nucleus shown here contains several chromosomes ▌. Chromosomal regions called NORs (nucleolar organizer regions) organize the formation of nucleoli (concentric rings of green, rose, and yellow). Nucleoli are biomolecular condensates that assemble the protein-making apparatus— ribosomes (subunits R and r)— around ribosomal RNA (rRNA).

The protein-making instructions are encoded in messenger RNA (mRNA), transcribed from chromosomal DNA ▌. The mRNA, along with the ribosome components, migrates from the nucleus to the cytoplasm via the nuclear pores ▓. There, the subunits are assembled into a ribosome (Rr), which attaches to the mRNA and builds the protein ▌ to order.

Endoplasmic Reticulum
Endoplasmic Reticulum
Endoplasmic Reticulum
Nuclear membrane
Nuclear membrane
Nuclear membrane
NOR
NOR
NOR
rRNA
mRNA

Chromosomes and nucleoli share tasks in the nucleus

In the early 19th century, a prominent organelle at the heart of plant cells drew the attention of botanists, among them Robert Brown, who named it the *nucleus* in 1831. Seen through the microscopes of the time, the nucleus appeared to house delicate filaments—later identified as chromosomes—and conspicuous granules that, like chromosomes, seemed to appear and vanish to the rhythm of cell division. These granules, the most visible nuclear structures, were named *nucleoli*, meaning "little nuts" in Latin. The ubiquity of both nuclei and nucleoli in plant and animal cells inspired Schleiden and Schwann, who formulated the cell theory *(see p. 14)*.

Within the nucleus of an animal cell, each chromosome occupies a distinct domain. The DNA double helix is intricately coiled into nucleosomes, which are themselves twisted and folded into the dense structure of a chromosome. For a gene to be expressed, its nucleosomes must loosen, exposing a portion of linear DNA that can then be transcribed into mRNA by the enzyme protein RNA polymerase. This polymerase, together with its accessory proteins, forms a nanomachine that faithfully copies the nucleotide sequence of DNA into the complementary sequence of mRNA *(see p. 57)*.

Nucleoli—the other major nuclear structures—vary in number and, depending on the cell type, may occupy up to a quarter of the nucleus.

Their function remained a mystery until the 1930s, when Barbara McClintock, through her pioneering work on the genetics of corn, identified specific chromosomal regions now known as NORs (<u>n</u>ucleolar <u>o</u>rganizer <u>r</u>egions) as essential to nucleolus formation. By the 1960s, it had become clear that nucleoli play a central role in the production of ribosomal RNA and the assembly of ribosomes—the indispensable nanomachines of protein synthesis, which are among the most abundant structures in the cell. More recently, in the past decade, a deeper understanding has emerged: Nucleoli are not solid granules, but dynamic biomolecular condensates formed through a phase transition called LLPS, for <u>l</u>iquid-<u>l</u>iquid <u>p</u>hase <u>s</u>eparation. Nucleoli are structured in three concentric domains, each one supporting a specific stage in the expression of ribosomal RNA. Within them, dozens of proteins assemble around the rRNA—which was transcribed from the chromosome's NOR—to assemble the small and large ribosomal subunits *(see p. 57)*. The physical state of the nucleolar condensate—more or less fluid—shifts throughout the cell cycle. Although nucleoli seem to disappear during mitosis, they do not vanish entirely. Instead, their components remain associated with the chromosomes. At the close of mitosis, they reassemble from the NOR regions within the daughter nuclei.

Beyond nucleoli, the nucleus contains several smaller condensates, including "nuclear speckles," which are involved in RNA editing and gene regulation. Our map of the nucleus—its architecture and functions—is still in its early stages, and each discovery reveals new layers of complexity.

View "Multiphase Liquid Behavior of the Nucleus" (HHMI iBiology)

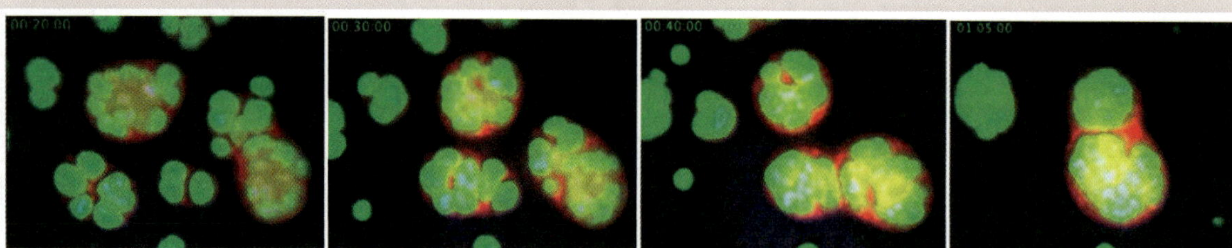

NUCLEOLI—DYNAMIC BIOMOLECULAR CONDENSATES
Large nucleoli (10 to 20 μm) are present in the voluminous oocyte nuclei of the toad *Xenopus laevis*. The images in this sequence reveal different domains within the nucleoli, and their coalescence via LLPS. The sequence was filmed frame by frame over two hours in the presence of a molecule that disrupts microfilaments, accelerating nucleolar remodeling and fusions.

STILLS FROM A CONFOCAL FLUORESCENCE MICROSCOPY VIDEO
© Courtesy Nilesh Vaidya and Clifford Brangwynne, Princeton University

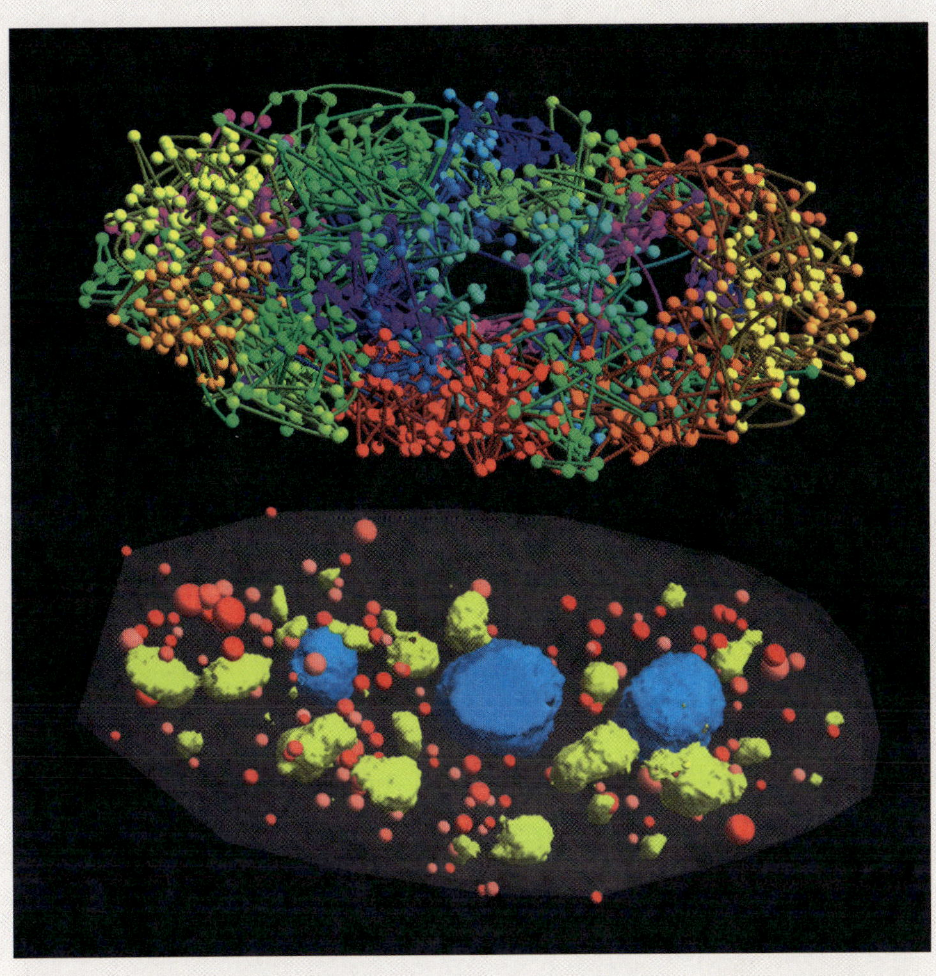

MAPPING A HUMAN CELL NUCLEUS—CHROMOSOMES, NUCLEOLI, AND MORE

Top: The 23 human chromosomes, represented by different colors, occupy distinct regions within the nucleus during interphase.

Bottom: Three nucleoli (*blue*), which transcribe DNA into ribosomal RNA and assemble the major components of ribosomes. Nuclear speckles (*yellow*) and other condensates (*red and pink*) also play a role in editing and expressing genes located on the chromosomes.

DIGITAL RECONSTRUCTION BASED ON LOCALIZATION OF GENES, RNA, AND PROTEINS WITH FLUORESCENT MOLECULES
© Courtesy Bogdan Bintu, University of California San Diego; and Xiaowei Zhuang, Harvard University

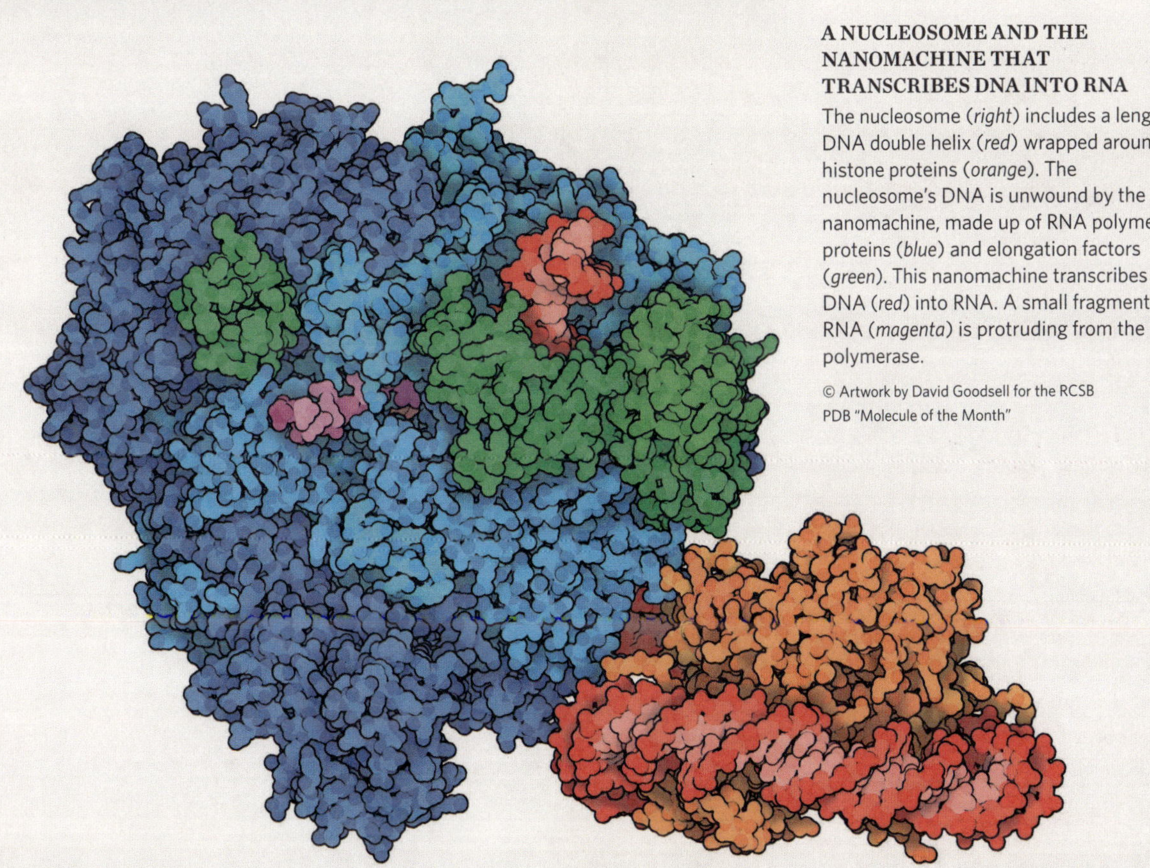

A NUCLEOSOME AND THE NANOMACHINE THAT TRANSCRIBES DNA INTO RNA

The nucleosome (*right*) includes a length of DNA double helix (*red*) wrapped around histone proteins (*orange*). The nucleosome's DNA is unwound by the nanomachine, made up of RNA polymerase proteins (*blue*) and elongation factors (*green*). This nanomachine transcribes the DNA (*red*) into RNA. A small fragment of RNA (*magenta*) is protruding from the RNA polymerase.

© Artwork by David Goodsell for the RCSB PDB "Molecule of the Month"

A DIGITAL REPRESENTATION OF AN ANIMAL CELL

In this cut-away view of a cell, the nucleus (*blue contents*) is at the bottom. Two nuclear pores (*yellow*) regulate exchanges between the nucleus and the cytoplasm. The pore on right has released components for protein-making: threadlike mRNA and ribosomes (*light blue*). Nearby, in the lower right corner, the nuclear membrane is continuous with that of the endoplasmic reticulum. On the left is a partial view of a mitochondrion (*pink contents*) with a microtubule (*grey*) just above. The cell's outer membrane, at top, is decorated with surface proteins. At top right, a junction connects it to a neighboring cell. This representation takes into account the true sizes and shapes of the cell's components, and their densities within the different spaces defined by the membranes.

"CELLULAR LANDSCAPE," DIGITAL ART USING MOLECULAR MAYA (MMAYA) SOFTWARE
© Courtesy Gaël McGill and Evan Ingersoll, Digizyme Inc., Brookline, Massachusetts

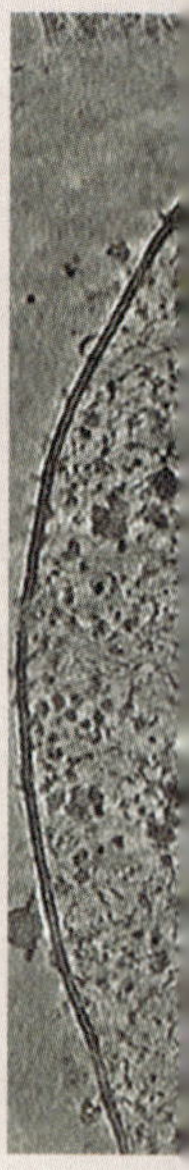

The protein factory

All cells, from bacterial to human, make proteins in roughly the same way: The genetic information contained in a DNA sequence is transcribed into an mRNA sequence, which is translated into an amino acid sequence, forming a specific protein. This two-step flow of information is directed by nanomachines—transcription by RNA polymerases and their partners, and translation by ribosomes and their associated factors. In prokaryotic cells that lack a nucleus, such as bacteria and archaea, all of these operations unfold seamlessly within the cytoplasm. There, a vast macromolecular complex known as an *expressome* exemplifies efficiency, coupling transcription and translation in real time, and allowing the newly synthesized mRNA to be immediately translated into protein—a powerful illustration of the genetic code in action.

Eukaryotic cells, by contrast, are compartmentalized. Within the membrane-bound nucleus, DNA is transcribed into various RNA species, and ribosomal components are assembled. The translation of mRNA into proteins, however, occurs outside the nucleus, in the cytoplasm. Separation is maintained by the nuclear envelope—a double membrane studded with nuclear pores. These pores are not mere openings, but sophisticated biomolecular condensates composed of hundreds of proteins and RNA molecules, which selectively regulate the traffic of molecules between the nucleus and cytoplasm.

To sustain this compartmentalized system, eukaryotic cells have evolved intricate mechanisms for transporting genetic messages and assembling the ribosomal machinery. Though energetically taxing, this division allows transcription and translation to be independently regulated, lending remarkable finesse to gene expression. Because eukaryotic cells are organized into multiple specialized compartments, proteins are typically synthesized close to where they are needed. The RNA and ribosomal components produced in the nucleus must be carefully packaged and exported through the nuclear pores. Once in the cytoplasm, mRNA—escorted by a suite of protein partners—is ferried to its destination by motor proteins moving along the "tracks" of the cytoskeleton, composed of microtubules and microfilaments. These destinations may include the cell periphery (the cortex), the membranes of the endoplasmic reticulum *(see p. 91)*, other organelles, or cytoplasmic condensates.

Once positioned, the threadlike mRNA is flocked by ribosomes, which assemble amino acids into polypeptide chains with astonishing efficiency. At a rate of roughly ten amino acids per second, most proteins are synthesized within just one or two minutes. Some of these proteins will be embedded in membranes or displayed on their surfaces; others will be directed into organelles or incorporated into dynamic condensates. Whatever their role, all proteins are subject to continual quality control. Their structural integrity and functionality are monitored, and when necessary, they are degraded and replaced. Depending on their structure and function, some proteins last for mere minutes; others remain active for hours or even days.

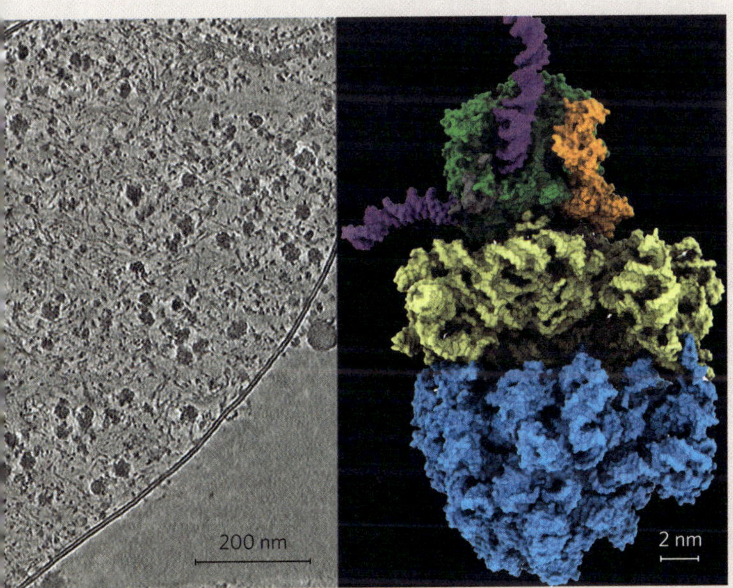

THE EXPRESSOME—THE NANOMACHINE THAT COUPLES TRANSCRIPTION AND TRANSLATION

An expressome performing simultaneous transcription and translation was recently visualized inside *Mycoplasma pneumoniae*, a parasitic bacterium measuring 0.1 μm.

Left: The tiny bacteria were rapidly frozen and examined by tomography with an electron microscope. In this tomographic section, expressomes appear as dark masses, seen from different angles. Mass spectroscopy revealed the relative positions of the proteins that make up the expressome.

Right: Computer reconstruction provides the 3D structure of an expressome at atomic resolution. We can see the DNA double helix (*purple*) being transcribed into RNA (*grey*) by RNA polymerase (*green*). Below are the small subunit (*light yellow*) and large subunit (*blue*) of a ribosome, which translates RNA into a protein. Other auxiliary proteins (*orange*) are also shown. This image, which captures the core principle of molecular biology, is a technological tour de force.

MASS SPECTROSCOPY AND CRYO-ELECTRON TOMOGRAPHIC IMAGING (CRYO-ET)
© Courtesy Liang Xue and Julia Mahamid, European Molecular Biology Laboratory, Heidelberg, Germany

Natural selection— from skin color to Darwin's moths

Skin tone variance across the human species is a remarkable example of the interplay between the environment, evolutionary pressure, and our genes. Our skin gets its color from melanins—pigments produced by specialized cells called *melanocytes*. Once synthesized, melanin is packaged into melanosomes and transferred to the outermost skin cells, where it forms a protective shield around the nuclei. This natural armor guards the DNA against damage from ultraviolet (UV) radiation, which can cause melanoma and other cancers. UV rays also cause the breakdown of folate—a key molecule the body needs for fetal development. But UV radiation offers a benefit as well: It supports the synthesis of vitamin D, which is essential for bone mineralization. The evolutionary balance between these harms and benefits has shaped human skin color. In darker skin, more melanin provides greater protection; while in lighter skin, less melanin allows UV rays to penetrate deeper, supplying vitamin D. Thus, darker pigmentation is advantageous in regions of intense sunlight, and lighter skin in regions with lower UV exposure.

Homo erectus, widely considered the ancestor of modern *Homo sapiens*, emerged in Africa over 1 million years ago. It is believed that these early hominins had light skin, similar to that of other primates. As they adapted to the African savannah, beginning to lose their thick body hair and regulate heat through sweating, darker skin emerged as a protective adaptation against solar radiation. Some 100,000 years ago, populations of *Homo sapiens* began migrating out of Africa, gradually settling across the globe. In regions with weaker sunlight, natural selection now favored lighter skin tones, shaped by changes in gene families that regulate melanin production, distribution, and composition.

This evolutionary dynamic is echoed across the animal kingdom, where melanin-rich pigmentation also adorns the skins, feathers, scales, and wings of various species. One of the most striking examples comes from the peppered moth, which offers strong evidence for Darwin's evolutionary theory. Typically, peppered moths have light wings that blend seamlessly with lichen-covered birch bark, offering them camouflage from predators. Dark-winged genetic variants, though present, were historically rarer—as they were more easily spotted against the pale bark and consumed by birds. But by the mid-19th century, industrial pollution began to blacken the

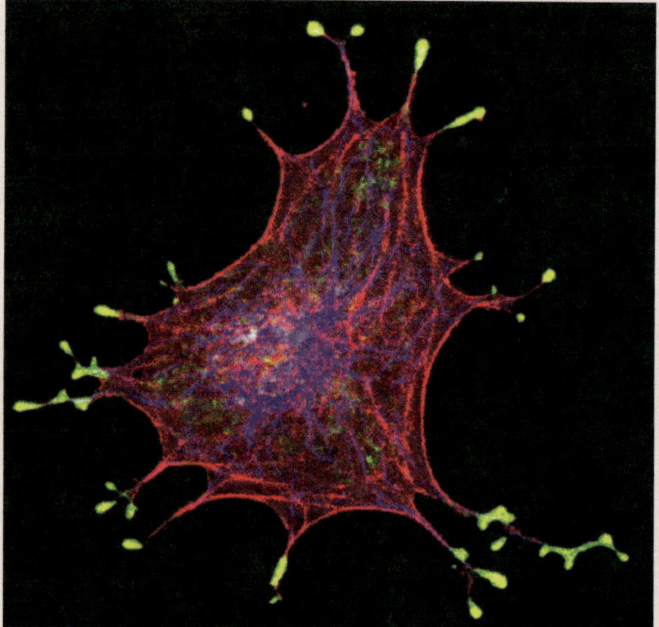

CELLS PRODUCE MELANIN PIGMENTS THAT PROTECT AGAINST UV RAYS

A melanocyte, the melanin-producing cell, packs melanin grains into sacs called *melanosomes* (*yellow*). The cytoskeleton concentrates the melanosomes in the cell's tips (*microfilaments are in red*). The melanosomes are then transferred to the keratinocyte cells of the epidermis.

FLUORESCENCE MICROSCOPY
© Courtesy Alistair Hume, University of Nottingham, Great Britain

PEPPERED MOTHS—NATURAL SELECTION IN ACTION

Resting on a lichen-covered birch trunk, a light-winged moth is less visible, and thus less susceptible to bird attacks, than a dark-winged moth of the same species (*Biston betularia*). This dynamic was temporarily reversed by industrial pollution.

trees of Great Britain. As a result, the once-advantaged light-winged moths declined, while the dark-winged form proliferated. This dramatic shift, known as *industrial melanism*, was traced to a genetic mutation: The insertion of a transposon—a mobile DNA element—into what's called the *cortex gene*. This gene, which influences cell division, also plays a pivotal role in determining the pigmentation of the scale cells that create vivid patterns on the wings of butterflies and moths.

Epigenetics—when the environment influences our genes

One need only observe the tireless ballet of worker bees around their languid queen to understand that DNA alone does not tell the whole story. Despite sharing an identical genetic blueprint, queens and worker bees diverge dramatically in form and function. The key lies not in their genes, but in how those genes are expressed—shaped by their respective diets. While future queens are nourished exclusively with royal jelly, workers are fed honey and pollen. This dietary influence on gene expression is a classic example of epigenetics, the realm of inheritance that operates "beyond the genes." This phenomenon extends to humans, as revealed by Swedish researchers who delved into the detailed historical records of Överkalix, a village in northern Sweden. They discovered that the grandsons of men who had experienced famine during puberty faced significantly higher mortality rates than the grandsons of men who did not. Strikingly, this intergenerational legacy appeared to pass through only grandfathers, and not grandmothers, suggesting an epigenetic imprint tied to paternal nutrition.

While human studies reach natural ethical limits, experiments on mice have shed light on the underlying molecular mechanism. In one such study, male mice were conditioned to fear a specific scent using mild electric shocks. Remarkably, this learned fear was transmitted to their offspring, even without further conditioning. The fear memory seemed to travel across generations through epigenetic information encoded in the sperm. The key messenger? Probably RNA—specifically, I call it epiRNA. When a sperm fertilizes an egg, it delivers not only paternal DNA but also a cargo of epiRNA molecules. But how do these molecules find their way into the sperm? Research points to the epididymis—a structure where sperm mature—as a likely source. There, cells release epiRNA via microscopic tunnels or tiny membrane vesicles known as *exosomes*, which shuttle the RNA molecules into developing sperm *(see pp. 200 and 205)*. Once delivered, epiRNA can influence how DNA is compacted within chromosomes or regulate the activity of enzyme proteins that write, read, or erase chemical marks on DNA and histone proteins. These markers— such as the addition or removal of methyl groups (CH_3)—can modulate gene expression. Intriguingly, even neurotransmitters like serotonin and dopamine, better known for mediating brain activity, can be grafted directly onto histone proteins, altering gene expression in ways linked to behavior and addiction. This emerging field opens new avenues for epigenetic therapies, or "epi-drugs," aimed at correcting the chemical tags associated with cancers, mental health conditions, or inherited behavioral traits.

Classical genetics still sits at the heart of heredity, but epigenetics has revealed a second layer of nuance—one that governs which genes are activated in response to the environment: temperature, light, nutrition, stress, parasites, microbiota, and more. Although we are only beginning to decipher its rules, many now speak of an epigenetic code—a language written in chemical markers and all sorts of RNA messages. This marks a quiet revolution in biology, echoing back to the once-marginalized ideas of Jean-Baptiste Lamarck, who long ago dared to imagine that acquired traits might be passed on.

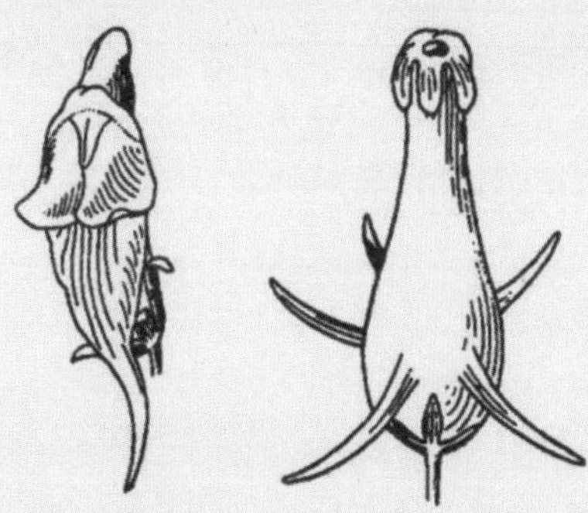

THE EPIGENETIC POWER OF PLANTS
It's an old story. It dates back to 1744, when Carl Linnaeus—the Swedish naturalist who first classified living beings—realized that the plant he named *Linaria vulgaris* had two different flower morphologies (the drawing is by the German poet Goethe). The common shape is a "dragon's mouth" with bilaterial or "mirror" symmetry (*left*). The other has radial symmetry, with five branches (*right*). This "peloric" form (from the Greek for "monstrous") is rare. When the plant is cultivated, the peloric form may disappear due to its rarity, to be replaced by the common form. This phenomenon is due to an epigenetic mutation of the Lcyc gene, involved in bilateral flower symmetry. The peloric form appears when the Lcyc gene is repressed by methylation marks, CH_3 groups that are deposited on the DNA. These marks, induced by environmental effects (temperature, light, humidity, stress, etc.), are transmitted epigenetically over several generations.

© Courtesy Eléa Héberlé, Le Plantoscope

View "Epigenetics" (BBC World Service)

CHAPTER X

REPRODUCE, AGE & DIE

Cancer & infectious agents

REPRODUCTION AND DEVELOPMENT

Reproduction is a vital necessity for all living beings. Bacteria, archaea, many protists, and some multicellular organisms reproduce with elegant simplicity. They replicate their chromosomes and components, then divide into two or more new organisms, which reproduce in turn. But many eukaryotes take a more complex approach, shaped by a remarkable innovation that emerged among our unicellular ancestors over a billion years ago—sexual reproduction. Over time, this mode of reproduction has evolved into a dazzling array of strategies, involving specialized organs and tissues and extremely complex mating behaviors.

At first glance, sex may appear to be an inefficient way to reproduce. While bacteria, archaea, and many protists effortlessly double their numbers with each division, sexual reproduction generally yields just one offspring from the union of two cells. It grants an evolutionary edge not through quantity of offspring, but through diversity—by enabling organisms to shuffle and share their genetic material. In animals, this takes place within and between the germ cells, or reproductive cells. Like somatic (nonreproductive) cells, germ cells are diploid, with two sets of chromosomes. But rather than undergo mitosis, which would produce two diploid daughter cells, germ cells perform meiosis to create two haploid gametes (oocytes or spermatozoa), each with a single set of chromosomes. Pre-division, the two sets of chromosomes swap material between them, so that each gamete is genetically unique.

When a spermatozoon fertilizes an oocyte, they create a zygote—a new, single cell endowed with two novel sets of chromosomes, one from each parent. As the zygote begins to develop, it divides again and again, becoming an embryo composed of somatic cell lineages and its own distinct line of germ cells. The genetic reshuffling of meiosis and fertilization confers a profound evolutionary advantage. Within a population, the individuals carrying more favorable gene combinations—and thus better equipped to face challenges such as predators, parasites, environmental shifts, and resource competition—are more likely to thrive and reproduce, which is the essence of natural selection.

From embryo to larva to adult, organisms develop through the coordinated differentiation of both somatic and germ cells. This complex ballet is governed by regulatory genes. In some species, such as the fruit fly *Drosophila*, maternal factors (proteins and RNA molecules known as *determinants*) positioned in the oocyte cortex guide development. In others, including mammals, the zygotic genome steers development—acting on each cell at the proper moment according to its precise location, in an exacting pattern that starts with fertilization. Thus, our many human cell types, including the vital germline, emerge from a plethora of signals exchanged among somatic cells, as they differentiate to form the embryonic tissues.

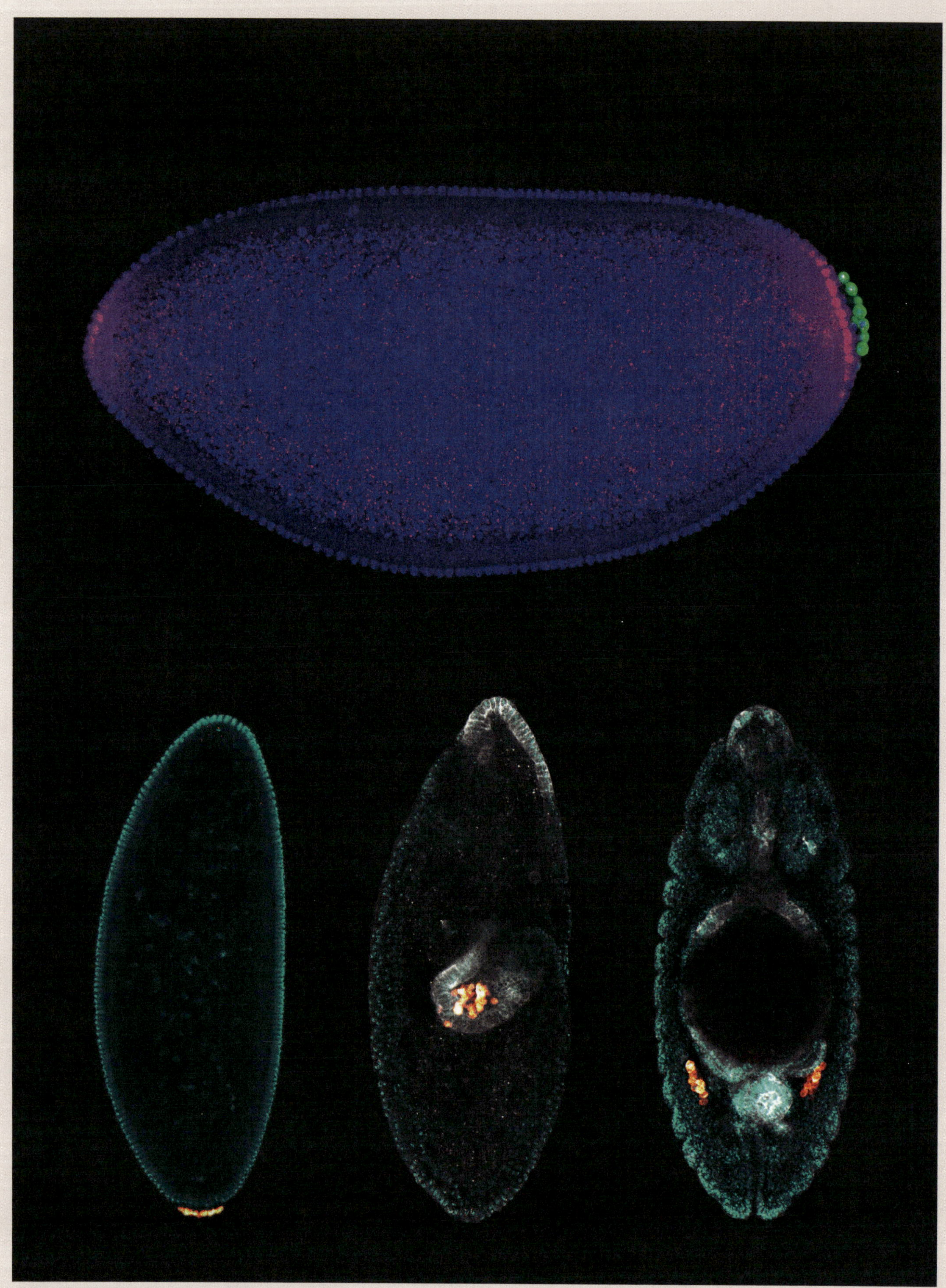

The germline— immortal cells

Oocytes and spermatozoa arise from germ cells—cells of the germline entrusted with transmitting hereditary information across generations. In this respect, germ cells possess a unique kind of immortality. Unlike somatic cells, which perish with the body and leave no trace in future generations, germ cells bridge past and future. In organisms such as the fruit fly *Drosophila melanogaster* or the nematode worm *Caenorhabditis elegans*, a small cluster of primordial germ cells emerges at one pole of the embryo soon after fertilization. This early specification is guided by "germ plasm" or "germ granules" present in the oocyte and early embryos of flies and nematodes. By contrast, most animals, including mammals, produce germ cells later in their development, prompted by complex signaling between somatic cells within the embryo *(see p. 138)*. These nascent germ cells then embark on a journey, migrating toward the gonads—the testes or ovaries—where they differentiate into male (spermatozoa) or female (oocyte) gametes.

Yet there is always room for variation. Creatures like planarians, hydras, and jellyfish can generate germ cells continuously into adulthood from their somatic cells. Plants also embody this regenerative principle. Their stem cells, nestled within meristems—special regions located at growing tips and nodes—allow plants to bud and bloom throughout their lifespans.

Despite their diversity, all animal germ cells share a common molecular hallmark—distinctive granules present in the cytoplasm, known as *germ granules* and made of germ plasm. The observation that such granules behave like fluid droplets led to the important discovery of biomolecular condensates in the nematode worm *C. elegans* *(see p. 138)*. In both *C. elegans* and in the fly *D. melanogaster*, analysis of the germ granules revealed that these essential biomolecular condensates—intricate assemblies of specific proteins and RNA—regulate gene expression with exquisite precision, directing the developing cells to assume their destiny as germ cells.

Remarkably, germ cells transmit not one, but two forms of hereditary information. First, there is the genetic information in the nucleus—the DNA housed within the chromosomes. Second, the germ granules, which reside in the germ cells' cytoplasm, also carry genetic information. Alongside the RNA and proteins that gave the germ cells their identity in the first place, they include short RNA strands (microRNA), carrying epigenetic information that helps regulate gene activity across generations. It is even suggested that these condensates act as sensors of environmental changes in germ cells.

Our understanding of germ and somatic cells is evolving rapidly. In species such as sea urchins, ascidians, and annelid worms, destruction of the germ cells does not doom future fertility, since somatic cells can generate new germline cells. An even more striking example of this kind of cellular plasticity comes from experiments performed on somatic cells from mice—specifically fibroblasts. Through the targeted expression of just a handful of master genes, fibroblasts can be reprogrammed into stem cells called IPS (induced pluripotent stem) cells. When cultured under the right conditions, IPS cells can be coaxed to differentiate into various cell types, including oocytes. And when these laboratory-generated oocytes are fertilized, they give rise to living, breathing mice—an extraordinary demonstration of the latent potential within each somatic cell.

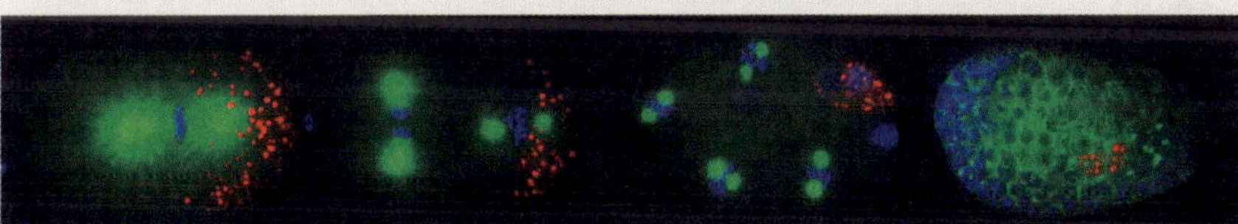

GERM CELLS IN A *DROSOPHILA MELANOGASTER* FLY EMBRYO

Left, top: Germ cells are produced at the posterior pole of the embryo. Here, fluorescent molecules reveal the germ cells by tagging their essential vasa proteins (*green*) and their nuclei (*blue*). The activity of the Torso gene (*magenta*), which regulates the formation of the head and tail regions of the embryo, is also shown.

Left, below: Three stages in the localization of germ cells (*red*). After they are produced at the posterior pole, the germ cells migrate into the developing gonads.

CONFOCAL FLUORESCENCE MICROSCOPY
© Courtesy Ruth Lehmann, Whitehead Institute, MIT, Cambridge, Massachusetts

GERM PLASM AND GERMLINE CELLS IN THE NEMATODE WORM *CAENORHABDITIS ELEGANS*

From the fertilized oocyte (*left*) to the embryo (*right*) of a nematode worm, the germ plasm (*red*)—as biomolecular condensates called P granules (*see p. 138*)—progressively assembles in the germ cells at the posterior pole. Also shown are the microtubules (*green*) and chromosomes (*blue*) of the dividing embryonic cells.

CONFOCAL FLUORESCENCE MICROSCOPY
© Courtesy Carsten Hoege and Tony Hyman, Max Planck Institute, Dresden, Germany

Sex—how and why

Sexual reproduction rests on a fundamental difference between the chromosomes of males and females—a discovery first made in the early 20th century by American biologist Nettie Stevens, through her work on insects. Stevens discovered the XX chromosome pair in females and the XY pair in males—a pattern shared by humans and other mammals (although not universal to all animals). In mammals, a single X chromosome bears around 800 genes, and the much smaller Y chromosome harbors roughly ten times fewer. Among these, the pivotal SRY gene (from <u>s</u>ex-determining <u>r</u>egion of <u>Y</u>) directs formation of the testes and thus initiates male development.

At the heart of sexual reproduction lies a defining moment: the fusion of gametes. In humans, the spermatozoon, carrying a single set of 23 paternal chromosomes, delivers its nucleus into the oocyte—a large cell equipped with rich nutrient reserves and a corresponding set of 23 maternal chromosomes. Their union results in a new, single cell—a zygote—equipped with the full complement of 46 chromosomes, including the two sex chromosomes. Barring exceptional cases, the zygote receives a single X chromosome from the mother's oocyte (as she is XX), and either an X or a Y chromosome from the father's spermatozoon (as he is XY). Thus, the spermatozoon determines whether the zygote is female (XX) or male (XY). This zygote then embarks on a journey of development, ultimately giving rise to a genetically unique individual.

This mechanism is far from universal. In birds and certain crustaceans, the maternal gamete dictates sex. In jellyfish and some reptiles, ambient temperature determines whether an individual becomes male or female. Other organisms—nematode worms, some insects, some lizards, and various plants—sidestep fertilization altogether through parthenogenesis: the ability to reproduce without spermatozoa. Meanwhile, many species, including snails, fish, tunicates, and the majority of flowering plants, are hermaphrodites—organisms with both male and female reproductive organs—and sometimes engage in self-fertilization. Sexual diversity, then, is not the exception but the rule.

Sexual reproduction is an ancient innovation. It first arose more than a billion years ago, among protists, our single-celled ancestors. Many protists still engage in sexual reproduction today. For instance, the unicellular green alga *Chlamydomonas* (see p. 103), a vegetal protist which usually reproduces by binary fission, will convert into sexed cells of equal size in nitrogen-deprived conditions. When two cells of opposite sex encounter one another, they fuse to form a zygote, which subsequently divides to create genetically diverse offspring.

The true evolutionary advantage of sexual reproduction lies in this genetic remixing. Through the mechanisms of meiosis and fertilization, genes are shuffled and recombined, producing offspring with novel genetic combinations. This genetic dynamism enhances a population's ability to adapt to changing environments and resist the persistent threat of parasitic organisms, which often wreak havoc on asexually reproducing populations, which lack genetic variation.

Another profound outcome of sex lies in maternal inheritance. Alongside the 23,000 genes encoded by our chromosomes, a human embryo inherits an abundance of mitochondria—each bearing its own set of genes—exclusively from the oocyte. These maternally derived mitochondria, numbering in the hundreds of thousands, power the cell's energy systems. If mutations are present, they can potentially cause metabolic disorders. Remarkably, our mitochondrial legacy—passed strictly from mother to offspring—has enabled scientists to trace the deep ancestry of humankind. Through this unbroken maternal line, we are led back to a single woman, an ancestral Eve who lived in Africa about 150,000 years ago.

THE DISCOVERY OF SEX CHROMOSOMES
In 1905, Nettie Maria Stevens, working with *Tenebrio molitor* mealworm larvae, observed the chromosomes during meiotic and mitotic division and identified the sex chromosomes. This discovery is co-attributed to her predecessor and mentor at Bryn Mawr College, Edmund Beecher Wilson. Although aware of Stevens's work, Wilson published before his female colleague. In the highly paternalistic atmosphere of the time, Stevens's contribution remained unrecognized for a long time. Nevertheless, she was the first female biologist to establish herself by publishing her discoveries under her own name.

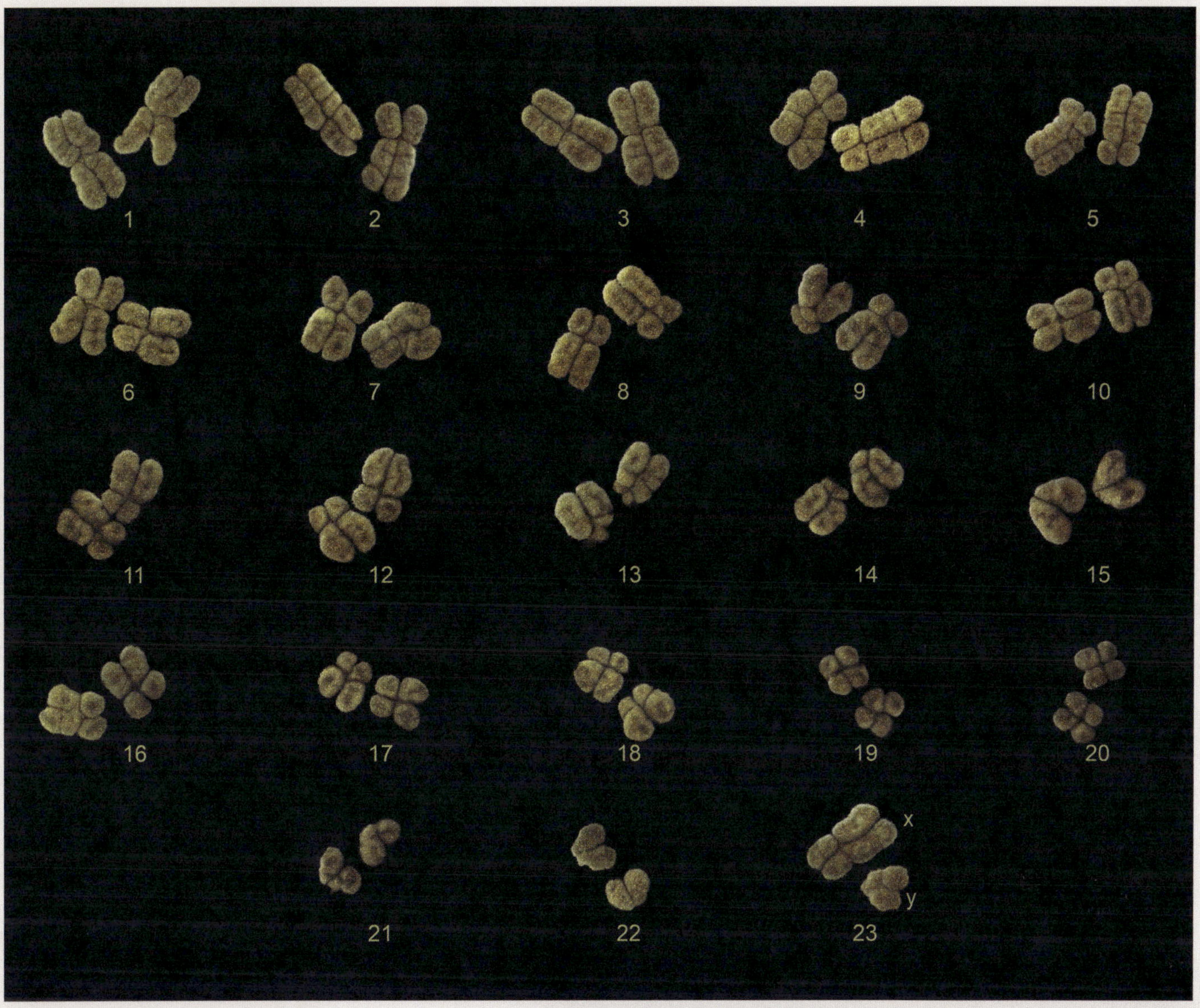

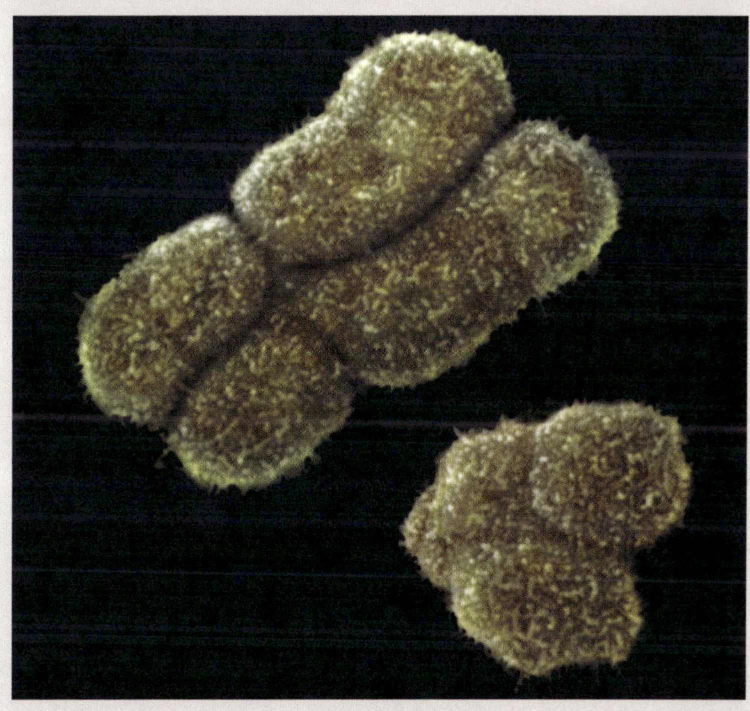

THE 23 PAIRS OF HUMAN CHROMOSOMES, INCLUDING THE X AND Y SEX CHROMOSOMES

Humans have 23 pairs of chromosomes (numbered 1 to 23). Each pair contains one chromosome from the mother and one from the father. These chromosome pairs are isolated at the metaphase stage of mitosis, during which the chromosomes are condensed into chromatids and the nuclear membrane is dispersed (see p. 152). The sex chromosomes (magnified at right) form pair 23, in this case consisting of a larger X chromosome and a smaller Y chromosome.

COLORIZED SCANNING ELECTRON MICROSCOPY

View Molly Webster's TED talk,
The Weird History of "Sex Chromosomes"

FROM FERTILIZATION TO EMBRYO

In all animals, the moment of fertilization is heralded by one or several calcium waves that sweep through and activate the oocyte. This is triggered by the fertilizing spermatozoon, which also contributes the paternal chromosomes and a centrosome *(see p. 144)*. The resulting zygote cell divides multiple times to form an embryo. This embryo then takes a path that unfolds along three developmental axes: first the anterior–posterior (head to tail) and dorsal–ventral (back to belly) axes, and later the left–right axis. These axes may be established before, during, or after fertilization, depending on the organism. For many years, the nature of the molecules guiding these axes remained a mystery—until the 1980s, when the fruit fly *Drosophila melanogaster*, long a cornerstone of genetics, provided a key insight. A fruit fly's elongated oocyte acquires its developmental axes within the ovary. This is achieved by maternal mRNA molecules, which travel to the cortex of the growing oocyte to concentrate at its anterior and posterior poles. These maternal "RNA determinants," along with other macromolecules and organelles, are transferred from interconnected "nurse cells" to the oocyte before fertilization occurs.

In jellyfish, ascidian, and amphibian embryos, axis determination also depends on maternal RNA determinants, which localize during oogenesis and/or meiosis. Yet this elegant mechanism is not universal. In the nematode worm *Caenorhabditis elegans*, for example, the site of fertilization itself dictates the anterior–posterior axis. Rather than RNA, "polarity proteins" at the cell cortex, at opposing ends of the embryo, establish its orientation and symmetry. Mammal embryos take yet another path. Their axes emerge only after fertilization, guided by intricate signaling networks and dynamic interactions among various cell types. Despite this diversity in timing and molecular machinery, a common developmental architecture soon emerges.

In nearly all animals, the embryo's early cell divisions produce a hollow sphere of cells—the blastula. Then, the cells begin to organize into three primary types: The epithelial cells of the ectoderm form a protective outer layer. Within, cells destined to become the mesoderm and endoderm interact with one another across the cavity. At gastrulation—a dramatic developmental turning point—the three germ layers reorganize, and the blastula folds in on itself, forming a "pocket" of cells that fills its interior. The ectoderm, mesoderm, and endoderm are now better positioned to communicate with each other, whether via molecular signals or intercellular contacts *(see p. 200)*. As the embryonic cells continue to divide and specialize, the germ layers begin to form the body tissues. The ectoderm gives rise to the nervous system and the skin; the mesoderm to muscle, blood, and immune cells; and the endoderm to the body's inner linings and organs such as the intestines, liver, and lungs. Depending on its species, the embryo may develop directly into a miniature adult, or pass through one or more larval stages before metamorphosis into its final form.

Top right: The threadlike spermatozoon contains a nucleus ● carrying the male chromosomes ▮, as well as mitochondria ● and microfilaments ▮. The spermatozoon swims toward the oocyte using the undulating motion of its flagellum, powered by microtubules ▮.

Top left: The oocyte is a voluminous cell with a nucleus ● enveloping the female chromosomes ▮. Its cytoplasm contains numerous mitochondria ●, various other organelles, and a cytoskeleton made up of microtubules ▮ and microfilaments ▮.

The spermatozoon attaches to the oocyte by its anterior part. At fertilization, the membranes ▮ of the spermatozoon and the oocyte fuse, giving rise to a zygote cell with both male and female chromosomes.

An embryo develops: The cells divide, forming a blastula, a ball of cells—each with its own nucleus ● and chromosomes ▮, and a cilium filled with microtubules ▮ and microfilaments ▮. Mitochondria ● and other organelles ● are also shown.

From oocytes to embryos—building the body axes

There are striking contrasts between the oocyte and the spermatozoon—two gametes with profoundly different forms and destinies. The oocyte, large and immobile, gradually accumulates nutrients as it matures within the gonad during oogensis, whereas spermatozoa are generally slender, motile, and produced in staggering numbers. Propelled by its flagellum, a sperm's main purpose is to reach and unite with an oocyte. In certain organisms, such as the fruit fly, the oocyte already exhibits a high degree of polarity by the end of oogenesis. At the time of fertilization, when the spermatozoon delivers its nucleus and centriole into the oocyte's cortical region, it triggers and enhances this pre-established polarity. In others, like the nematode worm *C. elegans*, the entry site of the fertilizing spermatozoon activates and determines the polarity of the embryo. As the fertilized oocyte cleaves and gives rise to multiple cells, it establishes its fundamental body axes—anterior–posterior (A-P), dorsal–ventral (D-V), and left–right (L-R). These axes function as molecular signposts, guiding cellular differentiation and the emergence of organized tissues. Embryogenesis unfolds as a symphony of gene expression, regulated according to time and place through intricate intercellular communication.

Until the 1980s, our understanding of axis specification was largely based on morphological studies and micro-manipulation experiments, such as tissue ablation (targeted cell removal or destruction) and transplantation. The field underwent a profound shift, however, with the advent of molecular genetics and the study of mutant embryos in the fruit fly *Drosophila melanogaster*. At the newly founded European Molecular Biology Laboratory in Heidelberg, Christiane Nüsslein-Volhard and Wolfgang Driever made a landmark discovery: In the fruit fly, the oocyte's A-P axis is determined within the gonad before fertilization, through the localization of specific messenger RNA (called *maternal RNA determinants*) at the anterior and posterior poles. These RNA messengers are synthesized by nurse cells and transferred into the oocyte, where they are transported and anchored to precise cortical locations. Following fertilization, they are translated into proteins—both activators and repressors—whose concentration gradients shape the expression of downstream genes. These maternal cues provide the embryo with a molecular blueprint, laying the foundation for tissue development along its major axes.

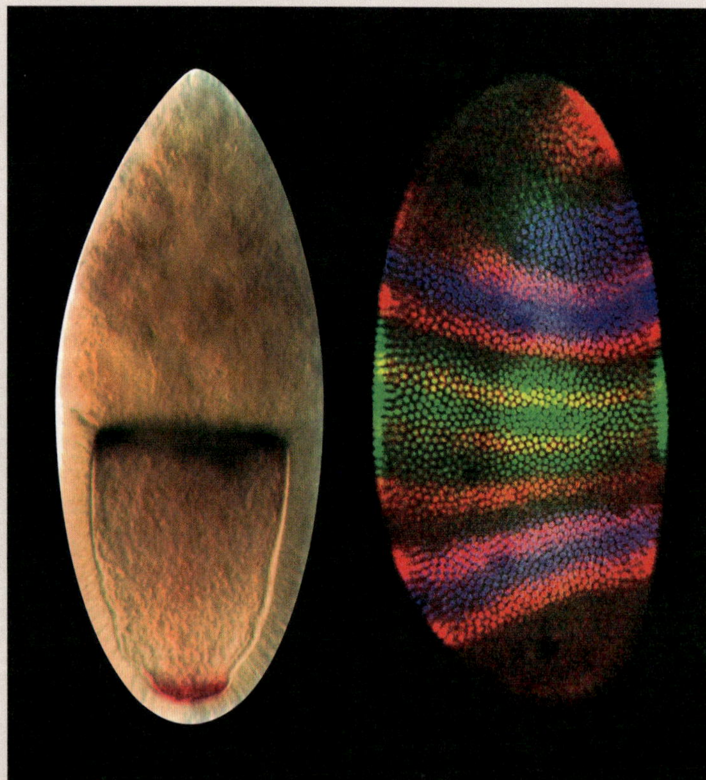

On left

DETERMINANT RNA IN A *DROSOPHILA* OOCYTE IN THE GONAD

This oocyte (*at left, lower half*) is topped by a mass of nurse cells. Where they meet, Bicoid mRNA (*black*), a maternal determinant, is expressed in the oocyte's cortex at its anterior pole. Oskar mRNA (*red*), a germ cell determinant, is localized in germ plasm—biomolecular condensates—at its posterior pole.

LIGHT MICROSCOPY/RNA LOCALIZATION USING *IN SITU* HYBRIDIZATION TECHNIQUES
© Courtesy Daniel St Johnston and Katie Smith-Litière, Cambridge University, Great Britain

On right

VISUALIZATION OF GENE EXPRESSION

The expression of three different genes—Hairy (*red*), Krüppel (*green*), and Giant (*blue*)—are visualized in the cells of a *Drosophila* embryo. Expression of these antagonistic genes defines different cell domains along the embryo's anterior–posterior (top to bottom) axis. The cell nuclei are visible as a multitude of small fluorescent dots.

CONFOCAL FLUORESCENCE MICROSCOPY
© Courtesy Jim Langeland, Stephen W. Paddock, and Sean Carroll, University of Wisconsin, Madison

Further axis determinants have been identified within the oocyte cortex of several key developmental model organisms, including the toad *Xenopus laevis*, the ascidian *Ciona intestinalis*, and the nematode worm *C. elegans*. These discoveries reveal the diverse mechanisms that underlie axis specification—mechanisms that rely on the spatial distribution of molecular signals, such as RNA or proteins, and on intricate interactions between cells.

But how do plants establish their developmental axes? It is worth recalling that fertilization was first discovered not in animals, but in plants. In 1854, Gustave-Adolphe Thuret observed fertilization in the brown alga *Fucus*, two decades before Hermann Fol described the process in starfish.

When a *Fucus* zygote forms in complete darkness, the sperm's site of entry into the oocyte determines the axis of development. However, when the oocyte is illuminated, light itself dictates the embryo's polarity. The illuminated side of the fertilized oocyte gives rise to the frond—the olive-green, photosynthetic part of the alga responsible for energy capture and reproduction. By contrast, the shaded side forms an extension that develops into rhizoids—fine, rootlike filaments that anchor the young alga to its rocky substrate. This exemplifies a striking interplay between intrinsic cues and environmental signals in the establishment of embryonic axes.

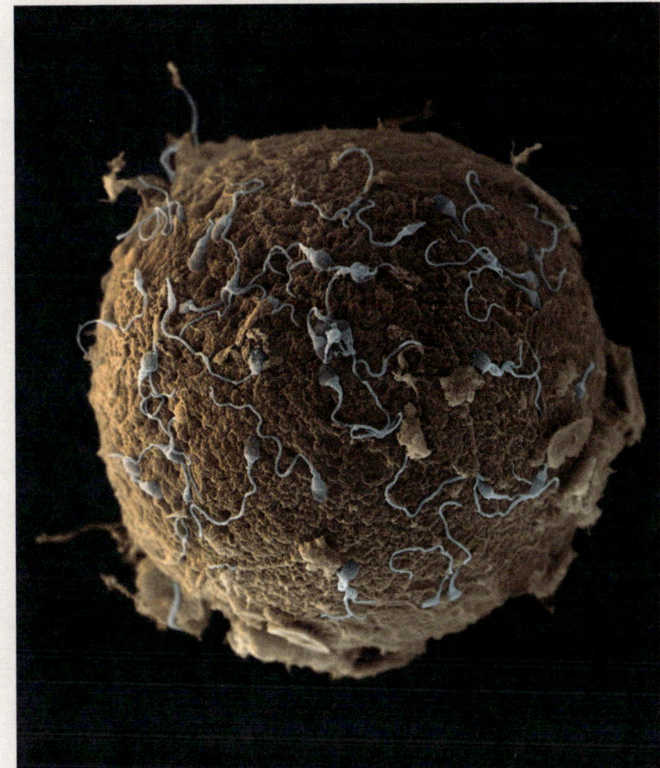

HUMAN OOCYTE WITH SPERMATOZOA

A human oocyte, measuring 0.1 mm, is covered with spermatozoa. Women produce one oocyte per month, while men produce millions of sperm. To fertilize the oocyte, a single sperm cell must penetrate its outer layer (called the *zona pellucida*). This is a surplus oocyte from *in vitro* fertilization.

COLORIZED SCANNING ELECTRON MICROSCOPY
© Courtesy Nicole Ottawa and Oliver Meckes, *eye of science*, Germany

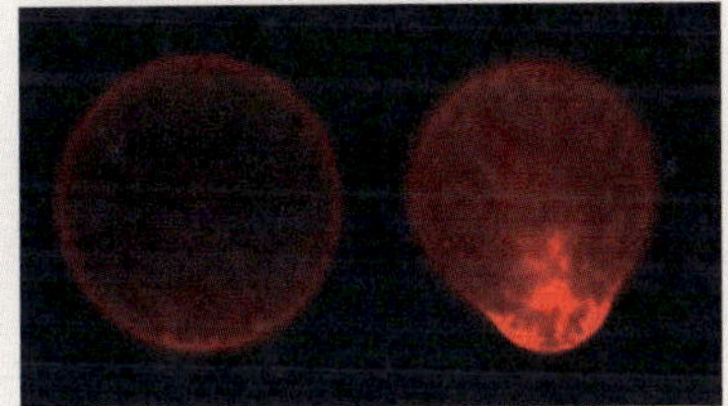

A FERTILIZED OOCYTE OF THE BROWN ALGA *FUCUS* IS POLARIZED BY LIGHT

A recently fertilized oocyte (*on left*) is exposed for a few hours to a light coming from above. This is enough to establish the axis of the developing embryo (*on right*).

The fertilized oocyte develops an outgrowth on the side opposite the light source. Its fluorescence is due to the accumulation of multiple secretory vesicles which give rise to the rhizoids—filaments that anchor *Fucus* alga (*right*) to rocks.

CONFOCAL FLUORESCENCE MICROSCOPY
© Courtesy Darryl Kropf, The University of Utah, Salt Lake City; and Whitney Hable, University of Massachusetts, Dartmouth

AGING AND LIFE EXPECTANCY

Like all living beings, people age—each at our own rhythm. So do our cells. When cells derived from healthy tissues are cultured *in vitro*, they undergo only a limited number of divisions before their progress is arrested. This phenomenon, known as *cellular senescence*, arises from a variety of causes. One of the most widely recognized is the gradual shortening of telomeres—the protective caps at the ends of chromosomes—over successive cell divisions. As the telomeres erode, they trigger built-in cellular checkpoints that ultimately halt proliferation. But telomere attrition is only part of the story. Other chromosomal alterations also contribute to senescence, including the slow accumulation of DNA mutations and epigenetic marks—chemical additions that subtly reshape gene expression without altering the DNA sequence itself. Among the most studied of these marks are methyl groups (CH_3) added directly to DNA. Intriguingly, the pattern and density of methylation can serve as a molecular clock, offering a glimpse into an individual's age. Their number and frequency can even be used to predict a person's future longevity. Aging also brings about widespread changes in gene expression. Genes coding for large proteins are expressed less frequently over time, while those coding for smaller proteins continue to function normally. This imbalance may help explain why tissues lose their capacity to heal as we grow older, and organs lose their ability to regenerate.

In addition to chromosomal changes, mitochondria—the cell's power-houses—play a central role in the aging process. With age, their capacity to fuse, divide, and repair diminishes. One consequence of this decline is the accumulation of abnormally large mitochondria in senescent cells. Moreover, mitochondrial performance wanes, particularly in the production of ATP (the cell's energy currency), while the super-oxidant molecules they produce, such as ROS *(see pp. 40–41)*, build up. These highly reactive molecules damage DNA and proteins, further accelerating cell aging.

The comparatively long lifespan of mammals—two years for mice and about seventy for humans—poses a practical challenge for aging research. To overcome this limitation, scientists have turned to a much smaller model: the nematode worm *Caenorhabditis elegans*. Although it lives only two to three weeks, this tiny organism still exhibits clear signs of aging, including pigment accumulation, reduced fertility, and declining motility. Crucially, this nematode worm can be genetically modified to produce mutants with significantly longer or shorter lifespans. Its flexibility allows researchers to dissect the gene networks involved in aging and explore the effects of diet. For instance, caloric restriction has been shown to extend lifespan, while nutrient- or glucose-rich diets accelerate aging. Current research is focused on the potential of anti-aging compounds—such as polyphenols—and their possible therapeutic applications. Remarkably, this tiny worm's metabolism and genetics closely resemble ours, offering valuable insights into human biology.

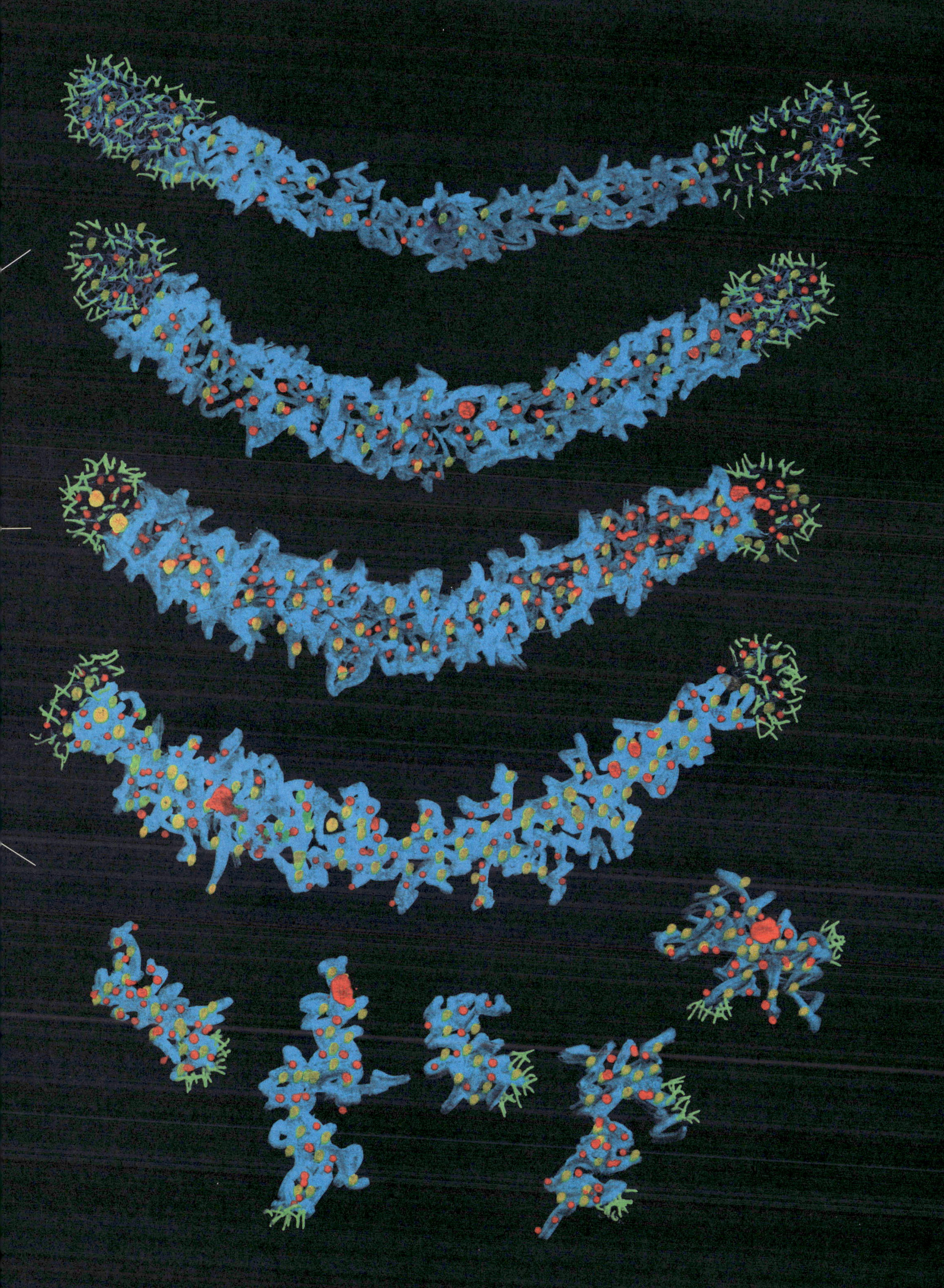

DEATH IS PROGRAMMED

Life is a delicate balance of birth and death, even at the cellular level. As new cells are created, others must die. This death can occur accidentally, or as a result of aging or disease. In 1842, the German biologist Carl Vogt made a novel observation: During the metamorphosis of a tadpole to a toad, groups of its cells died in a predictable pattern. Four decades later, Walther Flemming noticed that the first visible sign of impending cell death was fragmentation of the chromosomes—filamentous structures he had only recently observed. This process, known today as "programmed cell death," plays a crucial role in developing embryos. During metamorphosis, it sculpts the tissues with precision. Mammal embryos, for instance, form their fingers thanks to the systematic suicide of cells between the digits. Those same cells persist in web-footed animals like ducks and bats. Cell death also guides the closure of the neural tube and the formation of the heart. It even eliminates most of the millions of oocytes present in a female human fetus before birth.

Thus, development is not a story of cells solely dividing and specializing. It is equally a story of cells programmed to die—always the same ones, in the same places, at the same time—a sort of orderly ballet of destruction. A handful of genes orchestrate this cellular suicide, as shown with remarkable clarity in the tiny *C. elegans* worm. This elegant mechanism was named *apoptosis* in the 1970s by John Kerr, borrowing from the Greek word for "falling leaves." Like autumn foliage, cells undergoing apoptosis detach gently from their neighbors and fragment their chromosomes. Eventually, they break apart into small, decaying pieces. These fragments are swiftly engulfed and recycled by immune cells, particularly macrophages.

Apoptosis begins when specialized receptor proteins on the cell membrane are activated, or when the mitochondria become porous under stress, releasing internal distress signals. Either pathway activates protease proteins called *caspases* (gluttonous enzymes) as well as destructive nanomachines known as *proteasomes* (see p. 57), which chop up all the cellular structures made of proteins. However, apoptosis is just one form of programmed death. Molecular analysis of the dying process has revealed many other forms: autophagy, necroptosis, pyroptosis, and others, each with its own role and signature. Considering the stakes, it's essential that cell death be tightly regulated—to keep runaway cells, such as cancer cells, in check.

(see p. 57)

When cells are destined to die, they initiate a process known as *apoptosis*—programmed cell death.

Once apoptosis is triggered, protease proteins are activated, the mitochondria ● become porous, membranes ▌blister, and chromosomes ▌and the cytoskeleton (microtubules ▌and microfilaments ▐) begin to fragment.

The cell then breaks up into "apoptotic bodies" containing parts of all its cellular components. These cell fragments are engulfed and absorbed by the macrophages of the immune system.

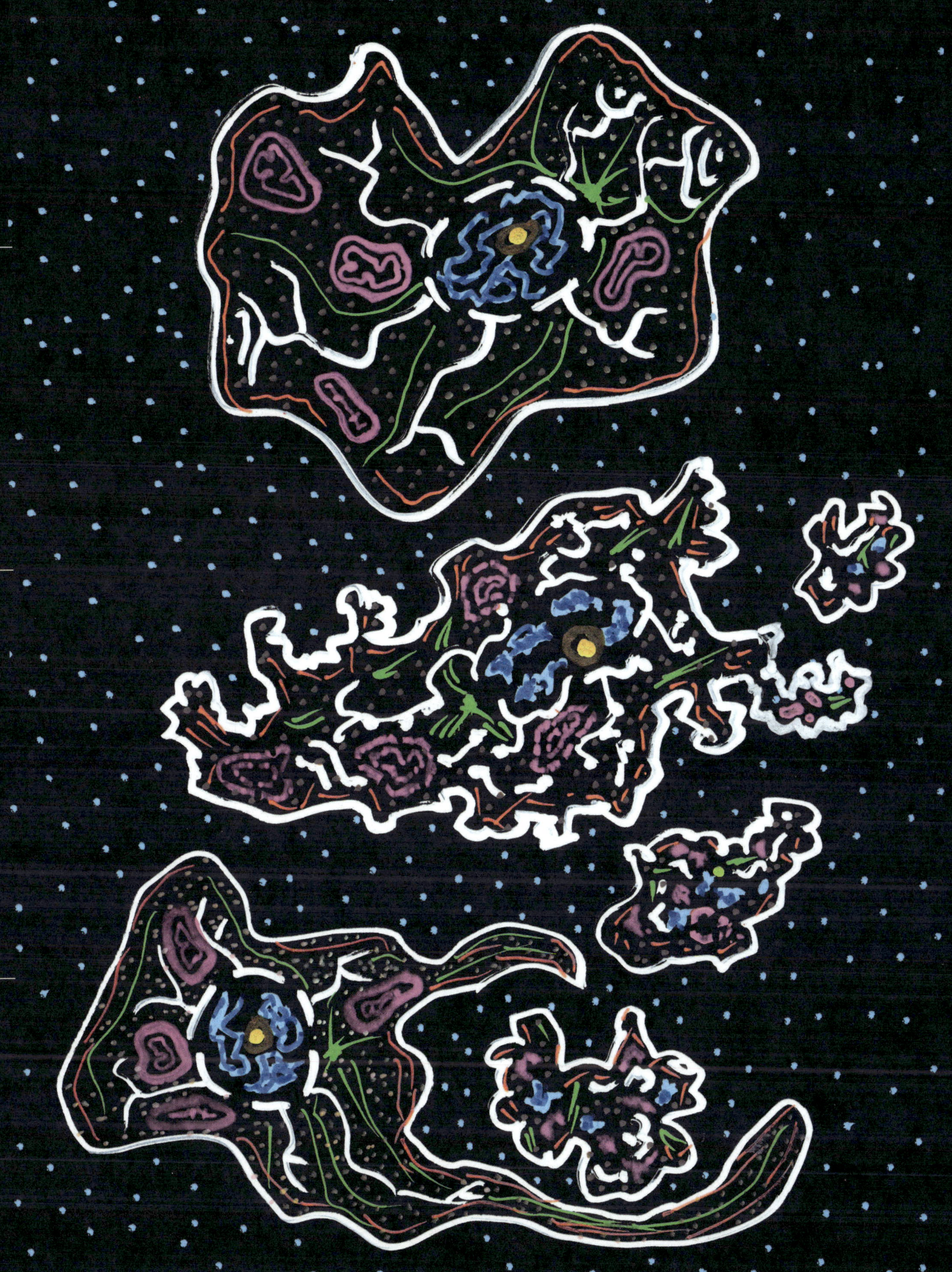

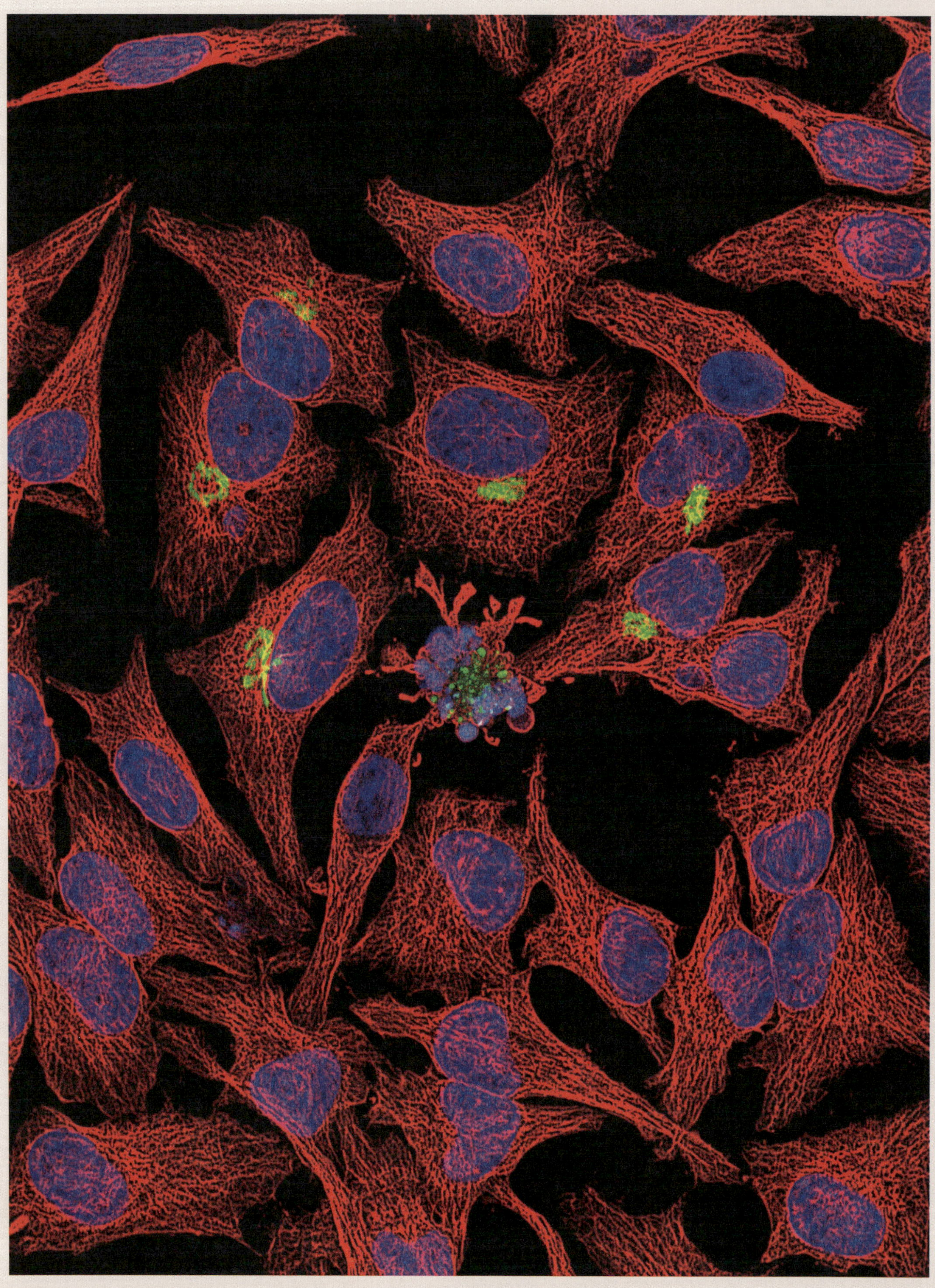

Life expectancy of human cells

The lifespan of human cells remained an enigma until the 1950s, when scientists finally mastered the art of cultivating them in petri dishes. Under ideal nutritional and environmental conditions, normal human cells divide between forty and sixty times before the end of their life. This boundary is now known as the *Hayflick limit* after Leonard Hayflick, who first observed it in the 1960s. Interestingly, this limit does not apply to cancer cells—a discovery that emerged from the observation of a line of tumor cells, later known as HeLa cells, which have been indefinitely cultured since the 1950s. Remarkably easy to grow, HeLa cells were quickly commercialized and distributed to laboratories across the globe, becoming a standard model in biomedical research—used to study cancer, viral infections, and vaccine development. Yet little thought was given to their origin. The truth was only popularized in 2010, with the release of science journalist Rebecca Skloot's bestselling book *The Immortal Life of Henrietta Lacks*. Skloot revealed that the popular HeLa cells are in fact those of Henrietta Lacks, an African-American woman who died of cervical cancer at age 31. Cultured and distributed without her consent, Henrietta Lacks's cells continue to proliferate *in vitro*. To the present day, about 800 billion HeLa cells have been produced in laboratories around the world.

A cell's destiny is intricately tied to that of its chromosomes, whose DNA is constantly repaired to preserve the integrity of its genetic information. To shield chromosomes from degradation or fusion with one another, their ends are capped by structures called *telomeres*—repetitive DNA sequences safeguarded by biomolecular condensates of specific proteins and RNA. Human telomeres shorten with each cell division. Once they become critically short, the chromosomes can no longer be replicated properly, leading cells into senescence—a kind of biological retirement marked by halted division and heightened inflammation. This gradual telomere erosion accelerates with age, stress, and disease, and is one of the primary explanations for the Hayflick limit. However, some cells escape this fate. Stem cells and cancer cells, which divide rapidly, maintain long telomeres thanks to an enzyme protein called *telomerase*—a sophisticated nanomachine composed of proteins and RNA. Telomerase attaches to the chromosomes' ends and extends their DNA sequences, effectively bypassing the countdown to senescence. In most human somatic cells, telomerase remains largely inactive. While it may seem tempting to activate it in the hope of slowing aging by preserving telomere length, this strategy is fraught with danger. Telomere shortening likely serves as a crucial safety mechanism, preventing unchecked cell proliferation that can lead to cancer.

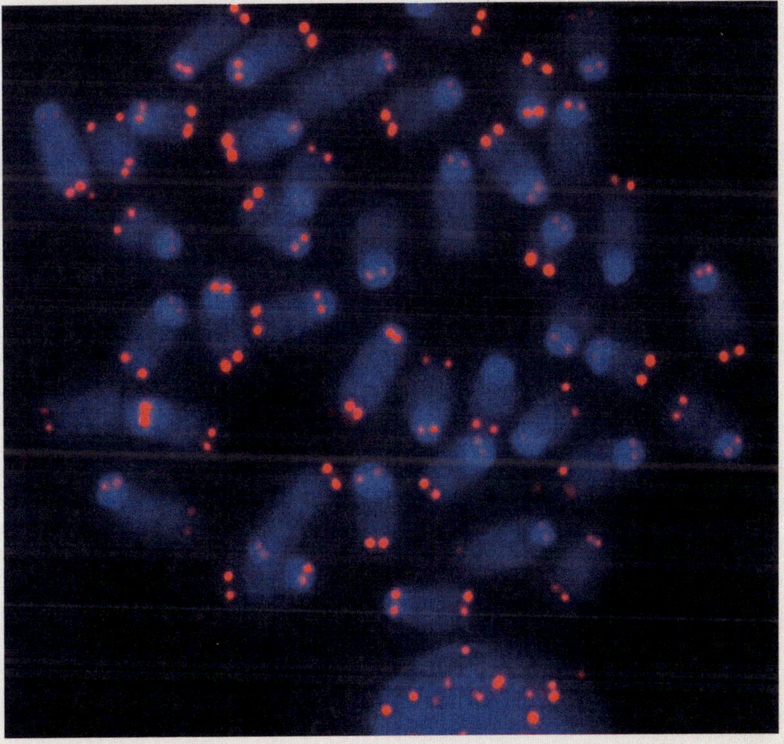

TELOMERES—THE ENDS OF HUMAN CHROMOSOMES

Telomeres, the cap-like structures at the ends of every pair of chromosome (*blue*), can be visualized using fluorescent proteins (*red*).

CONFOCAL FLUORESCENCE MICROSCOPY
© Courtesy Thomas Reid, NCI Center for Cancer Research, Bethesda

HELA CELLS IN CULTURE

HeLa cells (the first tumor cell line to divide indefinitely in culture) are imaged with fluorescent molecules, revealing microtubule networks (*red*), chromosomal DNA in the nuclei (*blue*), and Golgi apparatus organelles (*green*)—membrane organelles that specialize in the secretion of macromolecules. In the center, a cell dies by apoptosis.

CONFOCAL FLUORESCENCE MICROSCOPY
© Courtesy Thomas Deerinck and Mark Ellisman, NCMIR, University of California San Diego

From cell suicide to increased longevity

Embryos and larvae develop according to the choreography of cellular life and death. This became strikingly clear in the 1980s, when Sydney Brenner, John Sulston, and Robert Horvitz meticulously mapped the developmental journey of *Caenorhabditis elegans*, a transparent, soil-dwelling nematode. Their pioneering work, which earned them the Nobel Prize in 2002, revealed how the fate of cells—when they live and die—is built into the very architecture of life. Deliberately chosen by Brenner as a model organism, the nematode worm became the first animal to have its genome entirely sequenced in the early 2000s.

With around 20,000 genes—35 percent of which have homologues in humans—the nematode offers an ideal canvas for genetic manipulation and mutation analysis. The adult worm is composed of precisely 959 somatic cells, differentiated into various tissues such as skin, muscle, intestine, and neurons, alongside 1,000 to 2,000 germ cells producing the gametes—spermatozoa and oocytes. Notably, the species has no true females: Individuals are either male or hermaphrodite. The worm's anatomy is not only simple but invariant, enabling biologists to trace the lineage and fate of every single cell from the fertilized oocyte to the adult organism—which develops over the course of just three days. In observing its development, researchers found a striking pattern: 131 of the worm's cells die, in two predictable waves—always the same cells, at the same stages, at the same locations. This is no coincidence. As the cells die by apoptosis—programmed death by self-fragmentation—their remains are swiftly engulfed and digested by neighboring cells.

Just three genes orchestrate this coordinated cellular demise. If they are silenced by mutation, their targeted cells survive and continue to differentiate—often becoming neurons or muscle cells. If the genes are over-activated, the process intensifies and more cells perish. These genes encode key players in the apoptotic machinery, including caspase-like proteases, and enzyme proteins that cleave other proteins into fragments, effectively dismantling the cell from within.

Experiments on this humble worm have transformed our understanding of aging. Once thought to be the inevitable result of cellular wear and tear, aging is now recognized as a process that can be influenced—perhaps even modulated—by genetic and environmental factors. The lifespan of *C. elegans*, typically just two to three weeks, can be dramatically extended through genetic manipulation or dietary restriction. One particularly striking example involves the hormone insulin and its signaling pathway: When the genes regulating this pathway are altered, the worm can live from twice to even three times its normal lifespan. Though the mechanisms of this dramatic boost are not yet fully understood, the long-lived worms exhibit improved DNA repair and remarkable resistance to oxidative stress from UV rays or ROS (reactive oxygen species). This tiny model organism hints that the regulation of aging, far from being fixed, may be open to behavioral and environmental influence. Could it hold the secret to extending human life?

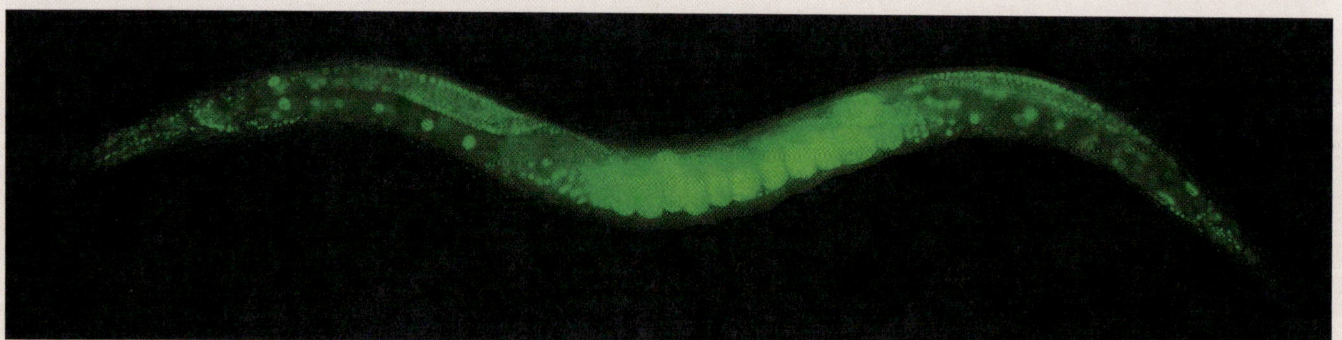

THE NEMATODE WORM *CAENORHABDITIS ELEGANS*

In this image of the hatched worm, the H3 histone proteins are labeled with the molecule GFP (green fluorescent protein), expressed in the cell nuclei thanks to genetic manipulation. These fluorescent nuclei appear as small green dots. The gonads, containing oocytes, are visible in the central region of the worm.

FLUORESCENCE MICROSCOPY

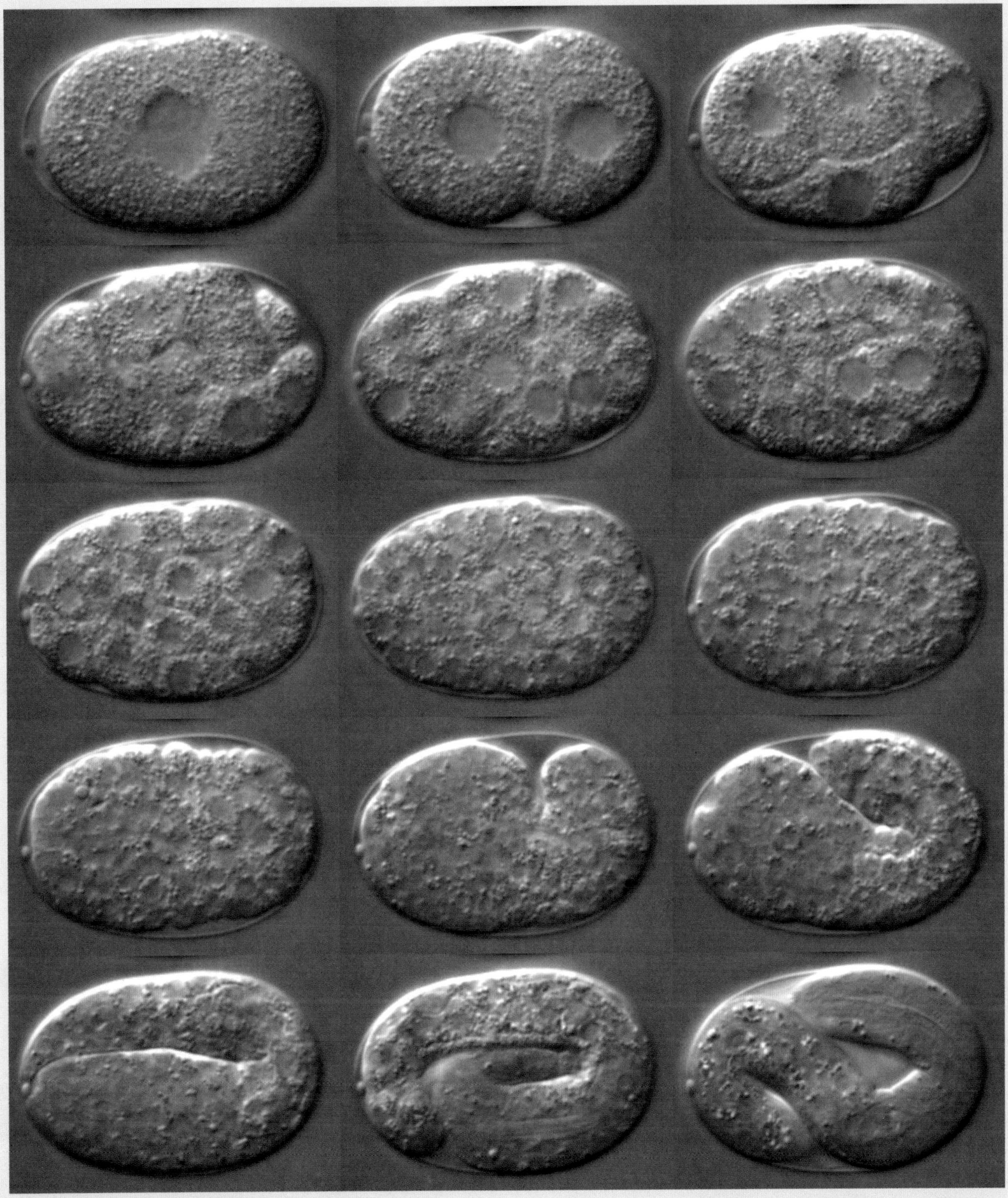

EMBRYONIC DEVELOPMENT OF THE NEMATODE WORM *CAENORHABDITIS ELEGANS*

Thanks to its transparency, *C. elegans*'s development can be easily observed, from the fertilized oocyte (*top left*) until the worm hatches in just one day (*bottom right*).

IMAGE SEQUENCE FROM A FILM USING PHASE-CONTRAST INTERFERENCE MICROSCOPY

© Courtesy Bob Goldstein and Daniel Dickinson, University of North Carolina, Chapel Hill

View timelapses of *C. elegans* development, compiled by the Goldstein Lab at the University of North Carolina, Chapel Hill

DISEASES—
AGGRESSIVE OR
FALTERING CELLS

As we grow older, our risk of serious illness—such as cancer or cardiovascular disease—rises steadily. These ailments often stem from internal malfunctions: tissues and organs that falter, or cells and molecules that malfunction or defy their roles, as with cancer cells that escape normal controls. Other diseases—such as malaria, tuberculosis, influenza, COVID-19, and AIDS—are caused by external invaders: infectious agents that infiltrate the body and disrupt its equilibrium. From smallest to largest, these invaders are prions, viruses, pathogenic bacteria, and parasitic eukaryotes. Transmission can occur directly between individuals or indirectly via animals—particularly insects like mosquitoes and ticks that act as biological syringes, injecting pathogens straight into the bloodstream.

The tiniest of these culprits, prions, came to public attention in the 1990s with the appearance of mad cow disease—a neurodegenerative bovine illness contracted from contaminated feed. Prions are simply misfolded proteins that, once inside the body, induce the same defect in other proteins they encounter. Prion-like proteins have also been implicated in Alzheimer's disease, where abnormal proteins aggregate in the neurons, leading to damage. Meanwhile, viruses, which carry DNA- or RNA-based genetic instructions, are behind some of history's most devastating pandemics—smallpox, polio, AIDS, and COVID-19—and continue to circulate among humans and animals *(see chap. VI)*.

Some unicellular organisms have long been infectious. Pathogenic bacteria are the agents of plague, cholera, tuberculosis, and numerous hospital-acquired infections. More complex are the single-celled eukaryotic parasites, like *Plasmodium*, responsible for spreading malaria, via mosquitos, in tropical regions. Larger still are the multicellular parasites, such as tapeworms, that infect both humans and animals.

Each of these diverse agents attacks the body in its own unique way. Prion proteins cluster into toxic plaques in nerve cells. Respiratory viruses like influenza and SARS-CoV-2 invade the airways, damaging epithelial cells and triggering inflammation and pneumonia. HIV transmits from person to person through bodily fluids and destroys immune cells, opening the door to countless pathologies such as cancers. The bacteria that spreads the plague, transmitted by flea bites, also attacks the immune system, and then all organs. By contrast, cholera bacteria secrete toxins that render the intestinal cells permeable, causing diarrhea and dehydration that can be fatal. Every infection is a battle between our cells and infectious cells, mediated by the armada of cells and molecules of the immune system.

Infectious agents—viruses, bacteria, and protists

The existence of major infectious agents (bacteria, viruses, and parasitic protists) responsible for contagious diseases was brought to light in the 19th century through the groundbreaking work of Louis Pasteur, Robert Koch, and Alphonse Laveran. For millennia, entire populations were decimated by epidemics, the true culprits behind which remained unknown until recently, when genetic analyses of ancient remains unveiled their identities. Among the most notorious historical pandemics are three caused by bacteria: plague, cholera, and tuberculosis. Other equally devastating diseases, such as smallpox, hepatitis, and influenza, are caused by viruses. All of these illnesses haunted the ancient world: In pharaonic Egypt and classical Greece, outbreaks of the plague and smallpox could eliminate up to half the population, often in agony. In more recent history, two global pandemics—AIDS and COVID-19—have swept across continents since the 1980s. Both are caused by RNA viruses, zoonoses that jumped from wild animals over to humans. Fortunately, our understanding of their RNA sequences, life cycles, and modes of transmission has allowed researchers to mount informed responses, mitigating their spread and severity *(see pp. 124–25)*.

Unlike bacterial and viral diseases, malaria stems from a parasitic protist: *Plasmodium falciparum*, a member of the Apicomplexa phylum *(see p. 96)*. This same group also includes *Toxoplasma gondii*. These microscopic protists are armed invaders with highly specialized cellular machinery, able to infiltrate host cells with precision. Once inside, they enclose themselves within a protective vacuole, fashioned from stolen fragments of the host's own membrane—a strategy that helps them evade the immune system.

Malaria is transmitted exclusively by female mosquitoes of the genus *Aphodele*, which harbor *Plasmodium* in their digestive tract and salivary glands. When a mosquito bites, it injects a form of the *Plasmodium* parasite into the bloodstream. From there, it rapidly travels to the liver, where it multiplies and morphs before launching its next assault on the red blood cells. These cycles of invasion and destruction trigger the hallmark symptoms of malaria—fever, chills, headaches, and profuse sweating, which recur every two or three days. Annually responsible for over 200 million cases and 400,000 deaths, malaria remains a formidable threat in many warm, tropical regions.

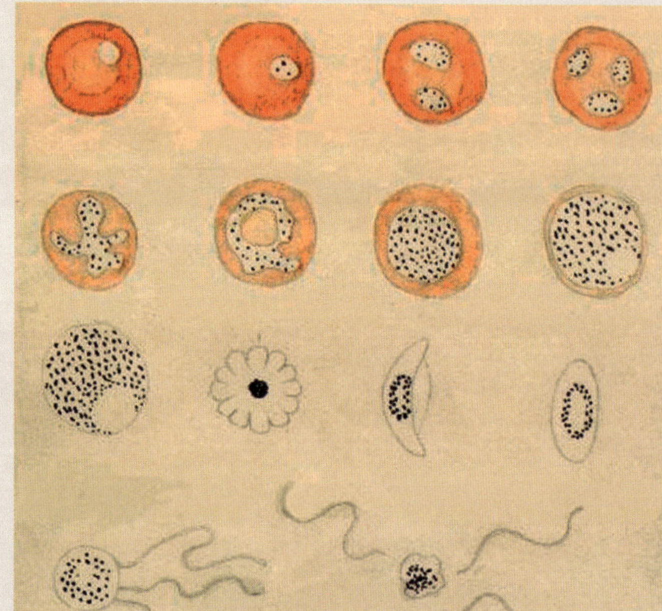

THE DISCOVERY OF *PLASMODIUM* BY ALPHONSE LAVERAN
Laveran, a military physician, was in Algeria in 1880 when he discovered this parasitic protist, which he called the "malaria hematozoan," in the blood of malaria patients. Here, his drawing includes red blood cells overtaken by multiple forms of the parasite (*black granules*).

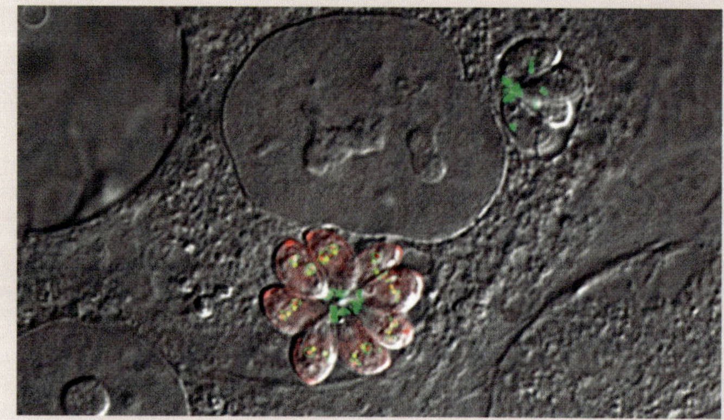

THE PARASITIC PROTIST BEHIND TOXOPLASMOSIS
Once *Toxoplasma gondii* has invaded its host, it forms multi-cell rosettes—surrounded by vacuoles—inside the host animal's cells. The eight *Toxoplasma* cells visualized here include organelles and cytoskeletal elements (*red and green*), which fluoresce in the anterior part of each protist. The cytoskeletal structure is designed to attack and penetrate other cells of the infected organism.

CONFOCAL FLUORESCENCE MICROSCOPY
© Courtesy John Murray and Ke Hu, Arizona State University, Tempe

Although less lethal than *Plasmodium*, the protist parasite *Toxoplasma* causes toxoplasmosis, a dangerous disease for immunocompromised individuals and pregnant women, transmitted primarily via domestic cats. *Toxoplasma* is capable of infecting nerve cells and altering behavior in mammals (including humans), provoking neurological and psychological changes related to risk-taking that continue to fascinate researchers *(see p. 96)*.

Cancer—cells run amok

The word *cancer* comes from the Greek *carcinos*, meaning "crab," the name Hippocrates gave to tumors. Traces of cancer can be found in the bones of prehistoric man and in Egyptian mummies. There are descriptions of tumors and surgical procedures on papyrus dating back 5,000 years! Cancer was thought to be contagious until the middle of the 19th century. But once biologists realized that every cell is born from another, by division, they understood that tumors are the result of uncontrolled proliferation. Whereas normal cell growth, division, and differentiation are regulated by the cell cycle *(see pp. 146 and 200)*, cancer cells break free from these constraints—rapidly dividing and proliferating as if they've regained the independence of single-celled protists.

Building on Rudolf Virchow's foundational idea that disease originates at the cellular level, Theodor Boveri proposed as early as 1910 that cancer might stem from chromosomal abnormalities—errors in their number or structure that could transmit aberrant instructions. In time, microscopic and genetic study of tumors revealed that certain genes—some inherited, some altered by external forces—are responsible for both the runaway proliferation of cancer cells and their invasive potential, which leads to metastases throughout the body. A groundbreaking moment came in 1911, when Peyton Rous discovered a virus capable of inducing connective tissue tumors (sarcomas) in chickens. The responsible gene, SRC (for <u>s</u>a<u>rc</u>oma), was isolated in 1970, marking the discovery of the first oncogene—a gene that can induce cancerous properties in a cell. Since then, a complex landscape of oncogenes, tumor suppressor genes, DNA repair mechanisms, and cell-death regulators has emerged. Cancer, we now understand, results when the precarious balance among these genetic and epigenetic factors—some inherited, others shaped by the environment—is disturbed. Carcinogenesis—the emergence of cancer cells—can be triggered by a vast array of external agents, such as ionizing radiation, pesticides, and infectious pathogens including HPV (<u>h</u>uman <u>p</u>apilloma <u>v</u>irus), the gastric bacterium *Helicobacter pylori*, and the parasitic *Schistosoma* worm.

A tumor is not a homogenous mass of identical cells, but a complex pseudo-organ, composed of both malignant and normal cells, complete with its own vascular network and supporting tissues. Tumors adapt to their surroundings with remarkable finesse, evading immune surveillance and often developing resistance to therapy. Once a tumor has metastasized (spread), cancer becomes even more difficult to

THE BOVERIS, CHROMOSOMES, AND CANCER

By observing sea urchin oocytes fertilized by more than one sperm cell, Theodor and Marcella Boveri came to understand that centrosomes (*a, b, c, d*) direct the distribution of chromosomes (*I, II, III, IV*). Fifty years before the discovery of the genetic code and DNA, the Boveris proposed that chromosomes carry hereditary traits, and that unequal distribution of chromosomes between daughter cells can cause cancer.

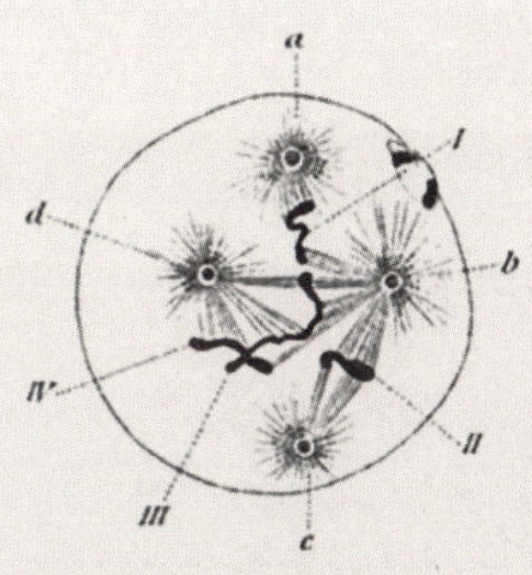

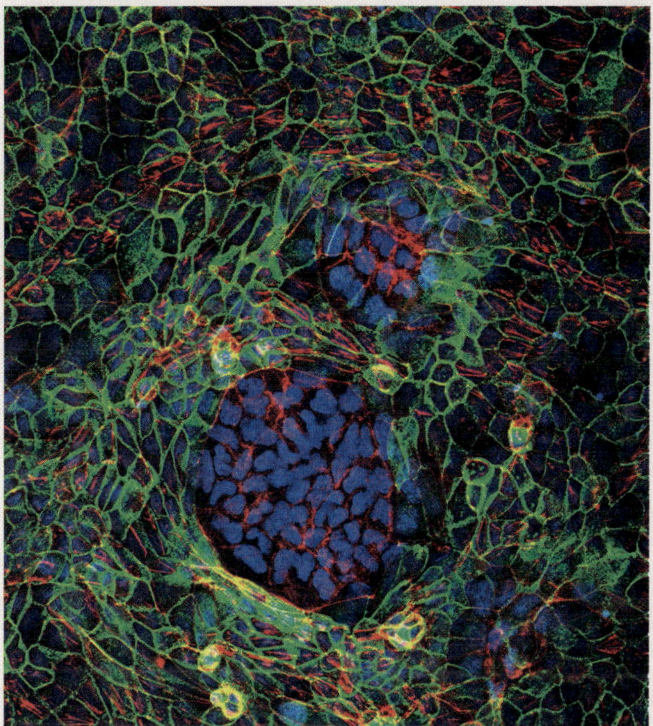

THE HETEROGENEITY OF TUMORS
Two islands of cancer cells (*blue nuclei*), surrounded by normal cells (*green membranes*), are observed by confocal fluorescence microscopy.

contain. In recent years, the metabolism of cancer cells has reemerged as a key area of research. As early as the 1920s, Otto Warburg observed that cancer cells exhibit a voracious appetite for glucose, a phenomenon known as the *Warburg effect*. The cancer cells rely on cytoplasmic glycolysis, accompanied by shifts in nitrogen metabolism, to fuel their growth. This metabolic "rewiring" reflects the heightened energy demands of tumor cells. Today, new insights on the metabolism, genetics, and epigenetics of cancer, as well as immunity, are converging to open new avenues for treatment—offering hope for more effective, personalized therapies to treat one of the most complex and persistent challenges in medicine.

WHAT LIFE IS

A social network of cells

TNT
GOLGI
GOLGI
RNA
RNA
mi RNA
MVB
EXOSOMES
mi RNA
RNA
RNA
RNA
RNA
RNA
MVB
EXOSOMES
RETICULUM
LYSOSOME
TNT
MVB
GOLGI
mi RNA
ECTOSOMES
GOLGI
GOLGI
RETICULUM

CELLULAR CONVERSATIONS

Cells are inherently social beings. They are in constant communication, whether by direct contact or sending messages across distances. Their conversations take many forms: electrical signals, exchanged molecules such as hormones, tiny membrane vesicles bearing messages, and even organelles, condensates, and nanomachines transferred through intercellular bridges and tunnels. Life itself can be seen as the sum of these cellular dialogues—a vast social network woven from the voices of individual cells, communicating within multi- and pluricellular organisms and across species and ecosystems.

Even the most ancient and seemingly simple forms of life—bacteria and archaea—are far from solitary. They sense their environment, respond to molecular signals emitted by neighbors, and adjust their behavior collectively. When their population reaches a critical threshold, these microbes begin to act as a coordinated group. This phenomenon, first discovered in glowing bacteria that live symbiotically in a tiny Hawaiian squid, is known as *quorum sensing*—a universal language of microbial communication. This social awareness enables prokaryotic cells to cooperate and synchronize their behavior, whether forming a resilient biofilm on virtually any surface (from rocks to human tissues) or invading a host. Pathogens like cholera bacteria, for example, "listen" for quorum signals before releasing their toxins—launching their coordinated strike only when their numbers are sufficient.

This microbial social network extends even further—to bacterial viruses known as *phages*. These viruses don't simply attack bacteria blindly; they "confer" before deciding whether to infect or lie low. Some phages communicate through short chains of amino acids called *arbitrium peptides*, effectively counting themselves before launching an infection. Remarkably, phages also serve as genetic couriers, transferring bits of DNA between bacteria—including genes for antibiotic resistance or virulence. In this way, phages become not just parasites, but actors speaking a viral language that shapes the behavior and evolution of their bacterial hosts.

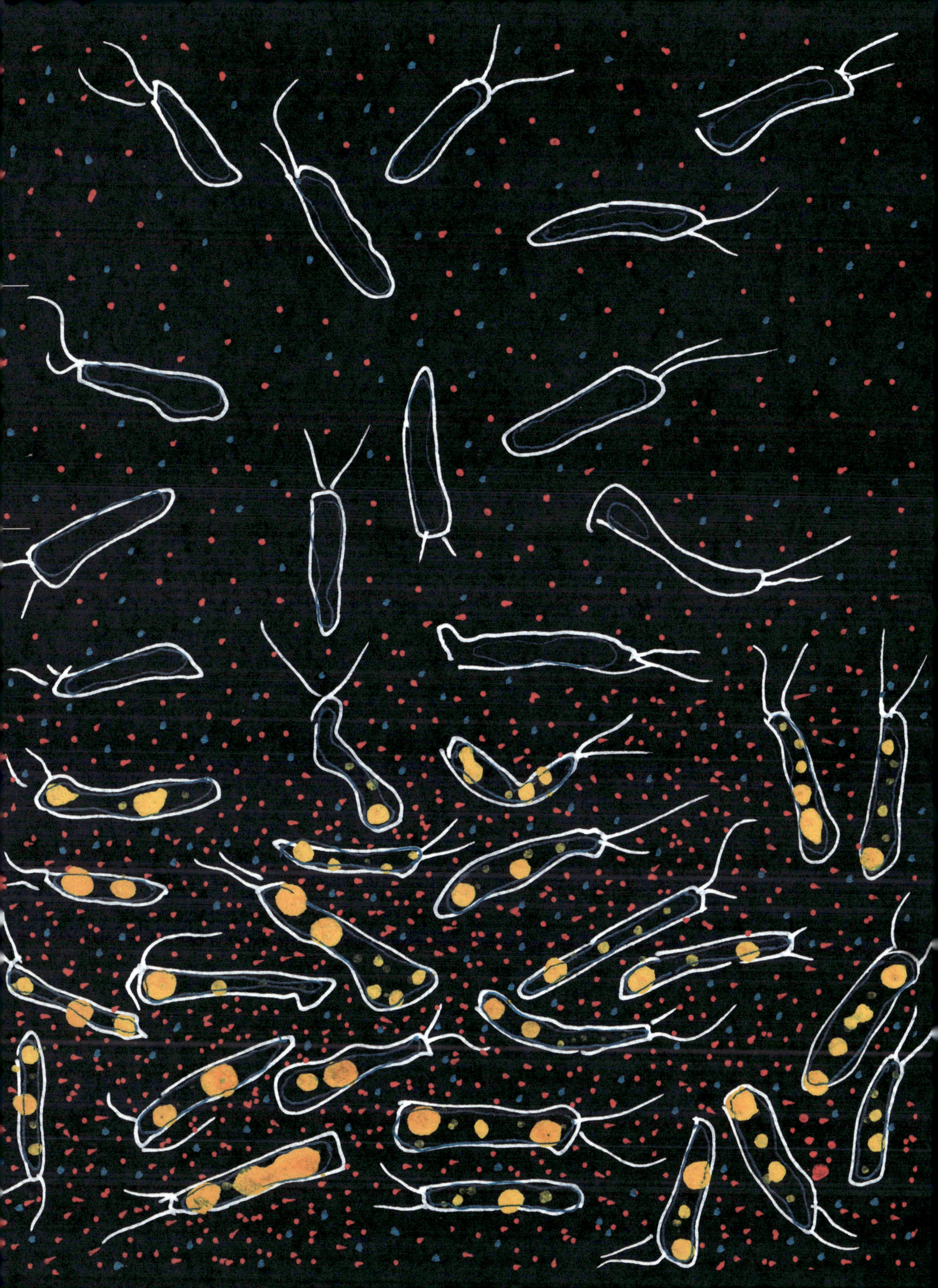

From gene transfer to phage chatter

Gene exchange is a natural mode of communication among unicellular organisms. This exchange is especially intense among prokaryotes—bacteria and archaea—through a process known as *horizontal gene transfer*. Japanese researchers first observed this phenomenon in the late 1950s, when different bacterial species were seen to transfer genes conferring antibiotic resistance. Bacteria not only exchange genes with one another, but

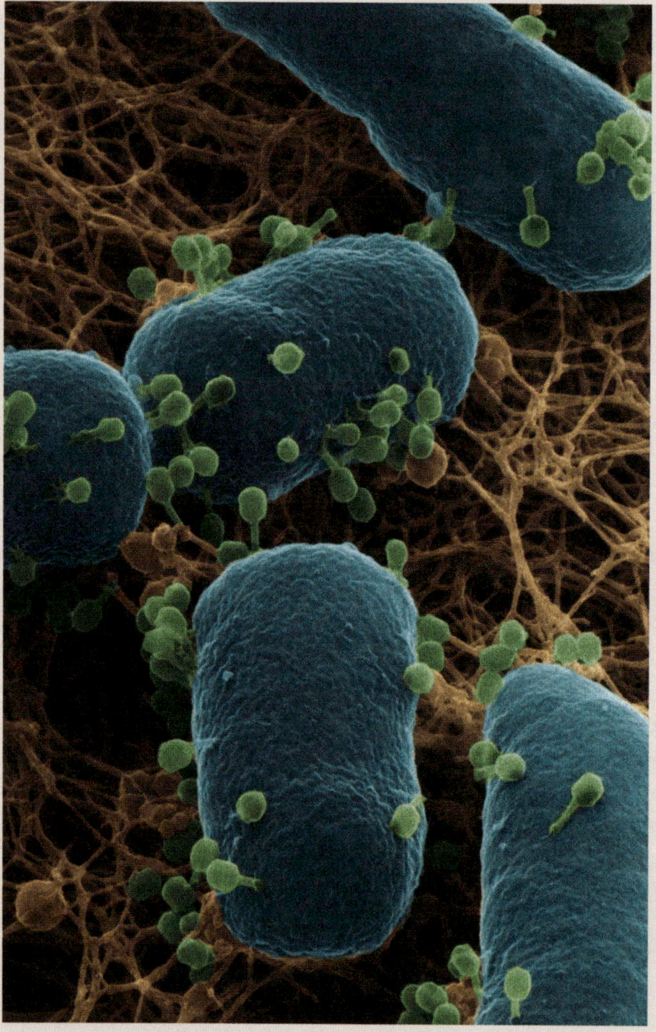

PHAGES INJECT THEIR GENETIC MATERIAL INTO BACTERIA

T4 phages (*green*) are anchored to the surface of bacteria (*blue*). The phage heads contain viral DNA which can be injected into the bacteria through the phage tails, appendages attached to the bacterial surface. Here, phages with shortened tails are in the process of injecting their DNA into the bacteria.

COLORIZED HELIUM ION SCANNING ELECTRON MICROSCOPY. PUBLISHED IN *ADVANCED BIOSYSTEMS*, VOL. 1, ISSUE 8, 03/07/2017

have shared genetic material with archaea since ancient times. In fact, up to 20 percent of the genes found in thermophilic bacteria—which thrive in hot springs—appear to have originated from archaea. Horizontal gene transfer also occurs between bacteria and archaea in our gut, facilitating the spread of antibiotic-resistant genes and enabling adaptation to the nutrient-rich intestinal environment.

How do these genetic exchanges occur? Some take place through transformation, where bacteria absorb free DNA from their surroundings. Others occur via conjugation, a more intimate process where bacteria physically connect and exchange DNA fragments through surface structures like pili and nanotubes *(see pp. 76 and 205)*. But horizontal gene transfer isn't limited to direct exchanges between bacteria—it also involves viruses. In transduction, phages—viruses that infect bacteria—pick up genes from one bacterial host and deliver them to another. The bacterium *Vibrio cholerae*, which causes cholera, offers a striking example. This bacterium becomes pathogenic only after being infected by the CTX phage, which inserts its own genes, including one that encodes the cholera toxin, into the bacterial genome. This toxin causes massive water loss into the intestine, and thus the sudden, severe diarrhea typical of cholera.

Phages even chat among themselves, coordinating their strategy based on the condition of their bacterial hosts. When resources are abundant, phages multiply rapidly within bacteria, causing them to burst—a violent process known as *phage lysis*. When resources are scarce, phages choose patience, integrating into the host genome via "lysogeny" and waiting for better conditions. Phage communication is governed by a simple yet effective system based on arbitrium peptides—short signaling molecules composed of a few amino acids. Phages induce their bacterial hosts to produce arbitrium, which accumulates in the environment. A low concentration of arbitrium signals that it is wiser to wait (and pursue lysogeny), while a high concentration indicates that conditions are favorable for an aggressive takeover (lysis). This minimalist yet remarkably ancient peptide-based language is finely tuned to aid the phages' survival. In addition, some phages can even eavesdrop on bacteria's quorum-sensing signals to inform their own lifecycle decisions *(see pp. 197 and 199)*.

THE HAWAIIAN SQUID
EUPRYMNA SCOLOPES

This small squid glows thanks to a light organ in its belly, made up of pockets filled with the luminescent bacteria *Aliivibrio fischeri*. In this juvenile squid, we can see the dark ink sacs that surround the light organ at the center of its transparent mantle. The structure of the luminous organ resembles that of an eye, with a lens and a diaphragm that control the light's intensity and focus.

When squid and bacteria cooperate

Every evening in the shallow waters of Hawaii, *Euprymna scolopes*, a squid no larger than a golf ball, hunts shrimp under the silver light of the moon. Its small size would make the squid an easy target for lurking predators, if not for an ingenious defense: It hides its own shadow by glowing. The light is not its own, but is produced by the luminous bacteria *Aliivibrio fischeri*, housed in a specialized light organ on the squid's belly. In exchange for this natural camouflage, the squid offers the bacteria a safe haven and nourishment, recruiting them from among the surrounding plankton, where they live freely. Each dawn, as sunlight returns, the squid expels the bacteria and buries itself in the sand—only to repeat the ritual again in the evening.

Biologists in Hawaii, curious about this phenomenon, realized that when these symbiotic bacteria are well fed by the squid, they multiply during the day, gradually accumulating in the light organ. By the evening, when they've reached a very high concentration, the bacteria emit light. It's as if the bacteria are aware of the number of other bacteria around them. The bacteria coordinate their glow through a remarkable form of communication known as *quorum sensing*. They release small messenger molecules called AHLs (acyl-homoserine lactones), which accumulate in the environment. Once they reach a critical concentration, these molecules bind to receptor proteins in the bacteria, activating genes that stimulate the production of luciferase—an enzyme protein that transforms chemical energy (ATP) into light *(see p. 55)*. A similar mechanism underlies the bioluminescence of fireflies and some deep-sea creatures.

This revelation from one little squid turned out to be a universal strategy. Many bacterial species use quorum sensing to measure their own density and synchronize their actions. During infections, for example, bacteria wait until they have reached sufficient numbers to launch an attack—secreting virulence factors and toxins in coordination as if they were one, multicellular entity. Understanding the molecular conversations of bacteria opens vast possibilities for controlling their behavior in agriculture, industry, and medicine.

But bacterial communication doesn't end with other bacteria. It also influences the organs and behaviors of their animal hosts. In the tiny Hawaiian squid, the light-producing bacteria not only aid its camouflage but also regulate its eye development and adaptations to day and night—its circadian rhythms. These discoveries also echo within us, as recent research reveals that the composition of our own gut microbiota plays a profound role in our health, metabolism, and even our behavior.

MESSAGES AND COMMUNICATION CHANNELS

THE MULTIPLE COMMUNICATION PATHWAYS BETWEEN CELLS

Animal and plant cells engage in long-distance communication through a rich array of secreted molecules, including hormones, growth factors, cytokines, and many others. In plants, hormones such as auxins orchestrate growth and development, while in animals, a handful of steroid hormones regulate metabolism and reproduction.

Among the most specialized animal cells are neurons, which communicate via messenger molecules—neurotransmitters like serotonin and dopamine—released at synaptic junctions, where two nerve cells meet. Whether hormone or neurotransmitter, each messenger molecule possesses distinct physical properties—size, hydrophilia, or hydrophobia—that influence its ability to cross membranes and enter cells *(see p. 48)*. When a molecule is too large or hydrophilic to traverse the membrane itself, it is recognized by a receptor protein on the cell surface, which relays its message within *(see p. 142)*.

To achieve long-distance signaling, cells have evolved sophisticated systems to package and dispatch their messages. Cells send vesicles—tiny membrane-bound sacs known as *exosomes* or *ectosomes*—laden with metabolites, proteins, and RNA—to convey regulatory or genetic information to other, distant cells. Inside the target cell, messenger molecules are ferried to their destinations by a network of carrier proteins.

Beyond secreted messenger molecules, cells also communicate by direct, physical contact. They exchange organelles (such as mitochondria), viral particles, condensates, and a variety of molecules and ions through fine, tubular membrane extensions known as *nanotubes*. Certain bacteria also employ these nanotubes to trade genes and molecular components in a form of microbial barter.

In parallel with these molecular exchanges, cells also exploit ultrafast, ionic communication pathways, governed by their membranes' selective permeability to ions. This form of electrical signaling is particularly crucial to coordinating the activities of tissues and developing embryos. Through junctions known as *gap* or *communicating junctions*—protein-based channels that open and close—small molecules and ions, especially calcium, pass directly from one cell to another. This mode of communication synchronizes the rhythmic contractions of cardiac muscle, the metabolic activities of epithelial cells, and the collective behavior of tissue cells. All told, cells deploy a remarkable diversity of communication strategies—chemical, electrical, and physical—to harmonize their actions and sustain life.

Some cells communicate via tunnels called **TNTs** (tunneling nanotubes), through which they can transfer cellular components, including organelles like mitochondria ●.

Another type of communication takes place via two kinds of extracellular vesicles: **exosomes,** which come from organelles called **MVBs** (multivesicular bodies) and **ectosomes,** which arise by budding from the plasma membrane.

Many cells communicate directly through openings in their membranes known as *gap* or *communicating junctions* (**CJ**), and/or by secreting messenger molecules such as hormones.

LYSOSOME
RETICULUM
RETICULUM
GOLGI
GOLGI
MVB
RNA
RNA
RNA
RNA
RNA
EXOSOMES
ECTOSOMES
TNT
TNT
mi RNA
mi RNA
mi RNA
RNA
EXOSOMES
MVB
MVB
GOLGI
GOLGI
GOLGI
RETICULUM
LYSOSOME
LYSOSOME
RETICULUM
RETICULUM
CJ CJ
CJ CJ
CJ CJ
CJ CJ
CJ CJ
HH
HH
HH
HH
HH

Messenger hormones

In humans, as in all multicellular organisms, cells communicate across long distances primarily using hormones. These vital messengers govern the development and functions of tissues and organs, as well as their responses to environmental cues.

The German physician Arnold Berthold intuited the existence of hormones in animals as early as the 19th century. In 1848, Berthold castrated the roosters in his barnyard and observed that they lost the vivid, red color of their crests and their instinct to chase hens. Through a simple yet brilliant experiment—transplanting a testicle into the abdomen of a castrated rooster—Berthold demonstrated that some active substance, originating from the testicle, restored the rooster's sexual vigor and coloration. A century later, researchers identified this substance—testosterone, a small molecule—after painstakingly extracting it from nearly thirty kilograms of ox testicles. Steroid hormones like testosterone were subsequently discovered to regulate metabolism and sexuality not only in mammals, but also in fish and insects.

Darwin was the first to predict the existence of a hormone-like messenger in plants. Shortly before his death, he demonstrated the existence of a factor enabling plants to orient themselves in relation to light—an ability called *phototropism*. In 1880, in their book *The Power of Movement in Plants*, Darwin and his son Francis explained how they prevented young grass shoots from turning toward a light source by cutting off the end of the blade, or putting a small cap on it. They deduced that an "influence" produced at the tip of the young shoot, when exposed to light, must propagate to its base, inducing the blade of grass to bend toward the light. These early manipulations led to the discovery, in the 1920s, of the hormone responsible for phototropism, which also regulates plant development and growth. Its name, *auxin*, is derived from the Greek word for "growth."

Testosterone and auxin are smaller hormones, hydrophobic molecules that pass easily through cellular membranes to act directly within the cell nucleus (*see chap. VIII*). Once inside the nucleus, they bind to specific receptor proteins that detect their presence and set off a cascade of events: Some genes and enzyme proteins are activated, while others are repressed—fine-tuning the cell's metabolic state. By contrast, hormones such as insulin or growth hormone are larger, hydrophilic molecules—peptides and small proteins—which cannot freely cross the cell's lipid-rich membrane. These larger hormones are recognized by receptor proteins embedded in the membrane (*see pp. 48 and 54*), which trigger signaling pathways inside the cell, conveying the hormone's message to its target genes and enzyme proteins. Cells use these intricate mechanisms of communication to coordinate their behavior, weaving a vast social network that sustains the function of tissues and organs.

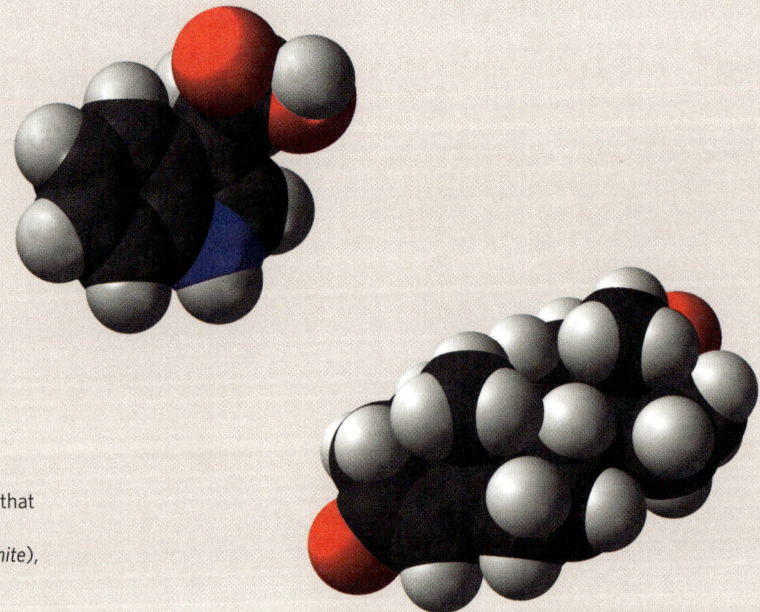

**AUXINS AND STEROIDS—
PLANT AND ANIMAL HORMONES**

Top: This auxin (indole-3-acetic acid) is a small, cyclic molecule composed of ten carbon atoms (*black*), eight hydrogens (*white*), two oxygens (*red*), and one nitrogen (*blue*). The auxin family of molecules controls plant phototropism, development, growth, and flowering.

Bottom: Steroids are also small, hydrophobic molecules that cross lipid membranes. Here, a testosterone molecule is composed of 19 carbon atoms (*black*), 28 hydrogens (*white*), and two oxygens (*red*).

THE MERISTEM OF THALE CRESS—A MODEL PLANT

Cell walls (*grey*) delimit cells in the meristem, at the tip of a stalk of thale cress, *Arabidopsis thaliana*. Here, we visualize the expression of two genes, which fluoresce in green and red, that define the future male and female regions of the plant. The plant hormone auxin affects the fate of these developing cells.

CONFOCAL FLUORESCENCE MICROSCOPY OF LIVE EMBRYOS

© Courtesy Nathanaël Prunet, Caltech, Pasadena, and UNC Chapel Hill

View "The Discovery of Auxin"
(Quick Biochemistry Basics)

Communicating junctions of the electrical heart

Our heart beats over 100,000 times each day, tirelessly pumping blood through the body thanks to 2 to 3 billion contractile cells—cardiomyocytes—whose countless cytoskeletal microfilaments contract in synchronized rhythm.

Among them, small clusters of specialized cardiomyocytes act as pacemakers, setting the rhythm of the cardiac contractions. Fifty to 100 times per minute, these self-exciting cells spontaneously generate an electrical signal, known as *action potential*, which is rapidly transmitted to neighboring cells. The electrical impulse propagates from one cardiomyocyte to another, triggering a coordinated wave of calcium ions to release within each cell. Activated by this surge of calcium, the microfilaments contract, producing the powerful, synchronized heartbeat. This extraordinarily fast and cohesive process is made possible by specialized structures called *communicating junctions* or *gap junctions*, which provide direct electrical couplings between cardiomyocytes. Discovered around sixty years ago, gap junctions are made of channel-forming proteins known as *connexins*, organized into rosette-like assemblies where neighboring cell membranes make contact. The connexin proteins span the adjacent membranes, forming gates that can open and close to regulate intercellular communication. When these gates are open, they permit ions, small molecules, and metabolites (such as ATP) to pass freely between cells. The gap junctions' opening and closing is finely controlled by kinase proteins, which add phosphate groups to the connexin proteins; and by

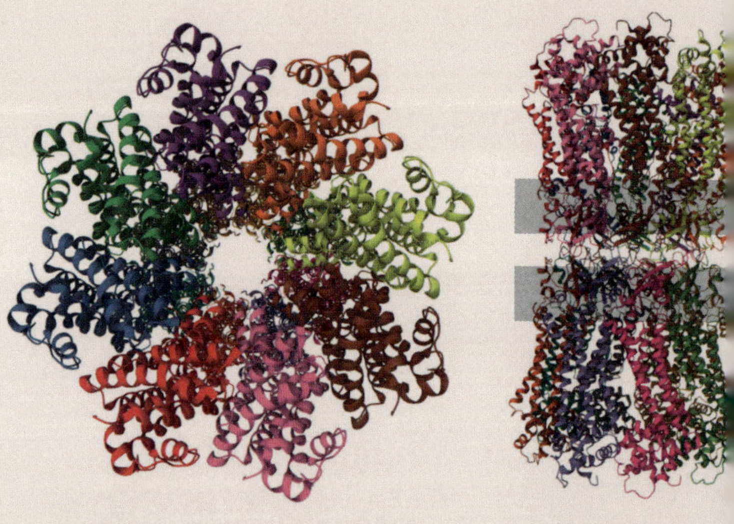

**COMMUNICATING/GAP JUNCTIONS
FORMED BY CONNEXIN PROTEINS**
Communicating/gap junctions connect adjacent cells via channels that open and close. A channel between adjacent membranes is established by a rosette-like arrangement (*front view, left*) of proteins called *connexins*, which face each other in the adjoined membranes of neighboring cells. The amino acid chains of the connexin proteins form helices, spanning the lipid bilayer of the membranes (*side view, right*). The open channel runs through both membranes of the two communicating cells (*grey horizontal bars*).

PROTEIN DATA BANK, 5H1R *C. ELEGANS* INX-6 GAP JUNCTION CHANNEL

phosphatase proteins, which remove them. The channels adjust their permeability in response to intracellular conditions, including concentrations of metabolites, protons (H^+), and calcium ions (Ca^{2+}). Notably, a sharp rise in calcium concentration within a cell prompts its gap junctions to close, shielding neighboring cells from calcium overload. This mechanism offers vital protection against the spread of harmful signals, such as those arising from apoptosis or viral infection.

Communicating/gap junctions are the most direct and rapid mode of cell communication, not only in the heart, but also in the epithelia, other tissues, and animal embryos. Communicating junctions between cells enable their metabolism to be coupled. Thus, when one cell receives a hormone signal and responds by producing a small messenger molecule in its cytoplasm, this molecule will pass through the communicating junctions, enabling neighboring cells to benefit from the hormonal signal.

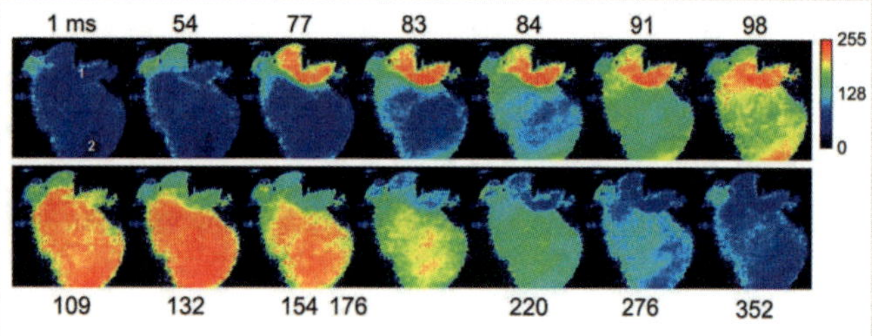

CALCIUM WAVES AND HEART CONTRACTION
This visualization captures a calcium wave propagating through a mouse embryo heart over about 350 milliseconds, causing it to contract. *Red* indicates the highest cellular concentrations of calcium, and *blue* indicates the lowest.

CONFOCAL FLUORESCENCE MICROSCOPY
© Courtesy Y. Tallini and M. I. Kotlikoff, Cornell University, New York

Tunnels, bubbles, and cellular kisses

Within our organs, tightly nestled-together cells communicate directly with their neighbors using communicating/gap junctions. But more spread-out cells can also communicate—even without membrane-to-membrane contact—by sending messages through tiny tunnels aptly dubbed TNTs (tunneling nanotubes). Recently discovered, these delicate conduits—mere tens to hundreds of nanometers in diameter—allow the transfer of ions, molecules, organelles, and even viruses between neighboring cells. For now, the rules governing these exchanges between animal cells remain largely mysterious.

Even plant cells, encased as they are in rigid cellulose walls, maintain direct communication. They achieve this through plasmodesmata—microscopic channels ten to twenty times wider than the communicating/gap junctions of animal cells *(see p. 106)*. Through these plasmodesmata, the cells' cytoplasm forms an almost continuous network, enabling proteins, nucleic acids, organelles, and viral particles to flow freely from one cell to another throughout the plant. Fungi, by contrast, are effectively networks of filament-like structures called *hyphae* which, all together, form the *mycelium*. Fungal cells share their cytoplasmic contents and organelles through septa—large passages, normally open, that only close in response to damage, preventing the loss of cellular material. Some soil-dwelling fungi even infiltrate plant tissues, exchanging minerals for sugars via junctions that, although not yet fully understood, hint at ancient symbiotic relationships *(see p. 212)*.

Bacteria, too, are masters of cellular exchange. They transfer macromolecules and genetic material—such as small fragments of DNA called *plasmids*—through tunnels with a similar, tubular structure to those at the base of their pili. Certain pathogenic bacteria in the intestine go even farther, using nanotubes like siphons to extract the contents of host cells, while simultaneously injecting molecules to shield themselves from the immune system's defenses. These exchanges between bacterial species, and between bacteria and eukaryotic cells, have evolved as a strategy for adaptation and survival.

In recent years, researchers have also uncovered cells' remarkable capacity to communicate via tiny, membrane-bound bubbles. These vesicles, known as *ectosomes*, bud directly from the plasma membrane, while others, called *exosomes*, form within intracellular compartments known as MVBs (multivesicular bodies).

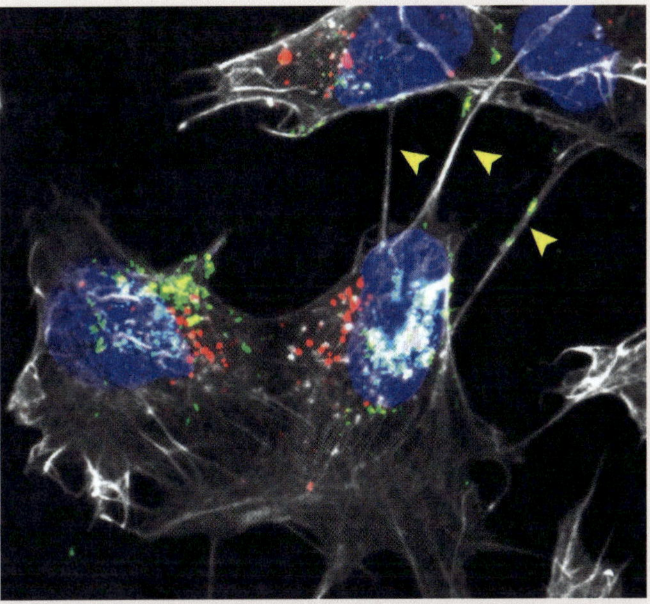

NANOTUBES BETWEEN HUMAN NERVE CELLS
Neurons communicate through thin tunnels called TNTs (tunneling nanotubes). Clusters of alpha-synuclein proteins (*red and green*) can be seen inside the cells, and passing from one cell to the next through the nanotubes. The nuclei (*blue*) are tagged with a fluorescent molecule.
CONFOCAL FLUORESCENCE MICROSCOPY
© Courtesy Anna Pepe and Chiara Zurzolo, Pasteur Institute, France

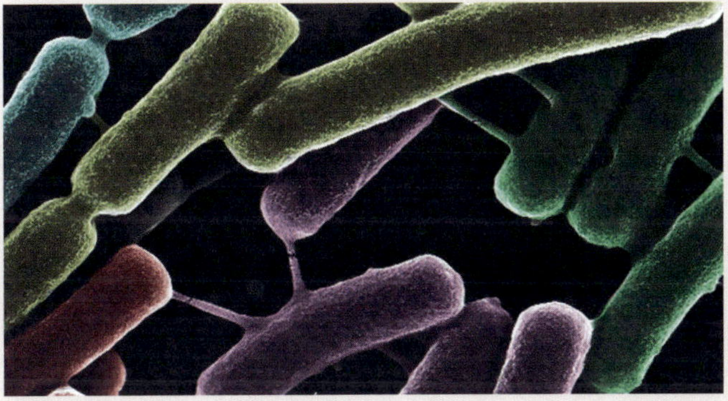

NANOTUBES BETWEEN BACTERIA
Bacteria of the *Bacillus subtilis* species communicate via nanotubes. These nanotubes act as conduits for the passage of metabolites, macromolecules, and small DNA fragments (plasmids) between cells.
COLORIZED SCANNING ELECTRON MICROSCOPY
© Courtesy Sigal Ben-Yehuda and Ilan Rosenshine, Hebrew University/Hadassah Medical Center, Israel

Exosomes are loaded with proteins and RNA destined for target cells. Some exosomes transfer short RNA strands known as *microRNA* (miRNA). As they carry genetic instructions, exosomes represent a novel and powerful means of cell-to-cell communication. It is becoming possible to manufacture synthetic exosomes to manipulate cells for therapeutic and industrial applications.

WE ARE ALL HOLOBIONTS WITH MICROBIOTA

Though the term *holobiont* may be unfamiliar, it describes a profound biological reality: the coexistence of an organism and its multiple partners, which are intricately involved with its formation and functions. Our human body, for instance, is composed of some 40 trillion human cells and an almost equal number of nonhuman cells, amounting to around 2 kg of mass. This vast collection of foreign, microscopic cells—our microbiota—primarily includes thousands of species of bacteria, archaea, and protists, along with their cohorts of viruses, as well as microscopic fungi and even tiny animals. Together, they represent a genetic reservoir of astonishing richness, harboring over 30 million genes—around 1,400 times more than our 22,000 human genes. Microbial communities inhabit every niche of our body: the digestive system, respiratory tract, skin, and orifices. Some species penetrate and live within our cells; some even lodge themselves within our organelles.

To describe these complex living entities—chimeras of interacting organisms—biologists initially turned to the term *super-organisms*. Yet this term, traditionally reserved for colonies of social insects like ants, bees, and termites, failed to capture the biological intimacy of these associations. Instead, they embraced the term *holobiont*, introduced by Lynn Margulis in the 1990s. *Holobiont* refers to the primary host—whether an animal, plant, fungus, or protist—closely entwined with a community of diverse microscopic species (the bionts) living both on and within the holobiont.

Coral is a quintessential example of a holobiont. Coral polyps harbor symbiotic microalgae known as *zooxanthellae*, a kind of phytoplankton (specifically, a photosynthetic dinoflagellate). These microalgae supply the coral with oxygen, glucose, and amino acids. The coral, in return, shelters them from predation. The coral holobiont is richer still, encompassing not only zooxanthellae but also other protist species, as well bacteria, archaea, and even animal species uniquely associated with specific corals.

Elsewhere in the natural world, many animal protists are, themselves, holobionts, in partnership with photosynthetic protists or cyanobacteria. Plants, too—from their roots to their leaves—nurture intricate relationships with a multitude of microbes, fungi, and specialized microfauna. And, of course, every holobiont has its parasites as well, with which it continuously co-evolves.

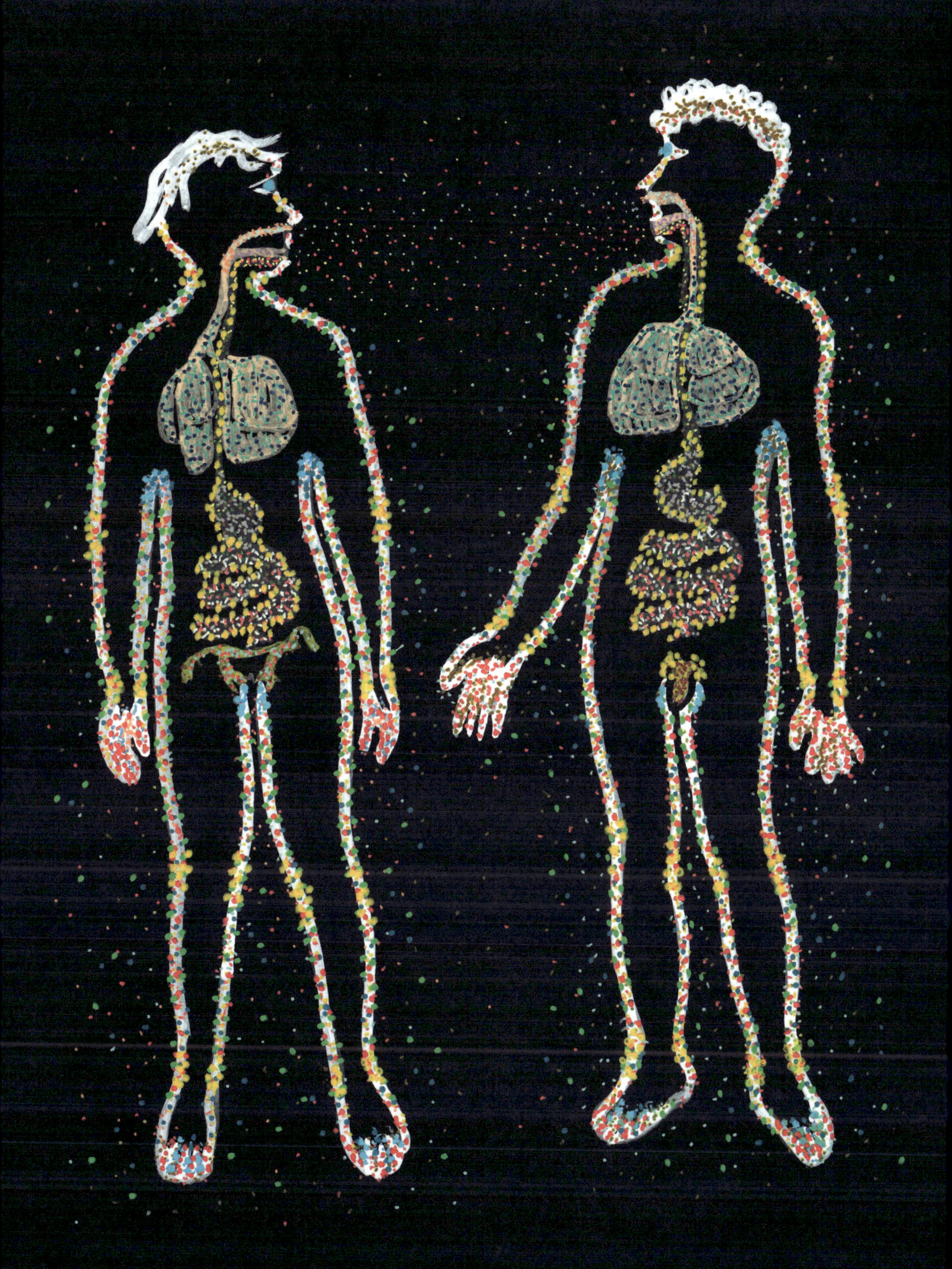

The human microbiota— from birth to death

During natural birth, as a baby passes through the vaginal canal and comes into contact with the perineal and anal regions, it acquires a rich collection of maternal microbes. This transfer of the mother's microorganisms and viruses forms the newborn's first microbiota. By contrast, babies delivered by Caesarean section inherit a microbiota composed primarily of skin bacteria from the mother and healthcare providers. To mimic natural conditions, some birth attendants now swab these newborns with the mother's vaginal secretions. Subsequent influences—such as breastfeeding, close contact with family members, and exposure to household pets—continue to shape and diversify the infant's microbiota.

Although the initial difference between babies born vaginally and by C-section tends to diminish by the end of the first year, studies suggest that C-section babies remain more vulnerable to inflammatory diseases, allergies, and obesity. This may be due to the microbiota's crucial role in stimulating the immune system through bacterial secretions that regulate its development. Beyond immunity, the microbiota appears to affect behavior as well. A Finnish study found that infants described as particularly cheerful harbored a gut microbiota rich in *Bifidobacterium*, a genus of beneficial bacteria. One explanation is that certain bacterial species produce metabolites—such as butyrate—that support tissue function and influence both immune responses and mental well-being.

Over time, the microbiota acquired at birth—whether vaginal or cutaneous—tends to fade. Each person gradually develops their own, distinctive microbial profile, shaped by heredity, diet, hygiene, illnesses, and exposure to medications such as antibiotics. This unique microbial signature, like a fingerprint, becomes an integral part of one's biological identity. Even short-term travel can temporarily alter the composition of an individual's intestinal flora. In fact, less than a year after immigrating to the United States, individuals from South Asia, Africa, or South America often acquire a less diverse microbiota, closely resembling that of established Americans with European descent—a shift with potentially lasting implications.

Throughout life, the microbiota evolves in a relatively predictable manner. So much so, in fact, that we can estimate a person's age by analyzing the composition of their gut bacteria. Microbiota diversity typically declines with age, which is believed to contribute to the rise in inflammatory and autoimmune diseases, as well as mood disorders. This has led to the idea that fecal transplants from young donors might offer therapeutic benefits to older individuals. Indeed, studies in mice have shown that such transplants can enhance immune function in elderly recipients. In humans, fecal microbiota transplants have already proven effective in treating stubborn intestinal infections caused by antibiotic-resistant *Clostridium difficile* bacteria. Today, both public institutions and private companies are investing heavily in research that explores the microbiota's potential for a wide range of therapeutic applications.

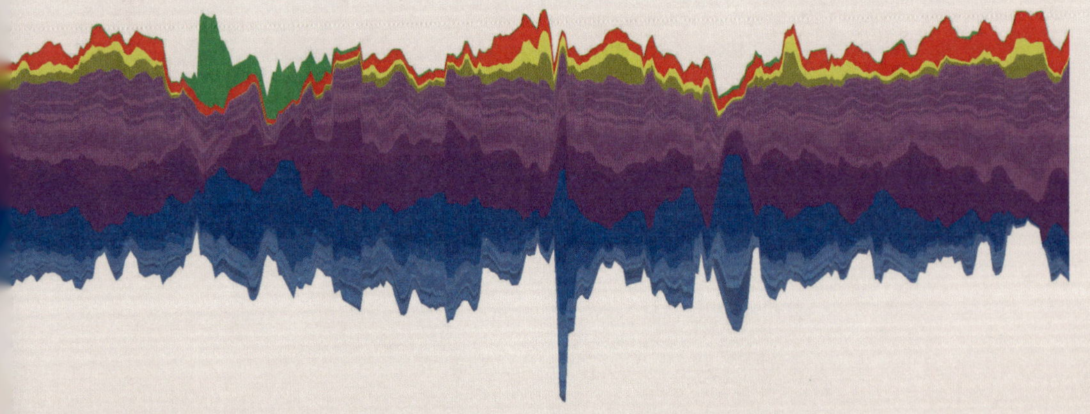

A YEAR IN THE LIFE OF A HUMAN MICROBIOTA

This graph illustrates the evolution of the main classes of bacteria in a person's intestinal microbiota, revealed by sampling and analyzing his stools for 365 days (*from left to right*). The relative stability of his microbiota was temporarily altered by a trip to Asia (around day 100), with the spectacular appearance of proteobacteria (*green*) and the virtual disappearance of Mycoplasmatota bacteria (*red*).

© Courtesy Lawrence David, Duke University, North Carolina

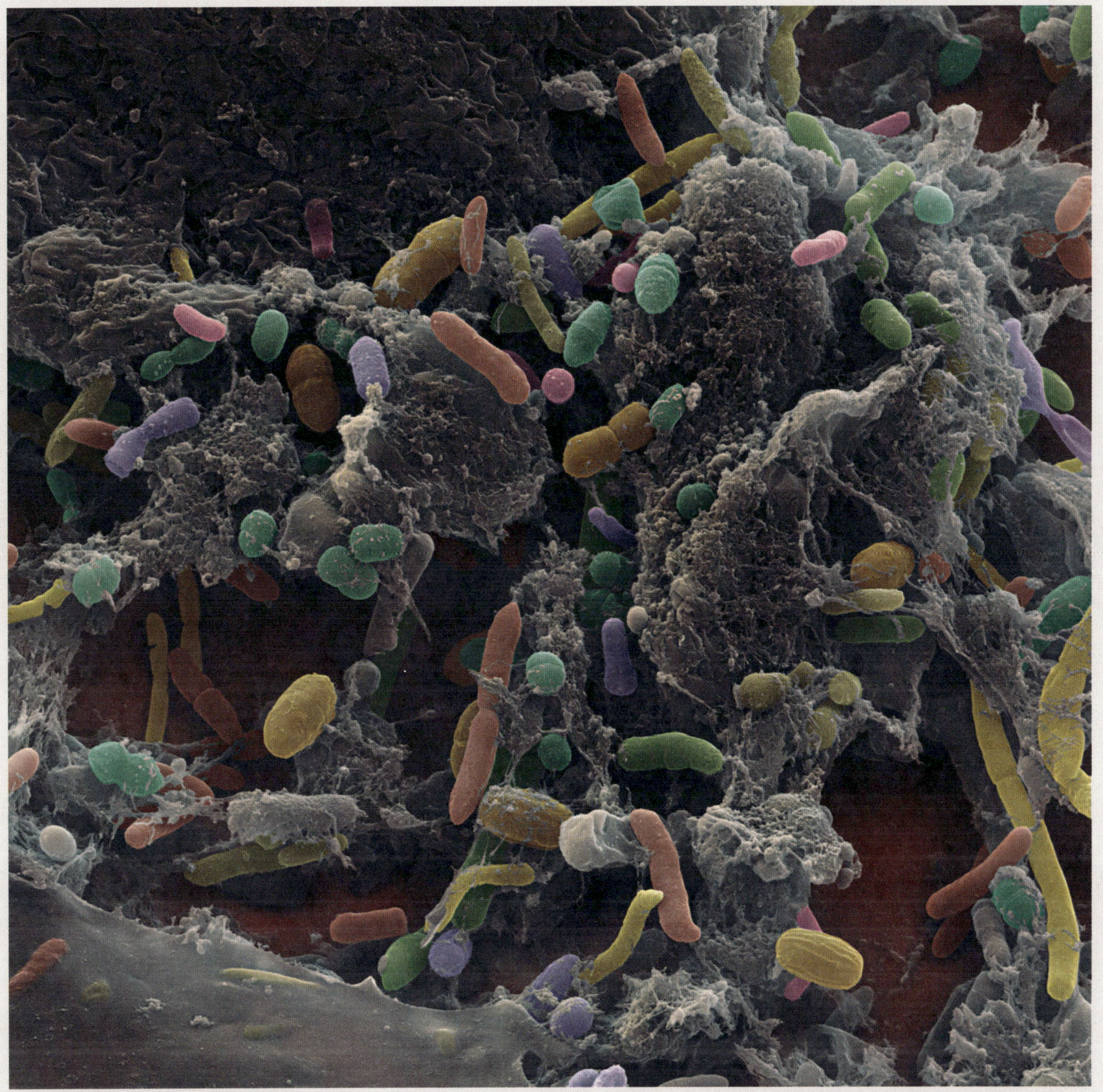

THE MICROBIOTA OF POOP

The stool microbiota is made up of thousands of species of microorganisms—bacteria, archaea, protists, and their viruses. These microorganisms digest food and secrete vitamins, among many other molecules. Microbes are expelled in the stool, along with the waste products of digestion. This microscopic image of a stool sample illustrates the diversity of these microorganisms.

COLORIZED SCANNING ELECTRON MICROSCOPY
© Courtesy Nicole Ottawa and Oliver Meckes, *eye of science*, Germany

ESSENTIAL SYMBIOSES

ALL HOLOBIONTS IN SYMBIOSIS

As human holobionts, we live in intimate symbiosis with vast communities of intestinal microbes. These microscopic partners help digest our food, synthesize essential vitamins and metabolites, and form a protective barrier against invading pathogens. Far from being mere passengers, these microbes—collectively known as our *microbiota*—play a fundamental role in shaping our health, immunity, and even behavior *(see p. 208)*.

An individual's microbiota is not just a random assemblage of microbes picked up from the environment. It's the product of multiple influences: how they were born, their genetic heritage, diet, and social and cultural practices. Over the course of life, this complex microbial ecosystem evolves, its composition shifting in response to age and experiences. Notably, disruptions or imbalances in the gut microbiota have been linked to a variety of chronic, age-related conditions, including obesity, cardiovascular disease, and neurodegenerative disorders.

Among nature's most striking examples of cooperative living are lichens. In fact, the term *symbiosis* was first coined in the 19th century to describe the extraordinary alliance between fungi and photosynthetic plant cells within lichens. These composite organisms, formed of interwoven fungal filaments and algal or cyanobacterial cells, thrive in some of the planet's most inhospitable environments—from bare rocks to icy mountain peaks and arid deserts. Not limited to just two partners, lichens also include consortia of bacteria and yeasts. Protists also engage in symbiosis, although in some cases the relationship teeters on the edge of exploitation, raising the question of whether it more closely resembles coercion or even "slavery."

Plants, too, harbor complex microbiota. Legumes, for example, have mastered the art of living in partnership with nitrogen-fixing bacteria in the soil—a symbiosis critical to their own survival and to soil fertility. In the animal kingdom, symbiotic relationships date back millions of years. Jellyfish, one of Earth's oldest creatures, have enlisted bacterial partners to perform photosynthesis, and thus provide both energy and nutrients. Even in the darkest depths of the ocean, entire ecosystems flourish through obligatory symbioses between invertebrate animals and chemosynthetic bacteria—organisms that draw energy not from the sun, but from chemical reactions involving metals and gases vented from Earth's core *(see p. 69)*.

The dynamic interplay—both cooperative and competitive—between the various bionts within a holobiont enables these composite beings to inhabit a wide range of ecological niches. Holobionts adapt by adjusting not only the types of organisms they host—including parasites—but also their relative abundance. This collective organization represents one of evolution's most powerful strategies for environmental adaptation and survival.

All protists, animals, plants, and fungi are holobionts, made up of different partner organisms (bionts) and, in particular, countless types and species of microorganisms that make up the microbiota.

Many bionts maintain complex relationships with each other and with the host: symbiotic, mutualist, or parasitic.

When plants, bacteria, and fungi cooperate

Beans, peas, soybeans, clover, and other legumes have nourished humanity and enriched soils since the dawn of agriculture. Long before the invention of synthetic fertilizers, these plants were celebrated as "green manure"—natural allies in maintaining soil fertility. For plants to thrive—particularly, to synthesize proteins and nucleic acids—they require nitrogen in a usable form, typically ammonia. Remarkably, for over 60 million years, *Rhizobium* bacteria in the soil have fulfilled this need by converting atmospheric nitrogen into ammonia compounds. To spark this ancient collaboration, legumes release signaling molecules from their roots into the surrounding soil. These attract *Rhizobium*—literally "root dwellers" in Greek—which respond by producing factors that prompt the roots' cortex cells to form nodules. These small, nublike growths serve as entry points that guide the bacteria from the root hairs into the root's deeper tissues. Once inside, the bacteria surround themselves with membranes to form organelle-like structures called *symbiosomes*. Once inside the symbiosomes, the bacteria can differentiate into "bacteroids" capable of fixing gaseous nitrogen (N_2) into ammonia (NH_3). Finally, this valuable nutrient is transported via the plant's sap to nourish its tissues.

The energy required for this transformation is provided by the plant itself. Through photosynthesis, the plant generates sugars and carbon compounds, which feed the bacteroids nestled within the nodules. Although the plant and bacterium are deeply integrated, they exhibit "facultative" symbiosis—meaning that, although they *can* live independently, they thrive together.

This kind of symbiosis has profound ecological significance. The rhizosphere—the shallow zone of soil surrounding plant roots—is a hub of biochemical dialogue between not only plants and bacteria but also fungi, protists, and soil-dwelling animals. The relationships between plants and fungi stand out for their complexity and importance. Roughly 90 percent of terrestrial plants acquire vital minerals from the soil via mycorrhizal symbioses—beneficial partnerships with fungi. A vast network of fungal filaments—the hyphae—either infiltrates the plant's root cells (as an endomycorrhiza) or externally envelops the roots (as an ectomycorrhiza) *(see p. 105)*. For every meter of root, over a kilometer of fungal hyphae weave through the soil, scavenging for dilute, mineral nutrients and delivering them to the plant. The plant repays the fungi with carbon-rich molecules—an offering that may represent anywhere from 10 to 50 percent of the plant's total photosynthetic output. This

ENDOMYCORRHIZA, SYMBIOSIS OF A PLANT AND A FUNGUS

In this cross-section of a grass root, we can see the cellulose cell walls, and the filamentous fungal hyphae (*grey*) that have infiltrated the root. The fungus provides minerals, drawn from the soil, and receives sugars and carbon molecules, produced by photosynthesis, from the plant.

COLORIZED SCANNING ELECTRON MICROSCOPY
© Courtesy Nicole Ottawa and Oliver Meckes, *eye of science*, Germany

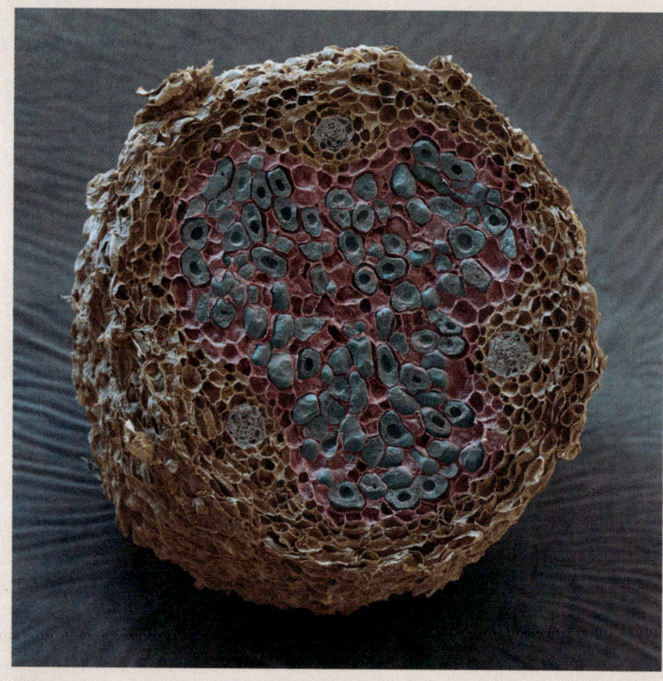

SYMBIOSIS BETWEEN LEGUMES AND BACTERIA

In this cross-section of a root nodule from a leguminous plant (clover), we can see *Rhizobium* bacteria (*blue*). The bacteria, attracted from the surrounding soil, have entered the root via its hairs. The root cells have divided and reorganized, creating a nodule to house the nitrogen-fixing bacteria. The pink color comes from hemoglobin, required for bacterial respiration.

COLORIZED SCANNING ELECTRON MICROSCOPY
© Courtesy Nicole Ottawa and Oliver Meckes, *eye of science*, Germany

ancient alliance between plants and fungi has profoundly influenced not only the structure of root systems, but also the development of soils and ecosystems across the globe.

The lichen holobiont—champion of symbioses

Lichens captivate with their vibrant colors, intricate forms, and extraordinary resilience. These slow-growing organisms colonize rocks, trees, even scorching deserts and the frozen expanses of the poles—and climb as high as 7,000 meters on the slopes of the Himalayas. Their remarkable hardiness stems from their hybrid nature: They are living amalgams of fungal and photosynthetic cells. Indeed, lichens inspired the term *symbiosis*, coined in 1875 from the Greek for "living together."

Until recently, lichens were considered to be a classic type of holobiont—with two principal allies in partnership: a fungal species (the mycobiont) and a photosynthetic one (the photobiont)—either a green microalga or a cyanobacterium.

Nearly 20,000 species of lichen have been identified, named after their fungal hosts (some 2,000 species), whose hyphal networks cradle and sustain photobionts from among 200 to 300 different species of cyanobacteria and microalgae. These photobionts provide the fungi with vital carbon compounds; cyanobacteria also contribute precious nitrogen. In exchange, the fungi draw minerals from the rock or bark, and shield the photobionts within a protective mesh of filaments, safeguarding them from predators and ultraviolet radiation. Ever adaptable, lichens dry out or soak up water in response to the surrounding humidity, thanks to layers of polysaccharides secreted by the fungal partner. When desiccated, their survival hinges on the microalgae, which produce polyalcohols (glycols) that bind to the polysaccharides, maintaining the lichen's internal moisture.

The traditional view of lichens as a simple duet of partners has recently evolved. In 2015, biologists in Montana made a striking discovery while investigating why two local lichen species—bright yellow *Bryoria tortuosa* and brown *Bryoria fremontii*—differ in color despite sharing identical symbionts. It turned out that *tortuosa*'s yellow hue comes from a dense layer of yeast cells, specifically basidiomycete yeasts, embedded in the lichen's cortical region. These yeasts stimulate the production of vulpinic acid, a toxic compound with yellow pigmentation. Since then, basidiomycete yeasts have been found in lichens across the globe, revealing that many lichens are in fact *ménages à trois*. And the story does not end there. Genetic analyses suggest that other hidden players, notably bacteria, contribute to the rich tapestry of the lichen holobiont.

Lichens stand among the most ancient examples of symbiotic life. Through their alliance with photosynthetic organisms, fungal mycelia ventured beyond the soil and onto exposed surfaces—including bark, rock, and even aerial habitats. Lichens and fungi were among the first pioneers to colonize terrestrial environments (as opposed to aquatic) over 400 million years ago. In so doing, they helped create the soil environments that nurtured the rise of terrestrial flora and fauna. Lichens continue their quiet but vital work even today, blanketing vast stretches of desert and tundra and absorbing nearly 7 percent of the carbon dioxide produced by human activities.

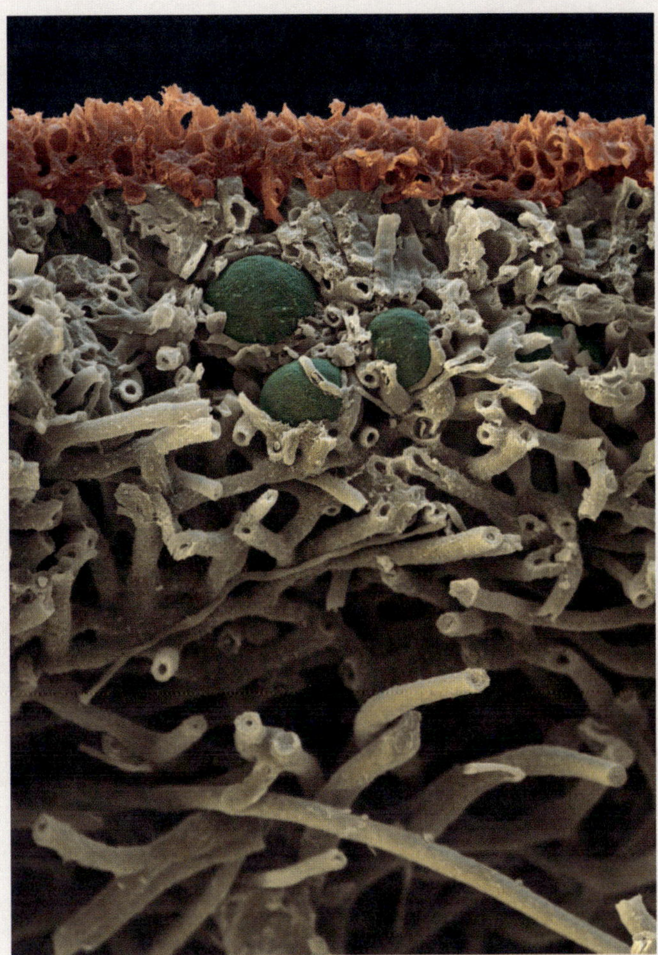

THE STRUCTURE OF A LICHEN'S TWO MAIN SYMBIOTIC PARTNERS

The two most visible partners of a lichen are a fungus (an ascomycete, or more rarely a basidiomycete) and a green microalga (Chlorophyta). Here, colorization reveals the fungal filaments, called *hyphae* (*beige*), the microalgae (*green*), and on top, the cortical region (*orange*), rich in polysaccharides, bacteria, and yeast.

COLORIZED SCANNING ELECTRON MICROSCOPY
© Courtesy Nicole Ottawa and Oliver Meckes, *eye of science*, Germany

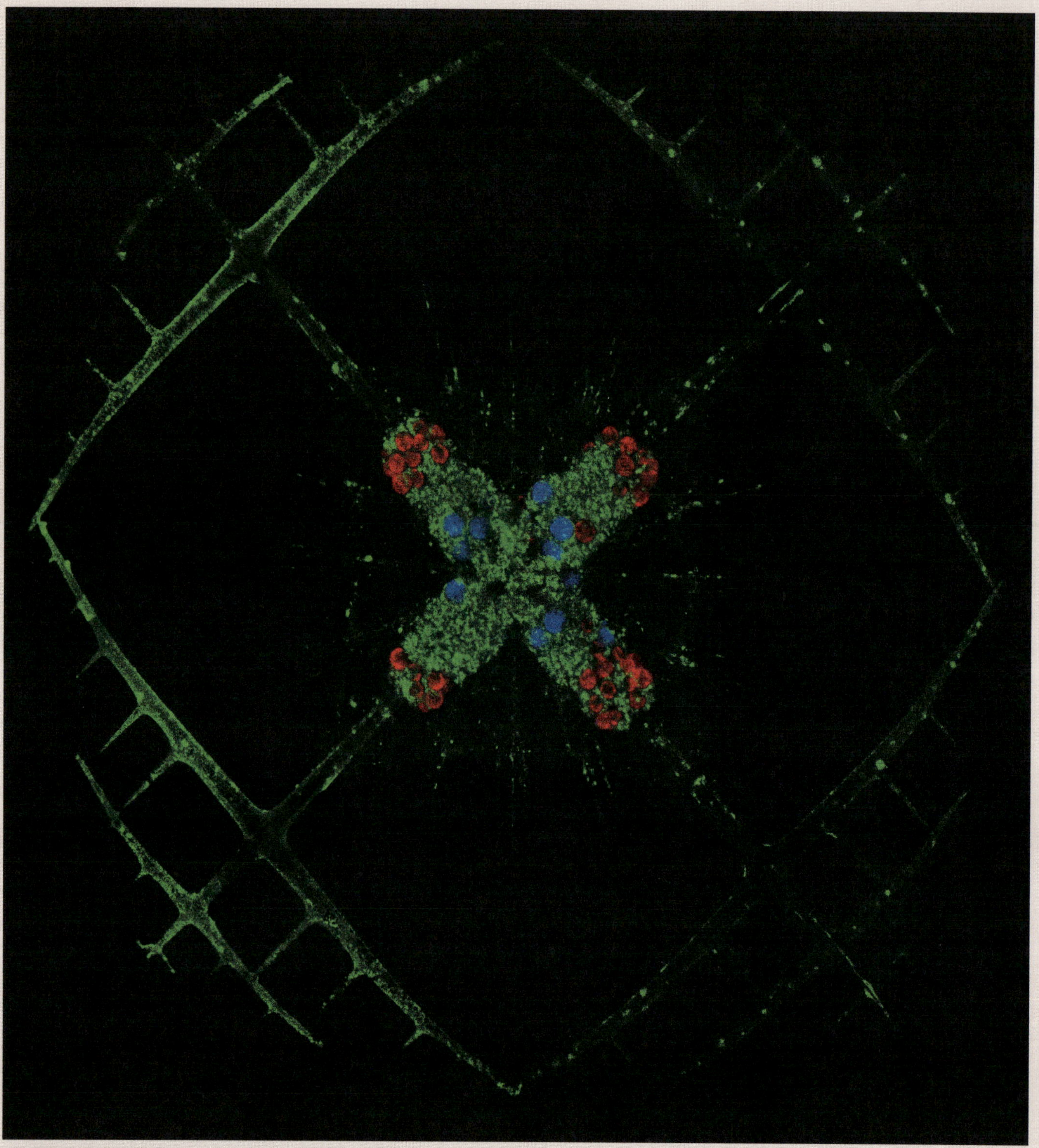

AN ACANTHARIAN WITH SYMBIOTIC MICROALGAE

The acantharian *Lithoptera fenestrata*, a planktonic protist, measures 0.1 to 0.5 mm depending on its age. *Fenestrata*, Latin for "windowed," refers to its delicately shaped skeleton, made of strontium sulfate. At its center, the cytoplasm (*green*) fills a cross-shaped region, with the symbiotic microalga *Phaeocystis cordata* present at the apexes of the four arms. The microalga's chloroplasts fluoresce (*red*), and the protist's nuclei (*blue*) are also visible.

CONFOCAL FLUORESCENCE MICROSCOPY

© Courtesy Johan Decelle and Sébastien Colin, Station Biologique de Roscoff, CNRS & Sorbonne University, France

View "Metamorphosis of the Microalga *Phaeocystis* in Symbiosis" (Decelle/Uwizeye)

Slaves and zombies in plankton

The vast realm of plankton is largely composed of unicellular organisms engaged in an intricate web of symbiotic relationships, spanning the entire spectrum from parasitism to mutualism. Among myriad species, *Lithoptera fenestrata*—an acantharian belonging to the actinopod family—stands out as one of the most delicate and graceful protists. Adorned with a windowed, starlike skeleton, *Lithoptera* drifts serenely with the ocean currents, feeding on bacteria and small protists. Among its prey are unicellular microalgae, particularly *Phaeocystis*, a genus of haptophyte (mineral-producing) protists that roam the waters propelled by two flagella. These tiny algae draw power from the respiration of a single mitochondrion, and from photosynthesis by their twin chloroplasts. At times, *Phaeocystis* blooms so prolifically that it carpets coastal waters with thick foam, releasing volatile sulfur compounds potent enough to influence cloud formation.

Yet the relationship between acantharia and microalgae extends beyond predation to a more intimate bond. *Phaeocystis* often live symbiotically inside *Lithoptera*, a relationship crucial to the survival of the host. Once ingested by *Lithoptera*, the planktonic microalgae are sequestered in special vacuoles, or symbiosomes, positioned at the four ends of *Lithoptera*'s cross-shaped cell body. This type of alliance—photosymbiosis, between a photosynthetic organism and an animal-like protist—is widespread among protists, including actinopods, radiolaria, and foraminifera, particularly in the nutrient-poor waters of the open oceans.

At first glance, the partnership appears mutually beneficial: The photosynthetic prowess of the microalgae nourishes the acantharian, while the host provides essential nutrients such as nitrogen, phosphorus, and minerals, as well as protection from predators. Yet, beneath this seemingly balanced exchange lies a more exploitative reality. Once inside, the microalgae are stripped of their autonomy. Their cell division is halted, and their size and number of chloroplasts increase dramatically. This transformation amplifies *Phaeocystis*'s capacity for photosynthesis and carbon fixation, benefiting the host cell at the microalgae's expense. Moreover, *Lithoptera* regulates *Phaeocystis*'s metabolism, supplying iron and sulfur to maximize their productivity. In essence, *Lithoptera* captures, cultivates, and exploits the *Phaeocystis* microalgae like enslaved workers, which it needs to survive!

The relationship between *Lithoptera* and *Phaeocystis* thus represents a curious reversal of parasitism, with the host acting as a slave-master to its symbionts. Elsewhere in the planktonic world, more traditional forms of parasitism are rampant. Among the most notorious players are dinoflagellates, some of which prey upon other microalgae—including their own kin. Their parasitic strategies are diverse and ingenious. To take one example, *Amoebophrya* invades other dinoflagellates and engulfs the nucleus to digest the chromosomes. Stripped of its genetic control center, the host dinoflagellate becomes a functional zombie, its metabolism co-opted to serve the parasite. While the zombified host continues to respire and photosynthesize, *Amoebophrya* feasts on its ill-gotten nutrients—phosphates and nitrogen from the dismantled nucleic acids and proteins. Ultimately, the parasite hijacks the host's cellular machinery, including the chloroplasts and mitochondria, to produce spores. Upon rupture of the host cell, the spores disperse into the water as new *Amoebophrya* parasites, ready to begin the cycle again.

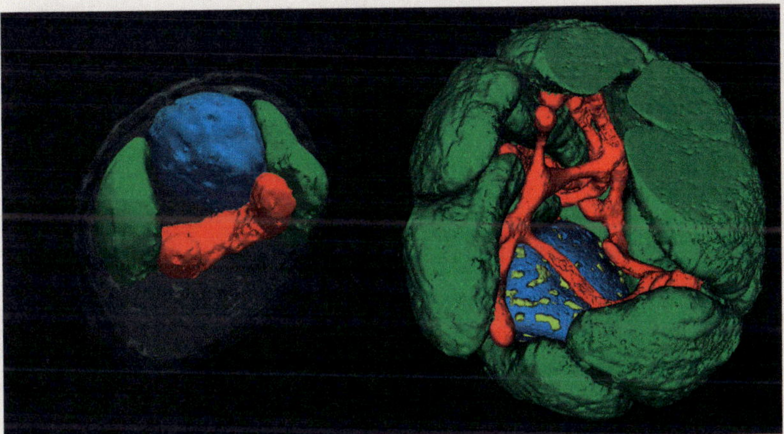

A MICROALGA TRANSFORMS INSIDE ITS HOST

On left: Swimming freely in the sea, the microalga *Phaeocystis*, a haptophyte protist, possesses a nucleus (*blue*) and two chloroplasts (*green*), in contact with a single mitochondrion (*red*).

On right: Once ingested by a *Lithoptera* acantharian, a *Phaeocystis* microalga increases its volume tenfold, and its chloroplasts multiply. The mitochondrion becomes threadlike, while maintaining contact with all the chloroplasts.

3D RECONSTRUCTION OF SEQUENTIAL ELECTRON MICROSCOPE SECTIONS
© Courtesy Clarisse Uwizeye, Johan Decelle, UMR 5168, CNRS/UGA/INRA/CEA, Grenoble, France

Animals with vegetals— from jellyfish to corals and mollusks

Plant cells, masters of photosynthesis, capture solar energy and atmospheric CO_2 to produce organic matter. This vital alchemy is performed by specialized organelles, the chloroplasts. Lacking such machinery, most animals must obtain their energy and carbon by feeding on plants or other organisms. But some creatures, including jellyfish and corals, have devised an extraordinary survival strategy: photosymbiosis—intimately hosting plant cells, microalgae, or cyanobacteria within their own tissues. In return for shelter and protection, the plant partners supply the animal host with sugars, amino acids, and precious metabolites produced through photosynthesis.

Among the most curious practitioners of this artful coexistence is the unusual jellyfish *Cassiopea*. It lives upside-down in shallow, warm, and sunlit waters, with its "umbrella" top resting on the seabed and its tentacles and mouth turned skyward. And why? This way, the jellyfish exposes the microalgae living inside its tissues, of genus *Symbiodinium*, to maximum light. The *Cassiopea*, while benefiting from photosymbiosis, continues to feed on its prey, small plankton that pass within reach. These jellyfish are easily cultivated in aquariums, but they lose weight and die when kept in the dark, or when "bleached" by a thermal shock that chases away their microalgae symbionts. A bleached *Cassiopea* can be recolonized if provided with a different species of *Symbiodinium*. As larvae, these jellyfish lack symbiotic microalgae, but after settling on the sea floor and transforming into polyps, they quickly acquire them. Polyps equipped with symbiotic microalgae grow more rapidly and are more likely to produce new *Cassiopea* babies through budding.

Reef-building corals are even more tightly bound to their photosynthetic partners, relying on an array of *Symbiodinium* species. In addition to these microalgae, corals harbor a rich microbiota of bacteria, archaea, fungi, and viruses, each contributing to the fragile equilibrium of the coral holobiont. Thriving in nutrient-poor waters, corals are paradoxically among the ocean's most productive organisms, thanks to their extensive photosymbiosis. However, when stressed by warming, pollution, or acidification, corals expel their symbionts and die. Although the oceans contain a wide diversity of free-living symbiotic algae from the *Symbiodinium* genus, it remains unclear how—or under what conditions—bleached corals might reacquire them and recover.

There are even stranger forms of photosymbiosis. One of the most exotic is kleptoplasty—chloroplast theft. Practiced by certain gastropod mollusks called *nudibranchs* (or commonly, sea slugs), notably the radiant *Elysia chlorotica*, kleptoplasty involves stealing chloroplasts from algae of the genus *Vaucheria*. The algal cells are digested, but their chloroplasts are preserved and integrated into the sea slug's epidermis, where they continue photosynthesizing, turning the mollusk a vivid green. Remarkably, this mollusk has evolved specific genes that prevent rejection of these foreign chloroplasts. Plant organelles, it seems, can also thrive inside animals. Perhaps we should call these mixed organisms "vegimals"!

CASSIOPEA, A JELLYFISH THAT LIVES UPSIDE-DOWN IN SYMBIOSIS WITH MICROALGAE

Cassiopea jellyfish often live statically along the edges of mangrove forests and play an important role in this fragile ecosystem. These jellyfish obtain a significant portion of their energy from symbiotic microalgae of the genus *Symbiodinium* that reside in their tissues.

© Courtesy Cheryl Ames, Tohoku University, Sendai, Japan

BLEACHED CORAL AND ZOOXANTHELLA MICROALGAE

Coral polyps of the *Pocillopora* sp. are partially bleached (*white area*) because they have lost their microalgal symbionts, *Symbiodinium*, under stress.

FLUORESCENCE MACROPHOTOGRAPHY
© Courtesy Pete West, Daniel Stoupin, BioQuest Studios, Queensland, Australia

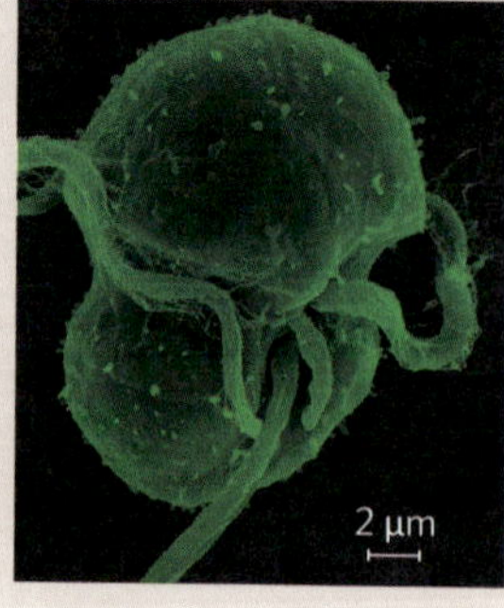

Symbiodinium microalgae are dinoflagellate protists with two flagella. Their chloroplasts give the coral holobiont its photosynthetic capabilities and green color.

COLORIZED SCANNING ELECTRON MICROSCOPY
© Courtesy Gert Hansen

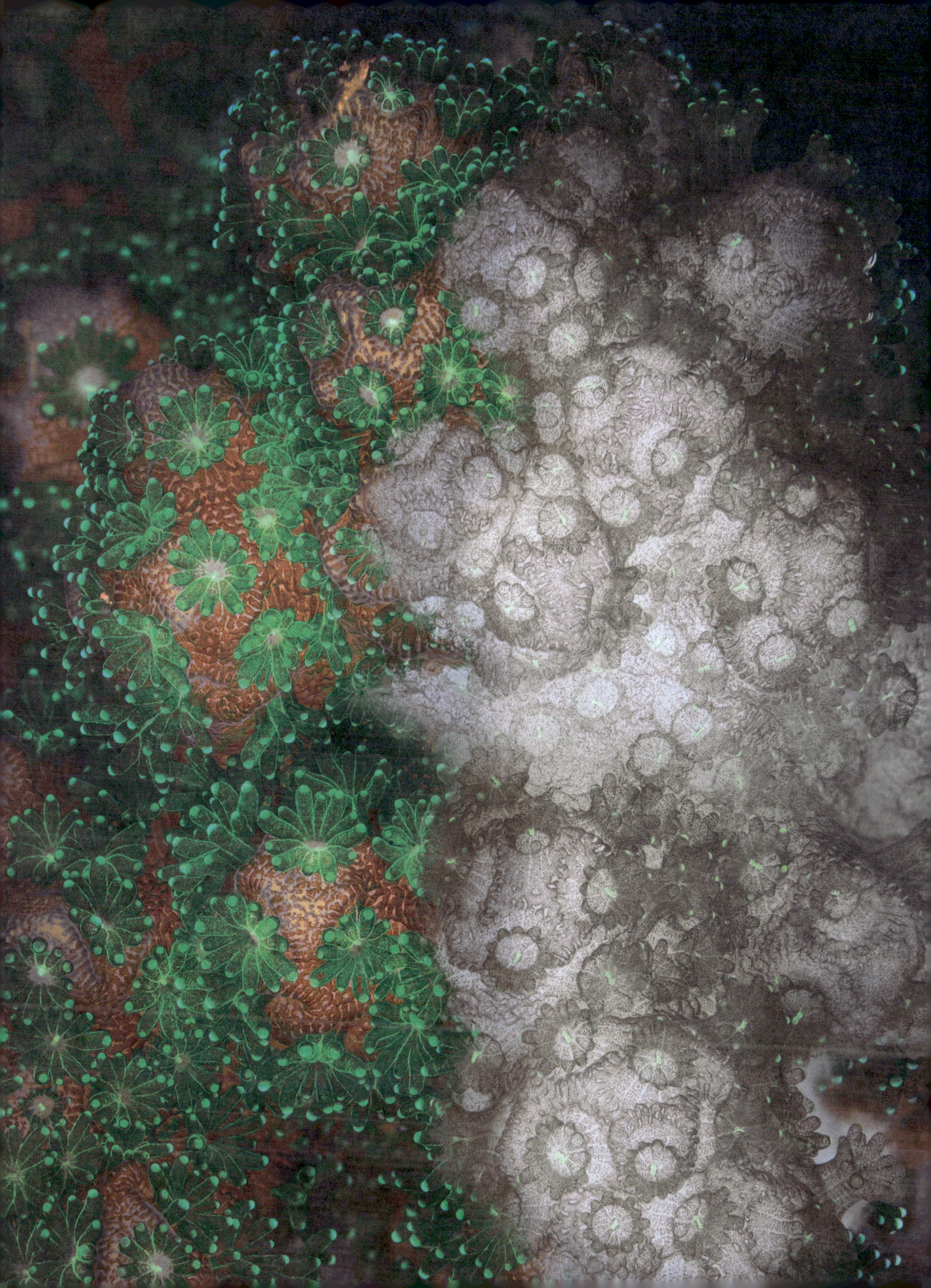

tubulin, 56, 148–49

U

unicellular, 95, 100–105

V

vacuole, 87, 192
vesicle, 48, 66, 96
virion, 116–17, 125
virocell, 118
virus, 26, 70, 73, 80, 84, 96, 106, 114–24
volvocale, 92

Y

yeast, 31, 95, 100, 104–5, 148, 210, 213

Z

zoonosis, 124, 192
zooxanthellae, 216
zygote, 144, 172–81

Species cited

A

Acanthamoeba, 117
Aliivibrio, 199
Amoebophrya, 215
Anabaena, 74–75
Arabidopsis, 110–11, 203
Ascaris, 158
Asterionellopsis, 97

B

Biston, 168
Bryoria, 213

C

Caenorhabditis, 31, 138, 149, 175, 189
Candida, 105
Cassiopea, 216
Chlamydomonas, 95, 103
Collozum, 97

D

Drosophila, 175, 178, 180

E

Escherichia, 77, 87, 150

F

Fucus, 181

H

Haloquadratum, 73
Homo, 6, 31, 156, 168

I

Ignicoccus, 73

K

Kylinxia, 23

L

Lauderia, 97
Lithoptera, 214–15

M

medaka, 145
Methanococcus, 73
Methanogenium, 73
Methanothermobacter, 73
Methanothermus, 73
Mus, 110

N

Nanoarchaeum, 73

P

Phaeocystis, 214–15
Physarum, 92, 94, 106
Plasmodium, 96, 190, 192
Promethearcheum, 86
Pyrobaculum, 73
Pyrodictium, 73

R

Rhizobium, 212
Riftia, 69

S

Saccharomyces, 95, 104–5, 148
Schistosoma, 193
Stentor, 92, 94
Symbiodinium, 216

T

Thiomargarita, 87, 90
Toxoplasma, 96, 192
Trichonympha, 149

V

Vaucheria, 216
Vibrio, 198

X

Xenopus, 148, 164, 181

People cited

A

Aristotle, 15
Avery, Oswald, 159

B

Bateson, William, 159
Beijerinck, Martinus, 116
Berthold, Arnold, 202
Boveri, Theodor, 148, 158, 193
Brangwynne, Clifford, 138
Brenner, Sydney, 134, 159, 188

C

Chatton, Édouard, 18, 90
Claude, Albert, 131
Crick, Francis, 134, 156, 159
Cuvier, Georges, 24

D

d'Hérelle, Félix, 116
Darwin, Charles, 10, 26, 28, 158, 168, 202
de Vries, Hugo, 158
Driever, Wolgang, 180

E

Epel, David, 144
Escherich, Theodor, 77

F

Fol, Hermann, 15, 158, 181
Fox, George, 30
Franklin, Rosalind, 134, 156, 159

G

Gosling, Raymond, 159
Gilkey, John, 144–45
Goethe, 169
Goodsell, David, 77

H

Haeckel, Ernst, 10, 26, 28–29, 80, 158
Hartsoeker, Nicolaas, 10, 12
Hayflick, Leonard, 187
Hertwig, Oscar, 158
Hooke, Robert, 12
Hyman, Anthony, 138

I

Ivanovski, Dmitri, 116
Iwasa, Janet, 66

J

Jaffe, Lionel, 144–45
Johannsen, Wilhelm, 159

K

Koch, Robert, 116, 192
Krebs, Hans, 131

L

Lacks, Henrietta, 187
Laveran, Alphonse, 192
Linnaeus, Carl, 28, 169
Lohmann, Karl, 131

M

Margulis, Lynn, 30, 206
Mendel, Gregor, 158–59
Meyerhof, Otto, 131
Miescher, Friedrich, 158
Mitchell, Peter, 132, 134
Morgan, Thomas Hunt, 159

N

Nüsslein-Volhard, Christiane, 180

O

O'Grady, Marcella, 158

P

Palade, George, 131
Parnas, Jakub Karol, 131
Pasteur, Louis, 10, 15, 95, 105, 116, 130, 192

R

Redi, Francesco, 15
Rous, Peyton, 193
Roux, Wilhelm, 158

S

Santorio, Santorio, 130
Schleiden, Matthias, 14, 158, 164
Schwann, Theodor, 14, 158, 164
Skloot, Rebecca, 187
Spallanzani, Lazzaro, 15
Stanley, Wendell, 116
Steinhardt, Richard, 144
Stevens, Nettie, 176
Sutton, Walter, 158
Szent-Györgyi, Albert, 131

T

Thuret, Gustave-Adolphe, 181
Twort, Frederick, 116

V

Van Leeuwenhoek, Antonie, 12
Virchow, Rudolf, 14, 193
Vogt, Carl, 184

W

Wallace, Alfred, 28
Warburg, Otto, 193
Watson, James, 134, 156, 159
Wilkins, Maurice, 159
Wilson, Edmund Beecher, 158, 176
Woese, Carl, 30

All illustrations by Christian Sardet, except:

© COURTESY Abderrazak El Albani, CNRS, University of Poitiers, France: p. 22 top. Diying Huang, Fangchen Zhao, Nanjing, Institute of Geology and Paleontology, China: p. 23. Jean Vannier, CNRS and University of Lyon 1, and Muriel Vidal, University of Brest, France: p. 24. Marc Vrakking, Max-Born Institute, Berlin: p. 39 top. Rosalind Rickaby, Oxford University: p. 39 bottom. Volker Brinkmann, Max Planck Institute for Infection Biology, Berlin: p. 41. Christian Sardet, Alex McDougall, Rémi Dumollard, Zoological Station, CNRS, Sorbonne University, Villefranche-sur-Mer: p. 42. David Goodsell for the RCSB PDB "Molecule of the Month": pp. 43, 54–57, 116 right, 135, 165 bottom. Andreas Koch, Laurent Larsonneur, Christian Sardet, Digital Studio, 1999: p. 48. David Goodsell, *The Machinery of Life*, Springer, 2009: pp. 52–53, 76, 134. Janet Iwasa, laboratory of Jack Szostak, Massachusetts General Hospital, Harvard Medical School: pp. 66–67. Debbie Kelley, Mitch Elend, University of Washington, Seattle: p. 68. Purificacion Lopez-Garcia, Laboratory of Ecology, Systems and Evolution, CNRS, Paris Saclay University, France: p. 69 top. Janet Iwasa, University of Utah, Salt Lake City: pp. 72, 73 left. Christian Sardet and the Macronauts, *Plankton Chronicles*, www.planktonchronicles.org: p. 74. Tagide deCarvalho, Keith R. Porter Imaging Facility, University of Maryland, Baltimore County: pp. 75, 103 top. David Goodsell, *Molecular Landscapes*, Scripps Research Institute, Protein Data Bank: p. 77. Hiroyuki Imachi, Masaru K. Nobu, JAMSTEC, Japan: p. 86. Tomas Tyml, The Regents of the University of California, Lawrence Berkeley National Laboratory: p. 87 top. Jean-Marie Volland, The Regents of the University of California, Lawrence Berkeley National Laboratory: p. 87 bottom. Christian and Dana Sardet: p. 89. Catherine Jessus, CNRS and Arago Laboratory Library, Sorbonne University, Banyuls, France: pp. 90–91 center. Andy Moore, Jennifer Lippincott-Schwartz, Advanced Imaging Center, Howard Hughes Medical Institute, Janelia Research Campus, Ashburn, Virginia: p. 91 top. Audrey Dussutour, CRCA Laboratory, CNRS, University of Toulouse and CNRS Images, France: p. 94 top. Igor Siwanowicz, Advanced Imaging Center, Howard Hughes Medical Institute, Janelia Research Campus, Ashburn, Virginia: p. 94 bottom. Eloïse Bertiaux, laboratory of Virginie Hamel and Paul Guichard, University of Geneva, Switzerland: p. 96 top. C. Sardet, excerpted from *Plankton: Wonders of the Drifting World*, University of Chicago Press, 2015: p. 97. Kayley Hake, Nicole King, HHMI, University of California, Berkeley: p. 102 left. Mark Dayel, www.dayel.com: p. 102 right. George Witman, UMass Chan Medical School;

Karl Lechtreck, University of Georgia: p. 103 bottom. Will Ratcliff, Anthony Burnetti, Ozan Bozdag, Georgia Institute of Technology: p. 104. Jonathan Nowak, Clément Ghigo, Marc Bajénoff, Marseille-Luminy Immunology Center & CNRS Images, France: p. 110. Yvon Jaillais, Plant Reproduction Laboratory, ENS Lyon & CNRS Images, France: p. 111. Karen-Beth G. Scholthof, Texas A&M University: p. 116 left. Hiroyuki Ogata, Kyoto University; Kazuyoshi Murata, National Institute for Physiological Sciences, Okazaki; Masaharu Takemura, Tokyo University of Science: p. 117. Willie Wilson, Keith Ryan, Marine Biological Association, Plymouth, Great Britain: p. 122 bottom. Timothy Vartanian, Yinghua Ma, Cornell University: p. 123 top. Simon Erlendsson, University of Copenhagen, with permission of *National Geographic*: p. 123 bottom. Richard Wheeler, Cambridge, Great Britain: p. 131. Anthony Hyman, Clifford Brangwynne, Max Planck Institute, Dresden. Germany; published in *Science*, vol. 324, pp. 1729–1732 (2009): p. 138. David Goodsell, Keren Lasker, RCSB Protein Data Bank, Scripps Research, doi: 10.2210/rcsb_pdb/goodsell-gallery-046: p. 139. Lionel Jaffe, coauthor of the original article "A free calcium wave traverses the activating egg of the medaka, *Oryzias latipes*," J. C. Gilkey, L. F. Jaffe, E. B. Ridgway, G. Reynolds, *J. Cell Biol.* 1978; vol. 76(2), pp. 448–466: pp. 144–45 top. A. McDougall, CNRS & Sorbonne University, IMEV, Villefranche-sur-Mer, France: pp. 144–45 bottom. Michael Shribak, Marine Biological Laboratory, University of Chicago; James R. LaFountain, University at Buffalo: p. 148 bottom. Pierre Gönczy, Swiss Federal Institute of Technology in Lausanne (EPFL), Swiss Institute for Experimental Cancer Research (ISREC): p. 149 left. Pierre Gönczy, EPFL/ISREC, Lausanne; Paul Guichard, Department of Cell Biology, University of Geneva, Switzerland: p. 149 right. Eric Clark, John Griffin, Nathan Claxton, Michael Davidson, Molecular Expressions, National High Magnetic Field Laboratory, Florida State University, Tallahassee: p. 153. Nilesh Vaidya, Clifford Brangwynne, Princeton University: p. 164. Bogdan Bintu, University of California San Diego; Xiaowei Zhuang, Harvard University: p. 165 top. Gaël McGill, Evan Ingersoll, Digizyme Inc., Brookline, Massachusetts: p. 166. Liang Xue, Julia Mahamid, European Molecular Biology Laboratory, Heidelberg, Germany: p. 167. Alistair Hume, University of Nottingham, Great Britain: p. 168 top. Eléa Héberlé, Le Plantoscope: p. 169. Ruth Lehmann, Whitehead Institute, MIT: p. 174. Carsten Hoege, Tony Hyman, Max Planck Institute, Dresden, Germany: p. 175. Daniel St Johnston, Katie Smith-Litière, Cambridge University, Great Britain: p. 180 left. Jim Langeland,

Stephen W. Paddock, Sean Carroll, University of Wisconsin, Madison: p. 180 right. Darryl Kropf, The University of Utah, Salt Lake City; Whitney Hable, University of Massachussetts, Dartmouth: p. 181 left. Thomas Deerinck, Mark Ellisman, NCMIR, University of California San Diego: p. 186. Thomas Reid, NCI Center for Cancer Research, Bethesda: p. 187. Bob Goldstein, Daniel Dickinson, University of North Carolina, Chapel Hill: pp. 188–89. John Murray, Ke Hu, Arizona State University, Tempe: p. 192 bottom. Ilari Maasilta, University of Jyvaskyla, Finland: p. 198. Nathanaël Prunet, Caltech, Pasadena, and UNC Chapel Hill: p. 203. A. Oshima, K. Tani, Y. Fujiyoshi, RCSB Protein Data Bank, doi: 10.2210/pdb5H1R/pdb: p. 204 top. Y. Tallini, M.I. Kotlikoff, Cornell University, New York: p. 204 bottom. Anna Pepe, Chiara Zurzolo, Pasteur Institute, France: p. 205 top. Sigal Ben-Yehuda, Ilan Rosenshine, Hebrew University/Hadassah Medical Center, Israel: p. 205 bottom. Lawrence David, Duke University, North Carolina: p. 208. Johan Decelle, Sébastien Colin, Station Biologique de Roscoff, CNRS & Sorbonne University, France: p. 214. Clarisse Uwizeye, Johan Decelle, UMR 5168, CNRS/UGA/INRA/CEA, Grenoble, France: p. 215. Cheryl Ames, Tohoku University, Sendai, Japan: p. 216 left. Pete West, Daniel Stoupin, BioQuest Studios, Queensland, Australia: p. 217.

IMAGE BANKS

Wikipedia/Wikimedia: pp. 12 top and bottom, 14, 15 top (© Library of the Faculty of Law and Labor Sciences, University of Seville), 22 bottom (© Paul Harrison, Reading, UK), 28–31, 40, 46–47 (ammonia © Frédéric Marbach), 69 bottom (© NOAA Okeanos Explorer Program, Galapagos Rift Expedition 2011), 73 right (©Andreas Klingl), 122 top (©Jacques Descloitres, MODIS Rapid Response Team, NASA/GSF), 130, 148 top, 159 right, 168 bottom (© Martinowsky; Chiswick Chap), 176, 193 top, 202 (© Ben Mills), 216 right (© Gert Hansen). **Science Photo**: pp. 13, 88, 96 bottom, 159 bottom, 177, 193 bottom. **Alamy**: pp. 25, 192 top. **ESA/Hubble**: p. 38. **Nicole Ottawa, Oliver Meckes**, *eye of science*, Germany: pp. 49, 95, 105, 181 top, 209, 212–13. **Adobe Stock**: p. 180 bottom. **Macroscopic Solutions**: p. 199 (© Mark Smith).

OTHER

Hermann Fol, *Recherches sur la Fécondation et le Commencement de l'Hénogénie Chez Divers Animaux*: p. 15 bottom. **Joseph Gall**, *A Pictorial History: Views of the Cell*, American Society for Cell Biology: pp. 158–59 top.

REFERENCES & SOURCES

This is a selection of books from my personal library that have most inspired me. With a few rare exceptions, I have not cited the hundreds of scientific articles I consulted in leading journals in the field (*Science, Nature, Cell, J. Cell Biology,* etc.). These publications are often behind costly paywalls and typically too technical for nonspecialists. By contrast, high-quality magazines intended for a broader audience (*New Scientist, Scientific American, Science News,* etc.) are available in print and online. Reliable websites (Phys.org, Live Science, Science Daily, Science News, etc.) are also good sources of short articles. Wikipedia is generally an excellent multilingual resource, accessible to all. And now AI-assisted searches can guide you to the latest information and sources on any topic.

BOOKS ON CELLS & LIFE

Curiously, many books devoted to cells omit the word from their titles, opting instead for broader references to "life" or living organisms. Perhaps the word *cell* carries academic overtones or an unsettling association with disease. But cells are the true focus of many books in my library, whatever their titles.

TEXTBOOKS & COFFEE TABLE BOOKS

There are several authoritative, essential textbooks on cells that have become "bibles" for biology students worldwide. Many are regularly updated, for example *Developmental Biology* is in its 10th edition. There are also beautifully illustrated books on biodiversity and evolution, including my previous book, *Plankton: Wonders of the Drifting World* (University of Chicago Press).

Christian Sardet's lifelong passion for cells and the history of life (1) was sparked by a childhood gift: a small microscope, through which he first observed the tiny creatures inhabiting the ponds of his village, Melle, France. Encouraged by a natural science teacher during secondary school, he was drawn to the life sciences and went on to earn a degree in biochemical engineering from the Institut National des Sciences Appliquées (INSA) in Lyon. In 1967, he moved to the United States, first joining a research team at the Wistar Institute in Philadelphia, then going to the University of California, Berkeley, to obtain a PhD in comparative biochemistry.

Back in France, Christian worked on protein structure as a postdoc at the Centre National de la Recherche Scientifique (CNRS) in Gif-sur-Yvette. In 1975, he joined the French Atomic Energy Commission (CEA) as a researcher at the Villefranche-sur-Mer Marine Station to work on fish osmoregulation.

The birth of his first child prompted Christian to investigate the mechanisms of fertilization and embryo development, leading him to join CNRS as a director of research. Affiliated with Sorbonne University and the CNRS, the Villefranche-sur-Mer Marine Station was an ideal setting to explore the exceptional diversity of marine and planktonic organisms. In 1983, together with a team of researchers and students, Christian founded the CNRS Marine Cell Biology team with the guiding principle: Ask the right question of the right organism. This led to collaborations worldwide and many publications (2). The research team evolved into the Laboratoire de Biologie du Développement (BioDev) at the Institut de la Mer de Villefranche (IMEV) where, as an emeritus researcher, Christian has pursued outreach activities during the past decade (3).

While the thrill of discovery and leading a team of thirty colleagues has been immensely rewarding, Christian also feels a desire to share his knowledge with students and a wider public. This sentiment gave rise to a series of audiovisual projects: films on fertilization in collaboration with Manfred Kage and Spektrum Heidelberg, animated shorts on the cell and the origin of life (4, 5), and the acclaimed 3D film *Voyage Inside the Cell* (6), produced with Digital Studio and shown in museums and at international symposiums. In 2000, with Véronique Kleiner, he made the DVD *Exploring the Living Cell* (7), which was distributed as bonus material with the leading textbook *Molecular Biology of the Cell*. With his son Noé, Christian launched Cinema of the Cell, a popular event at European Life Science Organization conferences.

In 2008, Christian Sardet joined Eric Karsenti and a few colleagues to launch the Tara Oceans expedition (8), a global exploration of plankton under the auspices of the Tara Oceans Foundation, fashion designer and philanthropist agnès b., CNRS, CEA, and the European Molecular Biology Laboratory (EMBL). In this context, Christian, in collaboration with the "Macronauts" (Noé Sardet and Sharif Mirshak), revealed the extraordinary beauty and diversity of plankton through *Plankton Chronicles* (9). This project includes a multilingual website, award-winning photos and films (10, 11, 12, 13, 14), conferences, workshops, and exhibitions. In 2013, Christian's art and science book *Plankton: Wonders of the Drifting World* (15), was published. The original French edition was followed by English, Japanese, German, and Chinese editions.

Cells: The Illustrated Story of Life is Christian's second book. The original French publication in 2023 was followed by an exhibition of his cell drawings in a New York City art gallery (16). Christian has been awarded the Grand Prix des Sciences de la Mer and the Prix Lequeux from the French Académie des Sciences, the Award for Communication in Life Sciences from EMBO, and the French Legion of Honor.

(1) Instagram: #cellsandplankton @cellsandplankton

(3) BioDev Laboratory, Villefranche-sur-Mer

(2) Christian Sardet's publications

(5) *O as Origin* with Y. Mahé, G. Macagno, M. C. Maurel, J. Iwasa

(4) *Marius Explores the Cell* with N. Sardet, Y. Mahé, G. Macagno

(7) *Exploring the Living Cell* with V. Kleiner, CNRS Images

(6) *Voyage Inside the Cell* with Digital Studio, Paris

(9) *Plankton Chronicles* with N. Sardet, S. Mirshak, Parafilms

(8) Tara Oceans expedition

(11) *How Life Begins in the Deep Ocean* with T. Thys, N. Sardet, S. Mirshak

(10) *The Secret Life of Plankton* with T. Thys, N. Sardet, S. Mirshak

(13) *The Smallest Solution to One of Our Biggest Problems* with T. Thys, E. Esteban

(12) *Plastic Vagabond* with T. Thys, N. Sardet, S. Mirshak, Tara Foundation, Parafilms

(15) *Plankton: Wonders of the Drifting World*, University of Chicago Press

(14) Oscar-nominated music video *Manta Ray* with J. Ralph, Anohni

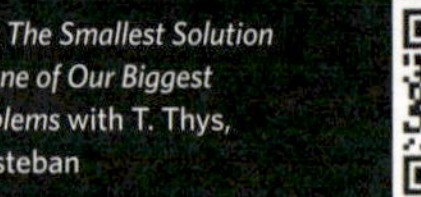

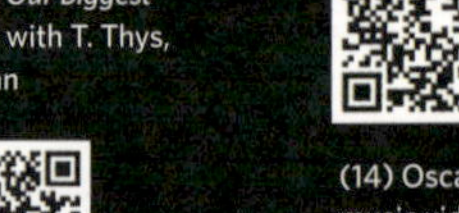

(16) Victoria Munroe Fine Art, New York